STUDIES IN THE THEORY OF RANDOM PROCESSES

by

A. V. SKOROKHOD

Translated from the Russian
by Scripta Technica, Inc.

DOVER PUBLICATIONS, INC., NEW YORK

Published in Canada by General Publishing Company, Ltd., 30 Lesmill Road, Don Mills, Toronto, Ontario.
Published in the United Kingdom by Constable and Company, Ltd., 10 Orange Street, London WC2H 7EG.

This Dover edition, first published in 1982, is an unabridged and unaltered republication of the work originally published in 1965 by Addison-Wesley Publishing Company, Inc., Reading, Mass., as part of their Adiwes International Series in Mathematics (A. J. Lohwater, Consulting Editor). The original Russian edition was published in 1961 by the Kiev University Press with the title *Issledovaniya po teorii sluchainykh protsessov.*
The present edition is published by special arrangement with Addison-Wesley Publishing Company, Inc.

International Standard Book Number: 0-486-64240-2

Manufactured in the United States of America
Dover Publications, Inc.
180 Varick Street
New York, N.Y. 10014

Library of Congress Cataloging in Publication Data

Skorokhod, A. V. (Anatoliĭ Vladimirovich), 1930-
Studies in the theory of random processes.
Translation of: Issledovaniı͡a po teorii sluchaĭnykh prot͡sessov.
Reprint. Originally published: Reading, Mass.: Addison-Wesley Pub. Co., 1965. (Adiwes international series in mathematics; 7021)
Bibliography: p.
Includes index.
1. Stochastic processes. I. Title. II. Series: Adiwes international series in mathematics; 7021. [QA274.S5813 1982] 519.2
81-17298 ISBN 0-486-64240-2 AACR2

PREFACE

The methods employed in the theory of probability, for all their diversity, can be divided into two rather distinct groups: analytic methods and probability methods. The basic distinction between these two consists in the fact that the first has to do only with the distributions of random processes and uses a varied analytical apparatus for the study of these distributions (generating functions, characteristic functions, differential equations, theory of one-parameter semigroups, etc.), whereas probability methods are based on operations with random variables themselves. Thus, to show that a sequence of distribution functions $F_n(x)$ converges to a particular distribution function $F(x)$ when we use analytical methods, we often proceed as follows: we show that $F_n(x)$ satisfies a certain equation $L_nF_n(x) = 0$ and that the operator L_n converges in some defined sense as $n \to \infty$ to an operator L such that F satisfies the relation $LF = 0$. (For example, the central limit theorem in the book *Asimptoticheskoe Zakony Teorii Veroyatnosteĭ* [*Asymptotic Laws of the Theory of Probability*], by A. Ya. Khinchine, is proved in this manner.) We would arrive at the same result if we constructed a sequence of variables ξ_n converging in probability to a variable ξ. Here the distribution function of ξ_n would be F_n, and that of the variable ξ would be $F(x)$. This type of proof of a fact has a probabilistic character.

A preference for one or the other group of methods is a meaningful question only in connection with a specific problem. The advantage of the analytic method lies in its generality; the advantage of probability methods lies in the unison with the subject matter. We must note that the vast majority of works on the theory of probability are mainly devoted to analytic methods. However, the increase in the number of works applying probability methods to the solution of purely analytic problems and also using the Monte Carlo method of approximate computation justifies in some measure (if a justification is necessary) the development of purely probabilistic methods. This book is devoted to the development of certain probabilistic methods in the specific field of stochastic differential equations and limit theorems for Markov processes.

The subject matter of the book falls into three parts. In the first (Chapters 1 and 2), the basic facts of the theory of random processes are introduced and the auxiliary apparatus of stochastic integrals is constructed. The elements of the theory of random processes are given without proof because the book is designed for readers already possessing a basic knowledge of this theory; in the footnotes at the end of the book there are references to sources containing the proofs. Results taken from the theory of martingales play a special role. For a detailed account of this theory, one should read the corresponding chapter of Doob's *Probability Processes.* The theory of stochastic integrals has not yet been presented in any book on the theory of random processes with the generality that is necessary for us; therefore, all proofs are presented in full here.

The second part of the book (Chapters 3, 4, and 5) contains the theory of stochastic differential equations, which permits us to construct a broad class of Markov processes on the basis of simple processes. This is an extremely large part of the theory of probability in which probability methods predominate.

The last part (Chapters 6 and 7) is devoted to various limit theorems connected with the convergence of a sequence of Markov chains to a Markov process with continuous time. In the proof of these limit theorems, we always use the probability method mentioned above by proving that a sequence of distribution functions $F_n(x)$ converges to $F(x)$. In this part, we also examine the probability method of estimating how fast the sequence converges in the limit theorems and how exact the limit theorems are. The results obtained by this technique cannot claim even a relative completeness, but can be justified by the fact that they are the first results in this field to be obtained by probability methods.

The remarks and the bibliography at the end of the book do not pretend to be complete.

I am very grateful to Professor Joseph Ilich Gikhman, with whom my conversations proved a great help in the clarification of many questions brought up in the book.

The Author.

CONTENTS

CHAPTER 1

CERTAIN FACTS FROM THE THEORY OF RANDOM PROCESSES

1. Basic definitions associated with the concept of a random process. We shall say that a probability field $\{\Omega, \mathbf{F}, \mathbf{P}\}$ is given if a σ-algebra $\mathbf{F}$ of subsets of the set Ω — $\mathbf{F}$ is defined and a measure $\mathbf{P}$ is defined on the σ-algebra such that $\mathbf{P}\{\Omega\} = 1$. We shall call the elements ω of the set Ω elementary events and the elements $\mathbf{A}, \mathbf{B}, \ldots$ of the σ-algebra $\mathbf{F}$ random events.

Let X be a metric space. If the function $\xi(\omega)$ is defined for almost all points relative to the measure $\mathbf{P}$, if ω assumes value from X, and if for every Borel set A in X, the set $\{\xi(\omega) \in A\}$ of those ω for which $\xi(\omega) \in \mathbf{A}$ belongs to $\mathbf{F}$, then $\xi(\omega)$ is called a *random variable with values in X*, or simply a *random variable in X*. If X is the m-dimensional Euclidean space $R^{(m)}$, the variable $\xi(\omega)$ is called an *m-dimensional random vector*. In this case, if $\xi(\omega)$ is a random variable in $R^{(m)}$, and if

$$\int |\xi(\omega)| \mathbf{P}\,(d\omega)$$

exists, then the vector $\int \xi(\omega)\mathbf{P}\,(d\omega)$ is called the *mathematical expectation* of the vector $\xi(\omega)$ and is denoted by $\mathbf{M}\,\xi(\omega)$.

A sequence of random variables $\xi_n(\omega)$ with values in X converges to the random variable $\xi_0(\omega)$

(a) almost everywhere (with probability 1), if $\xi_n(\omega) \to \xi_0(\omega)$ with respect to the measure $\mathbf{P}$ for almost all ω,

(b) in probability, if for every $\epsilon > 0$,

$$\lim_{n\to\infty} \mathbf{P}\{\rho(\xi_n(\omega), \xi_0(\omega)) > \epsilon\} = 0,$$

where $\rho(x, y)$ is the distance in X between x and y.

The function $\xi(t, \omega)$, defined for $t \in [t_0, T]$ and for every $t \in [t_0, T]$ for almost all ω relative to the measure $\mathbf{P}$, is called a *random process defined on* $[t_0, T]$ with values in X if for every $t \in [t_0, T]$, $\xi(t, \omega)$ is a random variable in X. A process $\xi(t, \omega)$ is said to be *stochastically continuous* at a point t_1 if for every $\epsilon > 0$,

$$\lim_{t\to t_1} \mathbf{P}\{\rho(\xi(t), \xi(t_1)) > \epsilon\} = 0;$$

a process is said to be *stochastically continuous from the right at the point* t_1 if for every $\epsilon > 0$,

$$\lim_{\substack{t \to t_1 \\ t > t_1}} \mathbf{P}\{\rho(\xi(t), \xi(t_1)) > \epsilon\} = 0.$$

Stochastic continuity from the left is defined in an analogous manner.

A process is *stochastically continuous* (from the right, from the left) if it is stochastically continuous at every point where it is defined.

Suppose that B_1 is a σ-algebra of Borel sets on the interval $[t_0, T]$, that $[t_0, T] \times \Omega$ is the set of pairs (t, ω) where $t \in [t_0, T]$, $\omega \in \Omega$, that $B_1 \times \mathbf{F}$ is the minimal σ-algebra of subsets of $[t_0, T] \times \Omega$ containing all subsets of the form $\Delta \times \mathbf{A}$, and that $\Delta \times \mathbf{A}$ is the set of pairs of points (t, ω) for which $t \in \Delta$, $\omega \in \mathbf{A}$, and $\Delta \in B_1$, $\mathbf{A} \in \mathbf{F}$. If the function $\xi(t, \omega)$ is measurable relative to the σ-algebra $B_1 \times \mathbf{F}$, then the process $\xi(t, \omega)$ is said to be *measurable.*

Let us examine the process $\xi(t, \omega)$ in $R^{(m)}$. Let Λ be a set that is everywhere dense on $[t_0, T]$. A process $\xi(t, \omega)$ is said to be Λ*-separable* (separable relative to the set Λ) if there exists an event $\mathbf{A}_1$ with probability 0 ($\mathbf{P}\{\mathbf{A}_1\} = 0$), such that for every closed set F in $R^{(m)}$ and (open) interval Δ, the event

$$\{\xi(t, \omega) \in F \qquad \text{for all} \qquad t \in \Lambda \cap \Delta\}$$

implies the event

$$\mathbf{A}_1 \cup \{\xi(t, \omega) \in F \qquad \text{for all} \qquad t \in \Delta\}.$$

If the process $\xi(t, \omega)$ is separable relative to some set, we shall call it *separable.*

The probabilities

$$\mathbf{P}_{t_1, t_2, \ldots, t_k}(A_1, A_2, \ldots, A_k) = \mathbf{P}\{\xi(t_1) \in A_1, \xi(t_2) \in A_2, \ldots, \xi(t_k) \in A_k\},$$

where $t_1, t_2, \ldots, t_k \in [t_0, T]$, and where $A_1, A_2, \ldots, A_k$ are Borel sets of the space on which the process is defined, are called *k-dimensional distributions of the process* $\xi(t, \omega)$. The collection of k-dimensional distributions with all possible values of k is called the set of *finite-dimensional distributions of the process* $\xi(t, \omega)$.

Two processes $\xi_1(t, \omega)$ and $\xi_2(t, \omega)$ defined on $[t_0, T]$ are said to be *stochastically equivalent* if for every $t \in [t_0, T]$,

$$\mathbf{P}\{\xi_1(t, \omega) = \xi_2(t, \omega)\} = 1.$$

Stochastically equivalent processes have identical finite-dimensional distributions.

THEOREM 1. *For every process $\xi(t, \omega)$ in X, there is a separable process stochastically equivalent to it. If a process $\xi(t, \omega)$ is defined on $[t_0, T]$ and is stochastically continuous at all points $t \in [t_0, T]$ with the possible exception of a finite number of points, then there exists a separable measurable process that is stochastically equivalent to the process $\xi(t, \omega)$.*

A process $\xi(t, \omega)$ defined on $[t_0, T]$ is said to be *continuous with probability 1* if for almost all $\omega \in \Omega$, $\xi(t, \omega)$ as a function of t is defined and continuous on $[t_0, T]$.

THEOREM 2 (KOLMOGOROV). *Let $\xi(t, \omega)$ be a separable process defined on $[t_0, T]$ and assuming values in $R^{(m)}$. If there exists $\alpha > 0$, $\beta > 0$, $C > 0$ such that for $t_1, t_2 \in [t_0, T]$,*

$$\mathbf{M}|\xi(t_2) - \xi(t_1)|^\alpha \leq C|t_2 - t_1|^{1+\beta},$$

then the process $\xi(t, \omega)$ is continuous with probability 1.

Let us examine a process $\xi(t, \omega)$ defined on $[t_0, T]$ with values in $R^{(m)}$. Let us designate by $\Phi^{(m)}[t_0, T]$ the space of all functions $x(t)$ defined on $[t_0, T]$ and assuming values in $R^{(m)}$. Let $C_{t_1}(A)$, where $t_1 \in (t_0, T]$ and A is a Borel set in $R^{(m)}$, be the set of all $x(t)$ for which $x(t_1) \in A$. A set formed by the intersection of a finite number of sets of the form $C_{t_1}(A)$ is called a *cylindrical set*. Let us indicate by $F^{(m)}[t_0, T]$ the minimal σ-algebra of subsets of $\Phi^{(m)}[t_0, T]$ containing all cylindrical sets. The measure $\mu(\mathbf{A})$, uniquely determined on $F^{(m)}[t_0, T]$ by the relations

$$\mu(C_{t_1}(A_1) \cap C_{t_2}(A_2) \cap \cdots \cap C_{t_k}(A_k)) = \mathbf{P}\{\xi(t_i, \omega) \in A_i, \qquad i = 1, 2, \ldots, k\}$$

for all k, $t_1, t_2, \ldots, t_k$ in $[t_0, T]$ and for all Borel sets $A_1, A_2, \ldots, A_k$ in $R^{(m)}$, will be called the *measure in the function space corresponding to the process* $\xi(t, \omega)$, or simply the *measure corresponding to the process* $\xi(t, \omega)$. Suppose that to every set $t_1, t_2, \ldots, t_k$ in $[t_0, T]$ and to all natural numbers k there corresponds a k-dimensional distribution

$$\mathbf{P}_{t_1, t_2, \ldots, t_k}(A_1, A_2, \ldots, A_k);$$

then for all $j \leq k$ and for the Borel sets $A_1, \ldots, A_{j-1}, A_{j+1}, \ldots, A_k$ in $R^{(m)}$, the relations

$$\mathbf{P}_{t_1, t_2, \ldots, t_{j-1}, t_j, t_{j+1}, \ldots, t_k}(A_1, A_2, \ldots, A_{j-1}, R^{(m)}, A_{j+1}, \ldots, A_k) = \mathbf{P}_{t_1, \ldots, t_{j-1}, \ldots, t_{j+1}, \ldots, t_k}(A_1, \ldots, A_{j-1}, A_{j+1}, \ldots, A_k),$$

$$\mathbf{P}_{t_1, \ldots, t_j, \ldots, t_k}(A_1, \ldots, A_j, \ldots, A_k) = \mathbf{P}_{t_1, \ldots, t_k, \ldots, t_j}(A_1, \ldots, A_k, \ldots, A_j)$$

hold. We shall call such a set of distributions *consistent* or *joint*.

THEOREM 3 (KOLMOGOROV). *If* $\mathbf{P}_{t_1,t_2,\ldots,t_k}(A_1, A_2, \ldots, A_k)$ *is a collection of joint distributions, then there exists a random process* $\xi(t, \omega)$ *for which these distributions will be finite-dimensional. To every joint collection of distributions, there corresponds a unique measure on* $F^{(m)}[t_0, T]$.

Suppose that $\xi(t, \omega)$ is a random process defined on $[t_0, T]$ and that $[t_1, t_2] \subset [t_0, T]$. Let us examine the minimal σ-algebra of events that contains all events of the form $\{\xi(s, \omega) \in A\}$ for $s \in [t_1, t_2]$ and for an arbitrary Borel set A in the space of values of the process. We shall henceforth designate this σ-algebra by $\mathbf{F}([t_1, t_2], \xi(t, \omega))$.

2. Conditional probabilities and mathematical expectations. Suppose that a probability field $\{\Omega, \mathbf{F}, \mathbf{P}\}$ is given and that $\mathbf{F}_1$ is a σ-algebra of events in $\mathbf{F}$, $\mathbf{F}_1 \subset \mathbf{F}$. If $\xi(\omega)$ is a certain random variable in $R^{(m)}$, then the random variable $\mathbf{M}(\xi(\omega)/\mathbf{F}_1)$ measured relative to $\mathbf{F}_1$ and satisfying the relation

$$\int_A \mathbf{M}(\xi(\omega)/\mathbf{F}_1)\mathbf{P}(d\omega) = \int_A \xi(\omega)\mathbf{P}(d\omega), \tag{2.1}$$

for all $\mathbf{A} \in \mathbf{F}_1$ for which

$$\int_A |\xi(\omega)|\mathbf{P}(d\omega) < \infty,$$

is called the *conditional mathematical expectation of the variable* $\xi(\omega)$ *relative to the* σ*-algebra* $\mathbf{F}_1$. If $\mathbf{M}|\xi(\omega)| < \infty$, then the conditional mathematical expectation exists and is uniquely determined to within a set of measure zero in the variable ω. In the case in which $\mathbf{F}_1$ is the minimal σ-algebra containing all events of the form

$$\{\xi_1(\omega) \in A_1\}, \quad \{\xi_2(\omega) \in A_2\}, \quad \ldots, \quad \{\xi_n(\omega) \in A_n\},$$

where $\xi_1(\omega), \xi_2(\omega), \ldots, \xi_n(\omega)$ are particular random variables and $A_1, A_2, \ldots, A_n$ are Borel sets in the space of their values, we write

$$\mathbf{M}(\xi(\omega)/\xi_1(\omega), \xi_2(\omega), \ldots, \xi_n(\omega))$$

instead of $\mathbf{M}(\xi(\omega)/\mathbf{F}_1)$. We shall call this quantity the *conditional mathematical expectation of the quantity* $\xi(\omega)$ *for fixed* $\xi_1(\omega), \xi_2(\omega), \ldots, \xi_n(\omega)$.

If the σ-algebra $\mathbf{F}_1$ coincides with the σ-algebra $\mathbf{F}([t_1, t_2], \xi(t, \omega))$, where $\xi(t, \omega)$ is a particular random process, then instead of $\mathbf{M}(\xi(\omega)/\mathbf{F}_1)$, we sometimes write $\mathbf{M}(\xi(\omega)/\xi(t, \omega), t \in [t_1, t_2])$. We shall call this quantity the *conditional mathematical expectation of the variable* $\xi(\omega)$ *for a fixed value of* $\xi(t, \omega)$ *on* $[t_1, t_2]$.

Let us denote by $\chi_A(x)$ the characteristic function of the Borel set A in $R^{(m)}$. Then the quantity

$$\mathbf{P}\{\xi(\omega) \in A/\mathbf{F}_1\} = \mathbf{M}(\chi_A(\xi(\omega))/\mathbf{F}_1)$$

is called the conditional probability of the event $\{\xi(\omega) \in A\}$ relative to $\mathbf{F}_1$. The quantities

$$\mathbf{P}\{\xi(\omega) \in A/\xi_1(\omega), \ldots, \xi_n(\omega)\}$$

and

$$\mathbf{P}\{\xi(\omega) \in A/\xi(t, \omega), t \in [t_1, t_2]\}$$

are defined analogously to the conditional mathematical expectations.

The collection of conditional probabilities $\mathbf{P}\{\xi(\omega) \in A/\mathbf{F}_1\}$ for all possible Borel sets A is called the *conditional distribution of the variable* $\xi(\omega)$ *relative to* $\mathbf{F}_1$.

We note that the relation

$$\int_{\mathbf{A}} \mathbf{P}\{\xi(\omega) \in A/\mathbf{F}_1\}\mathbf{P}(d\omega) = \mathbf{P}\{(\xi(\omega) \in A) \cap \mathbf{A}\} \tag{2.2}$$

follows from (2.1) for all $\mathbf{A}$ in $\mathbf{F}_1$.

From Formula (2.2), it is easy to deduce the following lemma:

LEMMA. *If for a sequence* $\xi_n(\omega)$ *of variables in* $R^{(m)}$ *and for some* $\mathbf{F}_1$,

$$\mathbf{M}(|\xi_n(\omega)|/\mathbf{F}_1) \to 0$$

in probability, then $\xi_n(\omega) \to 0$ *in probability.*

3. Processes with independent increments. A process $\xi(t)$ defined on $[t_0, T]$ and assuming values in $R^{(m)}$ is called a *process with independent increments* if, for any $t_1 < t_2 < \cdots < t_k$ in $[t_0, T]$, the variables

$$\xi(t_0),\ \xi(t_1) - \xi(t_0), \ldots, \xi(t_k) - \xi(t_{k-1})$$

are independent of each other, that is, if for all Borel sets $A_0, A_1, \ldots, A_k$ in $R^{(m)}$ the relation

$$\mathbf{P}\{\xi(t_0) \in A_0, \xi(t_1) - \xi(t_0) \in A_1, \ldots, \xi(t_k) - \xi(t_{k-1}) \in A_k\}$$
$$= \mathbf{P}\{\xi(t_0) \in A_0\} \prod_{i=1}^{k} \mathbf{P}\{\xi(t_i) - \xi(t_{i-1}) \in A_i\}$$

holds. It is obvious that finite-dimensional distributions of the process $\xi(t)$, and consequently the measure corresponding to the process $\xi(t)$, are completely determined in this case by the distribution of the quantity $\xi(t_0)$ and by the distributions $\xi(t_2) - \xi(t_1)$ for all possible values of t_1 and t_2 in $[t_0, T]$. The distribution of the quantity $\xi(t_0)$ can be arbitrary. For a stochastically continuous process with independent increments (and henceforth we shall consider only such processes with independent increments), the distribution of the variable $\xi(t_2) - \xi(t_1)$ for $t_1 < t_2$ is determined by its characteristic function:

$$\mathbf{M} \exp \{i(z, \xi(t_2) - \xi(t_1))\} = \exp \Big\{i(a(t_2) - a(t_1), z) - \tfrac{1}{2}([A(t_2) - A(t_1)]z, z) + \int_{t_1}^{t_2}\int_{|u|>1} (e^{i(z,u)} - 1)G\,(dt \times du) + \int_{t_1}^{t_2}\int_{|u|\le 1} [e^{i(z,u)} - 1 - i(z, u)]G\,(dt \times du)\Big\}, \quad (3.1)$$

where $a(t)$ is a continuous function on $[t_0, T]$ with values in $R^{(m)}$, $A(t)$ is a continuous function on $[t_0, T]$ whose values are nonnegative linear symmetrical operators in $R^{(m)}$, $A(t_2) - A(t_1)$ is also a nonnegative operator if $t_1 < t_2$, and, finally, G is the measure defined on the Borel sets of the space $[t_0, T] \times R^{(m)}$, i.e., of the space of pairs of points $(t; u)$, $t \in [t_0, T]$, $u \in R^{(m)}$, such that

$$\overline{\lim_{\epsilon\to 0}} \int_{t_0}^{T}\int_{|u|>\epsilon} \frac{(u, u)}{1 + (u, u)}\, G\,(dt \times du) < \infty.$$

[Here, (x, y) denotes the scalar product of the vectors x and y in $R^{(m)}$.]

A process with independent increments $\xi(t)$ that is defined for $t \ge 0$ and that satisfies the condition $\xi(0) = 0$ will be called a *homogeneous process with independent increments* if the distribution of the variable $\xi(t + h) - \xi(t)$ depends only on h. Finite-dimensional distributions of such a process are completely determined by the distributions of the variables $\xi(t)$ for all possible $t > 0$. For stochastically continuous homogeneous processes with independent increments, we can write the logarithm of the characteristic function in the form

$$\log \mathbf{M}e^{i(z,\xi(t))} = t\Big[i(z, \gamma) - \tfrac{1}{2}(Az, z) + \int_{|u|\le 1} (e^{i(z,u)} - 1 - i(z, u))M\,(du) + \int_{|u|>1} (e^{i(z,u)} - 1)M\,(du)\Big], \quad (3.2)$$

where $\gamma \in R^{(m)}$ and A is a linear symmetric nonnegative operator in $R^{(m)}$. M is a measure defined on the Borel sets of the space $R^{(m)}$ such that

$$\overline{\lim_{\epsilon\to 0}} \int_{|u|>\epsilon} \frac{(u, u)}{1 + (u, u)}\, M\,(du) < \infty.$$

Processes whose measure G (for homogeneous M) is identically equal to zero are called *normal processes.* If a process with independent increments and with probability 1 is continuous, it must necessarily be normal; the converse is true for stochastically continuous separable processes. A one-dimensional normal process (that is, one with numerical values) with independent increments $w(t)$ for which

$$\log \mathbf{M}e^{i\lambda(w(t+h)-w(t))} = -\frac{\lambda^2 h}{2}$$

is called a *process of Brownian motion.* We shall be encountering this process repeatedly in what follows.

A one-dimensional homogeneous process with independent increments $v(t)$ that assumes nonnegative integral values is called a *Poisson process* if for some $a > 0$,

$$\mathbf{P}\{v(t) = k\} = \frac{(at)^k}{k!} e^{-at}.$$

4. Markov processes. A process $\xi(t)$ defined on $[t_0, T]$ and assuming values in $R^{(m)}$ is called a *Markov process* if for every $s < t$ in $[t_0, T]$, with probability 1 the relation

$$\mathbf{P}\{\xi(t) \in A/\mathbf{F}([t_0, s], \xi(\tau))\} = \mathbf{P}\{\xi(t) \in A/\xi(s)\} \tag{4.1}$$

is fulfilled.

From (4.1), it follows that for every $t_1 < t_2 \cdots < t_k$ and for every Borel set A, with probability 1 the relation

$$\mathbf{P}\{\xi(t_k) \in A/\xi(t_1), \ldots, \xi(t_{k-1})\} = \mathbf{P}\{\xi(t_k) \in A/\xi(t_{k-1})\} \tag{4.2}$$

is fulfilled. If a function $\mathbf{P}(t_1, x, t_2, A)$ defined for all $t_1 < t_2, t_1, t_2 \in [t_0, T]$, $x \in R^{(m)}$ and all Borel sets A in $R^{(m)}$ is measurable with respect to x, and if it is a measure with respect to A such that with probability 1

$$\mathbf{P}\{\xi(t_2) \in A/\xi(t_1)\} = \mathbf{P}(t_1, \xi(t_1), t_2, A),$$

then such a function is called a *transition probability function of the Markov process* $\xi(t)$. If a Markov process $\xi(t)$ has a transition probability function $\mathbf{P}(t_1, x, t_2, A)$, then the finite-dimensional distributions of the process $\xi(t)$ for $t_1 < t_2 < \cdots < t_k$ are determined by the formula

$$\mathbf{P}\{\xi(t_1) \in A_1, \xi(t_2) \in A_2, \ldots, \xi(t_k) \in A_k\}$$
$$= \int m_0\,(dx_0) \int_{A_1} \mathbf{P}(t_0 x_0, t_1, dx_1) \ldots \int_{A_{k-1}} \mathbf{P}(t_{k-1}, x_{k-1}, t_k, A_k), \tag{4.3}$$

where $m_0(A)$ is the distribution of the variable $\xi(t_0)$.

A sequence of random variables $\xi_1, \xi_2, \ldots, \xi_n$ forms a *Markov chain*, if for every $k \leq n$ and every Borel set A in the region of values ξ_k, the relation

$$\mathbf{P}\{\xi_k \in A/\xi_1, \xi_2, \ldots, \xi_{k-1}\} = \mathbf{P}\{\xi_k \in A/\xi_{k-1}\}$$

holds with probability 1.

If there are functions $\mathbf{P}_k(x, A)$ that are measurable with respect to x and that are measures with respect to A such that with probability 1 the relation

$$\mathbf{P}\{\xi_k \in A_k/\xi_{k-1}\} = \mathbf{P}_k(\xi_{k-1}, A)$$

holds, then these functions are called *transition probabilities of the Markov chain* $\xi_1, \xi_2, \ldots, \xi_n$.

5. Martingales. A sequence $\xi_1, \xi_2, \ldots, \xi_n$ of random variables in $R^{(m)}$ is called a *martingale* if for every n, $\mathbf{M}|\xi_n| < \infty$, and with probability 1,

$$\mathbf{M}(\xi_n/\xi_1, \xi_2, \ldots, \xi_{n-1}) = \xi_{n-1}.$$

We note certain important properties of sequences of martingales:

1. For every $C > 0$,

$$\mathbf{P}\left\{\sup_{1 \leq k \leq n} |\xi_k| > C\right\} \leq \frac{1}{C}\mathbf{M}|\xi_n|.$$

2. If for some $\alpha > 1$, it is true for every $k = 1, 2, \ldots, n$ that

$$\mathbf{M}|\xi_k|^\alpha < \infty,$$

then

$$\mathbf{M}\sup_k |\xi_k|^\alpha \leq \left(\frac{\alpha}{\alpha - 1}\right)^\alpha \mathbf{M}|\xi_n|^\alpha.$$

A process $\xi(t)$ defined on $[t_0, T]$ and assuming values in $R^{(m)}$ is called a martingale if for every t, $\mathbf{M}|\xi(t)| < \infty$, and for $s < t$, with probability 1,

$$\mathbf{M}\big(\xi(t)/\xi(u), u \in [t_0, s]\big) = \xi(s).$$

Martingales—processes and sequences—will be used frequently in the following chapters; for this we shall need the following properties of martingale processes:

Suppose that $\xi(t)$ is a separable martingale defined on $[t_0, T]$:

3. As a function of t with probability 1, $\xi(t)$ cannot have discontinuities of the second kind.

4. For every $C > 0$, the inequality

$$\mathbf{P}\left\{\sup_{t_0 \leq t \leq T} |\xi(t)| > C\right\} \leq \frac{1}{C}\mathbf{M}|\xi(T)|$$

holds.

5. If for some $\alpha > 0$, it is true for all $\mathbf{M}|\xi(t)|^\alpha < \infty$ that $t \in [t_0, T]$, then

$$\mathbf{M}\left(\sup_{t_0 \leq t \leq T} |\xi(t)|^\alpha\right) \leq \left(\frac{\alpha}{\alpha - 1}\right)^\alpha \mathbf{M}|\xi(T)|^\alpha.$$

6. If $\xi(t)$ is a martingale with numerical values and $\xi^2(t) - t$ is also a martingale, then $\xi(t)$ is a Brownian motion process.

7. If $\xi(t)$ with probability 1 is a continuous martingale and $\tau(s)$, $s \in [\alpha, \beta]$ with probability 1 is a nondecreasing process for which the event $\{\tau(s) > t\}$ at every $s \in [\alpha, \beta]$ is contained in $\mathbf{F}([t_0, T], \xi(u))$, then the process $\xi(\tau(s))$ is also a martingale.

We also note that the definition of a martingale (process) given above is equivalent to the following: A process $\xi(t)$ defined for $t \in [t_0, T]$ with values in $R^{(m)}$ is called a martingale if to every $t \in [t_0, T]$ there corresponds a σ-algebra $\mathbf{F}_t$ relative to which all the quantities $\xi(s)$ are measurable for $s \leq t$ and which possesses the property that for $t_1 < t_2$,

$$\mathbf{M}(\xi(t_2)/\mathbf{F}_{t_1}) = \xi(t_1).$$

6. A limit theorem for random processes. In a given problem, we often need to deal with sequences of processes whose finite distributions converge to a particular limiting distribution. It turns out that under rather broad assumptions, we may treat the processes that we are examining as converging to some limiting process in probability. More precisely, we have the following theorem:

THEOREM. *Suppose that a sequence of random processes*

$$\xi_n(t), \quad n = 1, 2, \ldots,$$

defined for $t \in [t_0, T]$ and taking values in $R^{(m)}$, is stochastically continuous from the right at every point $t \in [t_0, T]$ such that for every k and $t_1, t_2, \ldots, t_k$ in $[t_0, T]$, the joint distribution of the values of $\xi_n(t_1), \xi_n(t_2), \ldots, \xi_n(t_k)$ converges weakly to a particular limiting distribution, and that for every $\epsilon > 0$,

$$\lim_{h \to 0} \lim_{n \to \infty} \sup_{|s_1 - s_2| \leq h} \mathbf{P}\{|\xi_n(s_1) - \xi_n(s_2)| > \epsilon\} = 0. \tag{6.1}$$

Then it is possible to construct a sequence of random processes $x_n(t, \omega')$, $n = 0, 1, 2, \ldots$, on the probability field $(\Omega', \mathbf{F}', \mathbf{P}')$, where Ω' is the interval $[0, 1]$, $\mathbf{F}'$ is a σ-algebra of Borel sets of the interval $[0, 1]$, and $\mathbf{P}'$ is the Lebesgue measure on $[0, 1]$, such that these processes $x_n(t, \omega')$ will have *the following properties:* $x_0(t, \omega')$ *will be stochastically continuous,* $x_n(t, \omega')$ *will converge for every* t *in probability to* $x_0(t, \omega')$, *and for* $n > 0$, *finite distributions of the processes* $x_n(t, \omega')$ *and* $\xi_n(t)$ *will coincide.*

To prove this theorem, we need a lemma.

LEMMA. *Let* R *be a separable complete metric space, let* $\rho(x, y)$ *be the distance between the elements* x *and* y *in* R, *and let* $\mathbf{B}$ *be a* σ*-algebra of Borel sets in the space* R. *If a sequence of measures* $\mu_n(A)$ *which are given on* $\mathbf{B}$ *and which satisfy the condition* $\mu_n(R) = 1$ *converges weakly to the measure* μ_0, *also defined on* $\mathbf{B}$, *then in the probability field* $(\Omega', \mathbf{F}', \mathbf{P}')$ *it is possible to construct a sequence of random variables* $x_n(\omega')$ *such that* $x_n(\omega') \to x_0(\omega')$ *in probability and for every* A *in* $\mathbf{B}$,

$$\mathbf{P}\{x_n(\omega) \in A\} = \mu_n(A).$$

Proof. We define the Borel sets $S_{i_1,i_2,\ldots,i_k}$ for all natural numbers $k, i_1, i_2, \ldots, i_k$ such that the following conditions are fulfilled:

1. $S_{i_1,i_2,\ldots,i_k}$ and $S_{i'_1,i'_2,\ldots,i'_k}$ do not have any common points with $i_k \neq i'_k$.
2. $\bigcup_{i_k=1}^{\infty} S_{i_1,i_2,\ldots,i_{k-1},i_k} = S_{i_1,i_2,\ldots,i_{k-1}}$; $\bigcup_{i=1}^{\infty} S_i = R$.
3. The diameter of $S_{i_1,i_2,\ldots,i_k}$ does not exceed $(\frac{1}{2})^k$.
4. For every $i_1, i_2, \ldots, i_k$,

$$\mu_0(\overline{S_{i_1,i_2,\ldots,i_k}} \cap (\overline{R \setminus S_{i_1,i_2,\ldots,i_k}})) = 0.$$

We also define subintervals of the interval $[0, 1]$, $\Delta^{(n)}_{i_1,i_2,\ldots,i_k}$ for all $n = 0, 1, 2, \ldots$ and for all natural numbers $k, i_1, i_2, \ldots, i_k$ such that the following conditions are fulfilled:

1. The intervals $\Delta^{(n)}_{i_1,i_2,\ldots,i_k}$ and $\Delta^{(n)}_{i'_1,i_2,\ldots,i'_k}$ do not intersect for $i_k \neq i'_k$.

2. The interval $\Delta^{(n)}_{i_1,i_2,\ldots,i_k}$ is to the left of the interval $\Delta^{(n)}_{i'_1,i'_2,\ldots,i'_k}$ if there exists an r such that $i_j = i'_j$ for $j < r$ and $i_r < i'_r$.

3. The length of the interval $\Delta^{(n)}_{i_1,i_2,\ldots,i_k}$ is equal to $\mu_n(S_{i_1,i_2,\ldots,i_k})$. In each of the sets $S_{i_1,i_2,\ldots,i_n}$, we choose one point $\bar{x}_{i_1,i_2,\ldots,i_k}$ and set

$$x_n^m(\omega') = \bar{x}_{i_1,i_2,\ldots,i_m} \qquad \text{for} \qquad \omega' \in \Delta^{(n)}_{i_1,i_2,\ldots,i_m}.$$

Then for all ω' with the exception of a countable set of endpoints of intervals $\Delta^{(n)}_{i_1,i_2,\ldots,i_m}$, the relationship

$$\rho(x_n^m(\omega'), x_n^{m+p}(\omega')) \leq (\tfrac{1}{2})^m$$

holds. Since R is a complete space, the limit $\lim_{m\to\infty} x_n^m(\omega')$ exists for almost every ω'. We designate this limit by $x_n(\omega')$. Since the length of the interval $\Delta^{(n)}_{i_1,i_2,\ldots,i_m}$ approaches the length of the interval $\Delta^{(0)}_{i_1,i_2,\ldots,i_m}$ as $n \to \infty$, it follows that for $\omega' \in \Delta^{(0)}_{i_1,i_2,\ldots,i_m}$,

$$\lim_{n\to\infty} \rho(x_n^m(\omega'), x_0^m(\omega')) = 0,$$

which means that no matter what m is,

$$\lim_{n\to\infty} \rho(x_n(\omega'), x_0(\omega')) \leq \frac{1}{2^{m-1}}.$$

Thus $x_n(\omega') \to x_0(\omega')$ with probability 1 and hence in probability. It remains to show that

$$\mathbf{P}\{x_n(\omega') \in A\} = \mu_n(A).$$

For this, it is sufficient to establish the relationship indicated for those A for which $\mu_n(\overline{A} \cap \overline{R\backslash A}) = 0$, where $\overline{A}$ denotes the closure of the set A.

Let C_m indicate the set of points for which the distance from $\overline{A} \cap \overline{R\backslash A}$ does not exceed $(\frac{1}{2})^m$. Then, indicating by A_m the sum of all the sets $S_{i_1,i_2,\ldots,i_m}$ entirely contained in $A\backslash C_{m+1}$, we get $A \supset A_m \supset A\backslash C_m$,

$$\mathbf{P}\{x_n(\omega') \in A\} \geq \mathbf{P}\{x_n^{(m+1)}(\omega') \in A_m\}$$
$$= \mu_n(A_m) \geq \mu_n(A) - \mu_n(C_m).$$

Since $\lim_{m\to\infty} \mu_n(C_m) = 0$, we have $\mathbf{P}\{x_n(\omega') \in A\} \geq \mu_n(A)$. Analogously,

$$\mathbf{P}\{x_n(\omega') \in R\backslash A\} \geq \mu_n(R\backslash A).$$

But

$$(\mathbf{P}\{x_n(\omega') \in A\} - \mu_n(A)) + (\mathbf{P}\{x_n(\omega') \in R\backslash A\} - \mu_n(R\backslash A)) = 1 - 1 = 0.$$

Consequently

$$(\mathbf{P}\{x_n(\omega') \in A\} = \mu_n(A)).$$

This proves the lemma.

We now proceed to prove the theorem. We choose a sequence of points $t_1, t_2, \ldots$ that is dense everywhere on $[t_0\ T]$ and that includes the point T. Let R be the metric space of sequences $x = \{x_1, x_2, \ldots\}$, where $x_k \in R^m$, with the metric

$$\rho(x, y) = \sum_{k=1}^{\infty} \frac{1}{k!} (1 - e^{-|x_k - y_k|}),$$

if $x = \{x_1, x_2, \ldots\}$, $y = \{y_1, y_2, \ldots\}$.

Let μ_n be a measure on the Borel sets of R such that no matter what the Borel sets $A_1, A_2, \ldots, A_k$ in $R^{(m)}$ may be, the set $C \subset R$ will contain those and only those points $x = \{x_1, x_2, \ldots\}$ for which the following relationships with $x_1 \in A_1, x_2 \in A_2, \ldots x_k \in A_k$, are satisfied:

$$\mu_n(C) = \mathbf{P}\{\xi_n(t_1) \in A_1, \ldots, \xi_n(t_k) \in A_k\}$$

and

$$\mu_0(C) = F_{t_1,t_2,\ldots,t_k}(A_1, A_2, \ldots, A_k),$$

where $F_{t_1,t_2,\ldots,t_k}(A_1, A_2, \ldots, A_k)$ is the limiting distribution to which the simultaneous distribution of the variables $\xi_n(t_1), \ldots, \xi_n(t_k)$ converges. It is easy to see that the measures μ_n satisfy the conditions of the lemma. Therefore it is possible to construct on the probability field $\{\Omega', \mathbf{F}', \mathbf{P}'\}$ a sequence of quantities $x_n(\omega') = \{x_1^{(n)}(\omega'), x_2^{(n)}(\omega'), \ldots,\}$ $n = 0, 1, 2, \ldots,$ such that $x_n(\omega') \to x_0(\omega')$ in probability [that is, $x_k^{(n)}(\omega') \to x_k^{(0)}(\omega')$ in probability for every k], and the measures μ_n will serve as distributions of the variables $x_n(\omega')$; that is, for every n and all Borel sets $A_1, A_2, \ldots, A_k$ in $R^{(m)}$, the equation

$$\mathbf{P}\{x_1^{(n)}(\omega') \in A_1, \ldots, x_k^{(n)}(\omega') \in A_k\} = \mathbf{P}\{\xi_n(t_1) \in A_1, \ldots, \xi_n(t_k) \in A_k\}$$

is valid. Let $t_{k_r} \to t$, $t_{k_r} > t$; then $\xi_n(t_{k_r}) \to \xi_n(t)$ in probability; this means that $x_{k_r}^{(n)}(\omega')$ will also converge in probability to some particular value. If t is not an element of the sequence $t_1, t_2, \ldots,$ we shall indicate this limit by $x^{(n)}(t, \omega')$; if $t = t_k$, we obtain $x^{(n)}(t_k, \omega') = x_k^{(n)}(\omega')$. Thus finite-dimensional distributions of a process $x^{(n)}(t, \omega')$ set up in this manner will coincide with the finite-dimensional distributions of the process $\xi_n(t)$. Since

$$\mathbf{P}\{|\xi_n(t_k) - \xi_n(t_j)| > \epsilon\} = \mathbf{P}\{|x_k^{(n)}(\omega') - x_j^{(n)}(\omega')| > \epsilon\},$$

$$\mathbf{P}\{|x_k^{(0)} - x_j^{(0)}| > 2\epsilon\} \leq \overline{\lim_{n\to\infty}}\, \mathbf{P}\{|x_k^{(n)}(\omega') - x_j^{(n)}(\omega')| > \epsilon\},$$

according to (6.1), as $t_{k_r} \to t$ the sequence $x_{k_r}^{(0)}(\omega')$ converges in probability to some particular random variable. If we denote this limit by $x^{(0)}(t, \omega')$, then by again making use of (6.1) we see that $x^{(0)}(t, \omega')$ is a stochastically continuous process and that $\mathbf{P}\{x^{(0)}(t_k, \omega')\} = x_k^{(0)}(\omega')\} = 1$. We shall show that for all t, $x^{(n)}(t, \omega')$ converges in probability to $x^0(t, \omega')$. In fact,

$$\begin{aligned}\overline{\lim_{n\to\infty}}\, \mathbf{P}\{|x^{(0)}(t, \omega') - x^{(n)}(t, \omega')| > \epsilon\} &\leq \mathbf{P}\left\{|x^{(0)}(t, \omega') - x^{(0)}(t_k, \omega') > \frac{\epsilon}{3}\right\} \\ &\quad + \overline{\lim_{n\to\infty}}\, \mathbf{P}\left\{|x^{(0)}(t_k, \omega') - x^{(n)}(t_k, \omega')| > \frac{\epsilon}{3}\right\} \\ &\quad + \overline{\lim_{n\to\infty}}\, \mathbf{P}\left\{|x^{(n)}(t, \omega') - x^{(n)}(t_k, \omega')| > \frac{\epsilon}{3}\right\} \\ &= \mathbf{P}\left\{|x^{(0)}(t, \omega') - x^{(0)}(t_k, \omega')| > \frac{\epsilon}{3}\right\} \\ &\quad + \overline{\lim_{n\to\infty}}\, \mathbf{P}\left\{|x^{(n)}(t, \omega') - x^{(n)}(t_k, \omega')| > \frac{\epsilon}{3}\right\}\end{aligned}$$

We can make the right-hand side of the last inequality as small as we like by a sufficiently small choice of $t_k - t$ by using (6.1) and the stochastic continuity of $x^{(0)}(t, \omega')$. This proves the theorem.

Remark 1. In the hypothesis of the theorem, we can replace stochastic continuity from the right with stochastic continuity from the left.

Remark 2. Let there be a sequence of processes $\xi_n(t)$ that are stochastically continuous from the right (left) and that satisfy the conditions (6.1). Suppose that

$$\lim_{C\to\infty} \lim_{n\to\infty} \sup_{t_0 \le t \le T} \mathbf{P}\{|\xi_n(t)| > C\} = 0. \tag{6.2}$$

Then it is possible to exhibit a sequence n_k such that the limiting distributions for finite-dimensional distributions of the processes $\xi_{n_k}(t)$ exist when $k \to \infty$. In fact, the conditions (6.2) imply compactness of the sequence of distributions of the quantities $\xi_n(t_1), \ldots, \xi_n(t_l)$ no matter what $t_1, t_2, \ldots, t_l$ and l are. By using the diagonal method of Cantor, it is possible, for an arbitrary sequence, to construct a sequence n_k such that the common distribution of the variables $\xi_{n_k}(t_1), \ldots, \xi_{n_k}(t_l)$ for arbitrary l will converge to a certain limiting distribution. Making use of the stochastic continuity of the processes $\xi_n(t)$ (from the right or left) and (6.1), it is possible to show that for all $t_1', t_2', \ldots, t_r'$ in (t_0, T), the joint distribution of the variables $\xi_{n_k}(t_1'), \ldots, \xi_{n_k}(t_r')$ will converge to some particular limiting distribution.

COROLLARY 1. *If the processes $\xi_n(t)$ satisfy the conditions of Remark 2, then for some sequence n_k it will be possible to set up processes $x_{n_k}(t, \omega')$ on the probability field $\{\Omega, \mathbf{F}', \mathbf{P}'\}$ having the same finite-dimensional distributions as $\xi_{n_k}(t)$ and converging in probability to some particular process $x_0(t, \omega')$ as $k \to \infty$.*

COROLLARY 2. *If there are r sequences of processes $\xi_n^{(1)}(t), \ldots, \xi_n^{(r)}(t)$ and if each sequence satisfies the conditions of Remark 2, then for every sequence n_k it is possible to define processes $x_{n_k}^{(1)}(t, \omega'), \ldots, x_{n_k}^{(r)}(t, \omega')$ on the probability field $\{\Omega', \mathbf{F}', \mathbf{P}'\}$ the joint finite-dimensional distributions of which are the same as those of the processes $\xi_{n_k}^{(1)}(t), \ldots, \xi_{n_k}^{(r)}(t)$ and each of the sequences $\xi_{n_k}^{(1)}(t), \ldots, \xi_{n_k}^{(r)}(t)$ will converge in probability to a certain limit.*

This assertion is obtained as a consequence of the preceding by examining the composite process $\{\xi_n^{(1)}(t), \ldots, \xi_n^{(r)}(t)\}$, that is, a process in the space of points of the form $\{x_1, x_2, \ldots, x_r\}$, where x_i is a point in the space of the values of $\xi_i(t)$.

In conclusion, we shall make certain observations concerning the use of the material of this chapter in subsequent chapters.

The probability field will everywhere be considered fixed (except for those cases in which the theorem of Section 6 and the field $\{\Omega', \mathbf{F}', \mathbf{P}'\}$ examined in Section 6 are involved); therefore the dependence on the elementary event ω of random variables and processes will not be indicated.

All the processes that we shall examine will be assumed to be measurable and separable. If values of a process $\xi(t)$ are determined for every t by means of some limiting transition, insofar as $\xi(t)$ is determined for every t up to a set of measure zero in the variable ω, the process $\xi(t)$ will be determined up to stochastic equivalence. Therefore we may always assume that in the results of the limiting transition by means of which $\xi(t)$ is defined a separable and measurable process will be obtained. (Measurability will be a consequence of the fact that all processes that we shall examine are stochastically continuous everywhere except possibly at a finite number of points.) The properties of measurability and separability will not be expressly stipulated but we shall always use them.

CHAPTER 2

STOCHASTIC INTEGRALS

1. The definition of a stochastic integral with respect to a Brownian motion. Let $w(t)$ be a Brownian motion defined on $[t_0, T]$. In this section, we shall define for an extremely broad class of functions the integral

$$\int_{t_0}^{T} f(t)\, dw(t), \tag{1.1}$$

which we shall call the stochastic integral with respect to a Brownian motion. We note that though a process $w(t)$ is continuous with probability 1, it is of unbounded variation with probability 1 on an arbitrary interval, and therefore the integral (1.1) cannot be taken as an ordinary Stieltjes integral.

Let us examine the set of σ-algebras $\mathbf{F}_t$ defined for all $[t_0, T]$ and satisfying the conditions:

(a) for $t_1 < t_2$, $\mathbf{F}_{t_1} \subset \mathbf{F}_{t_2}$;

(b) for every t, $w(t)$ is measurable with respect to $\mathbf{F}_t$;

(c) no matter what $t \in [t_0, T]$ and the event $\mathbf{A} \in \mathbf{F}_t$ may be, the variables $w(s_1) - w(t), \ldots, w(s_k) - w(t)$ do not depend on $\mathbf{A}$ (in the probability sense) if $s_1, \ldots, s_k \in [t, T]$. [Conditions (a), (b), and (c) will be fulfilled in the case where $\mathbf{F}_t$ is the minimal σ-algebra containing all events in $\mathbf{F}([t_0, t], w(s))$ and in some σ-algebra $\mathbf{F}$ such that for arbitrary $s_1, s_2, \ldots, s_k$ in $[t_0, T]$ the variables $w(s_1), \ldots, w(s_k)$ do not depend on every event in $\mathbf{F}$.]

We indicate by $M(\mathbf{F}_t)$ the set of functions $f(t)$ for which, for every $t \in [t_0, T]$, $f(t)$ is measurable with respect to $\mathbf{F}_t$. (In this chapter we shall call random processes functions.)

A function $f(t)$ will be called a step function if there exist points $t_0 < t_1 < \cdots < t_n = T$ such that $f(t) = f(t_i)$ for $t \in (t_i, t_{i+1})$. The class of all step functions belonging to $M(\mathbf{F}_t)$ will be denoted by $M_0(\mathbf{F}_t)$.

If $f(t) \in M(\mathbf{F}_t)$ and

$$\int_{t_0}^{T} \mathbf{M}|f(t)|^2\, dt < \infty,$$

we shall relate $f(t)$ to the class $M_1(\mathbf{F}_t)$. If $f(t)$ is square-integrable with probability 1 as a function of t on an interval $[t_0, T]$, that is, if

$$\mathbf{P}\left\{\int_{t_0}^{T} |f(t)|^2\,dt < \infty\right\} = 1,$$

then we shall say that $f(t)$ is in the class $M_2(\mathbf{F}_t)$.

Our purpose is to define the integral (1.1) for all $f(t) \in M_2(\mathbf{F}_t)$. Assume that $f(t) \in M_0(\mathbf{F}_t)$, that is, that there exist points $t_0 < t_1 < \cdots < t_n = T$ such that $f(t) = f(t_i)$ for $t_i \le t < t_{i+1}$. Then it is natural to set

$$\int_{t_0}^{T} f(t)\,dw(t) = \sum_{j=0}^{n-1} f(t_j)\big(w(t_{j+1}) - w(t_j)\big). \tag{1.2}$$

The following conditions are satisfied for the integral (1.2):

1. If

$$f_1(t) \in M_0(\mathbf{F}_t) \qquad \text{and} \qquad f_2(t) \in M_0(\mathbf{F}_t),$$

then

$$\int_{t_0}^{T} [\alpha f_1(t) + \beta f_2(t)]\,dw(t) = \alpha\int_{t_0}^{T} f_1(t)\,dw(t) + \beta\int_{t_0}^{T} f_2(t)\,dw(t), \tag{1.3}$$

no matter what the real numbers α and β may be.

2. Assume that $f(t) \in M_0(\mathbf{F}_t)$. Define $\psi(t) = 0$ if $\max_{t_0 \le s \le t} |f(s)| = 0$ and $\psi(t) = 1$ if $\max_{t_0 \le s \le t} |f(s)| > 0$. Then

$$\left|\int_{t_0}^{T} f(t)\,dw(t)\right| \le \psi(T)\left|\int_{t_0}^{T} f(t)\,dw(t)\right|. \tag{1.4}$$

3. If

$$f(t) \in M_0(\mathbf{F}_t) \cap M_1(\mathbf{F}_t),$$

then

$$\mathbf{M}\int_{t_0}^{T} f(t)\,dw(t) = 0, \tag{1.5}$$

and

$$\mathbf{M}\left|\int_{t_0}^{T} f(t)\,dw(t)\right|^2 = \int_{t_0}^{T} \mathbf{M}|f(t)|^2\,dt. \tag{1.6}$$

The proof of the last two formulas follows from the facts that $w(t_{j+1}) - w(t_j)$ does not depend upon $f(t_j)$, $w(t_{k+1}) - w(t_k)$, $f(t_k)$ for $k < j$, and that $\mathbf{M}\big(w(t_{j+1}) - w(t_j)\big) = 0$, $\mathbf{M}\big(w(t_{j+1}) - w(t_j)\big)^2 = t_{j+1} - t_j$.

Let us show that for every $f(t) \in M_1(\mathbf{F}_t)$ and $\epsilon > 0$, there exists $\bar{f}(t) \in M_0(\mathbf{F}_t) \cap M_1(\mathbf{F}_t)$ such that

$$\int_{t_0}^{T} \mathbf{M}|f(t) - \bar{f}(t)|^2\,dt < \epsilon. \tag{1.7}$$

Let us set $g_N(x) = 1$ for $|x| \leq N$ and $g_N(x) = 0$ for $|x| > N$. Then for $f(t) \in M_1(\mathbf{F}_t)$, the relation

$$\lim_{N\to\infty} \int_{t_0}^{T} \mathbf{M}|f(t) - f(t)g_N(f(t))|^2\, dt = 0 \tag{1.8}$$

is satisfied, because $|f(t) - f(t)g_N(f(t))| \leq |f(t)|$, Lebesgue's theorem on interchanging limit and integral being applicable.

From (1.8), we see that it is sufficient to prove (1.7) for bounded functions in $M_1(\mathbf{F}_t)$. Let $f(t)$ be a bounded function in $M_1(\mathbf{F}_t)$ and let $\alpha(t)$ be a real function that is bounded, nonnegative, continuous, equal to zero for $t \geq 0$, and that satisfies the relation

$$\int_{-\infty}^{\infty} \alpha(t)\, dt = 1.$$

Assume also that $f(t)$ is equal to $f(t_0)$ for $t \leq t_0$ and let

$$f_n(t) = n\int_{-\infty}^{\infty} \alpha(n(t - s))f(s)\, ds.$$

The $f_n(t)$ are bounded by the same constant as $f(t)$ and are continuous with probability 1. Also, since $f(t)$ is measurable with respect to t for almost all ω, for almost all $\omega \in \Omega f_n(t) \to f(t)$ and $t \in [t_0, T]$ [for all t that are points of asymptotic continuity of $f(t)$ as a function of t for given ω]. Therefore, on the basis of the theorem of Lebesgue mentioned above,

$$\lim_{n\to\infty} \int_{t_0}^{T} \mathbf{M}|f_n(t) - f(t)|^2\, dt = 0.$$

Thus it is sufficient to establish (1.7) for those $f(t)$ in $M_1(\mathbf{F}_t)$ that are bounded and continuous with probability 1. But for such functions $f(t)$, the step functions $\bar{f}_n(t) = f\left(t_0 + \frac{1}{n}[n(t - t_0)]\right)$, where $[\alpha]$ is the integral part of α, belong to $M_0(\mathbf{F}_t) \cap M_1(\mathbf{F}_t)$ and

$$\lim_{n\to\infty} \int_{t_0}^{T} \mathbf{M}|\bar{f}_n(t) - f(t)|^2\, dt = 0$$

(since $\bar{f}_n(t) \to f(t)$ with probability 1 for all $t \in [t_0, T]$). Consequently, if we take $\bar{f}(t)$ equal to $f_n(t)$ for sufficiently large n, the inequality (1.7) will be satisfied.

From (1.7) it follows that for every $f(t) \in M_1(\mathbf{F}_t)$, there exists a sequence $f_n(t) \in M_0(\mathbf{F}_t) \cap M_1(\mathbf{F}_t)$ such that

$$\lim_{n\to\infty} \int_{t_0}^{T} \mathbf{M}|f(t) - f_n(t)|^2\, dt = 0. \tag{1.9}$$

If a sequence $f_n(t)$ satisfies (1.9), it follows from (1.3) and (1.6) that

$$\lim_{\substack{n\to\infty \\ m\to\infty}} \mathbf{M}\left|\int_{t_0}^{T} f_n(t)\, dw(t) - \int_{t_0}^{T} f_m(t)\, dw(t)\right|^2 = 0.$$

Thus a sequence of random variables $\int_{t_0}^{T} f_n(t)\, dw(t)$ will converge in probability to some particular random variable. We shall denote this random variable by $\int_{t_0}^{T} f(t)\, dw(t)$. This integral is uniquely determined up to a set of measure zero in the variable ω. In fact, if a sequence $f_n'(t)$ also satisfies (1.9), then

$$\lim_{n\to\infty} \int_{t_0}^{T} \mathbf{M}|f_n'(t) - f_n(t)|^2\, dt = 0,$$

and consequently the limits of the sequences

$$\int_{t_0}^{T} f_n'(t)\, dw(t) \qquad \text{and} \qquad \int_{t_0}^{T} f_n(t)\, dw(t)$$

coincide with probability 1. For $\int_{t_0}^{T} f(t)\, dw(t)$, if $f(t) \in M_1(\mathbf{F}_t)$, Properties 1 to 3 which were established for the case in which $f(t) \in M_0(\mathbf{F}_t)$ are preserved. (By passing to the limit, we can easily show this in the corresponding equalities and inequalities.) From Property 1, we obtain

4. $$\mathbf{P}\left\{\left|\int_{t_0}^{T} f(t)\, dw(t)\right| > 0\right\} \le \mathbf{P}\left\{\sup_{t_0 \le t \le T} |f(t)| > 0\right\}. \tag{1.10}$$

Let us now extend the definition of the integral (1.1) to all functions in $M_2(\mathbf{F}_t)$. Suppose as above that $g_N(x) = 1$ for $|x| \le N$ and $g_N|x| = 0$ for $|x| > N$. Then if $f(t) \in M_2(\mathbf{F}_t)$, the function

$$f_N(t) = f(t) g_N\left(\int_{t_0}^{T} |f(s)|^2\, ds\right)$$

belongs to $M_1(\mathbf{F}_t)$, since

$$\int_{t_0}^{T} \mathbf{M}|f_N(t)|^2\, dt \le \mathbf{M}\int_{t_0}^{T} |f_N(t)|^2\, dt \le N.$$

We note also that with $N' > N$

$$\sup_{t_0 \le t \le T} |f_N(t) - f_{N'}(t)| = 0$$

if

$$\int_{t_0}^{T} |f(s)|^2\, ds < N.$$

Therefore, on the basis of (1.10), we can write

$$\mathbf{P}\left\{\left|\int_{t_0}^{T} f_N(t)\, dw(t) - \int_{t_0}^{T} f_{N'}(t)\, dw(t)\right| > 0\right\} \le \mathbf{P}\left\{\int_{t_0}^{T} |f(t)|^2\, dt \ge N\right\}.$$

Since $\int_{t_0}^{T} |f(t)|^2\, dt$ is a well-defined random variable,

$$\lim_{N\to\infty} \mathbf{P}\left\{\int_{t_0}^{T} |f(t)|^2\, dt \ge N\right\} = 0.$$

This means that $\int_{t_0}^{T} f_N(t)\, dw(t)$ converges in probability to some random variable which we shall denote by $\int_{t_0}^{T} f(t)\, dw(t)$. By use of the definition of $\int_{t_0}^{T} f(t)\, dw(t)$, it is easy to establish that Properties 1 and 4 are preserved for this integral. Furthermore, for this integral it is true that:

5. If $f_n(t) \in M_2(\mathbf{F}_t)$ and $f_n(t) \to f(t)$ in probability for almost all $t \in [t_0, T]$, and if there exists a $\varphi(t) \in M_2(\mathbf{F}_t)$ for which $|f_n(t)|^2 \le |\varphi(t)|^2$, then

$$\int_{t_0}^{T} f_n(t)\, dw(t) \to \int_{t_0}^{T} f(t)\, dw(t)$$

in probability.

Proof. We note that for every N,

$$f_n(t) g_N\left(\int_{t_0}^{t} |\varphi(s)|^2\, ds\right) \quad \text{and} \quad f(t) g_N\left(\int_{t_0}^{t} |\varphi(s)|^2\, ds\right)$$

belong to $M_1(\mathbf{F}_t)$; therefore

$$\mathbf{M}\left|\int_{t_0}^{T} f_n(t) g_N\left(\int_{t_0}^{t} |\varphi(s)|^2\, ds\right) dw(t) - \int_{t_0}^{T} f(t) g_N\left(\int_{t_0}^{t} |\varphi(s)|^2\, ds\right) dw(t)\right|^2$$

$$\le M\int_{t_0}^{T} |f_n(t) - f(t)|^2 g_N\left(\int_{t_0}^{t} |\varphi(s)|^2\, ds\right) dt \to 0,$$

on the basis of the same theorem of Lebesgue, because

$$|f_n(t) - f(t)|^2 g_N\left(\int_{t_0}^{t} |\varphi(s)|^2\, ds\right) \to 0$$

with respect to the measure that is the product of the Lebesgue measure on $[t_0, T]$ and the measure $\mathbf{P}$, and because

$$|f_n(t) - f(t)|^2 g_N\left(\int_{t_0}^{t} |\varphi(s)|^2\, ds\right) \le 4|\varphi(t)|^2 g_N\left(\int_{t_0}^{t} |\varphi(s)|^2\, ds\right),$$

but

$$\mathbf{M}\int_{t_0}^{T} |\varphi(t)|^2 g_N\left(\int_{t}^{t} |\varphi(s)|^2\, ds\right) dt \le N.$$

Also, $$\mathbf{P}\left\{\left|\int_{t_0}^{T} f(t)\,dw(t) - \int_{t_0}^{T} f_n(t)g_N\left(\int_{t_0}^{t} |\varphi(s)|^2\,ds\right) dw(t)\right| > 0\right\} \le \mathbf{P}\left\{\int_{t_0}^{T} |\varphi(s)|^2\,ds \ge N\right\},$$

$$\mathbf{P}\left\{\left|\int_{t_0}^{T} f_n(t)\,dw(t) - \int_{t_0}^{T} f_n(t)g_N\left(\int_{t_0}^{T} |\varphi(s)|^2 ds\right) dw(t)\right| > 0\right\} \le \mathbf{P}\left\{\int_{t_0}^{T} |\varphi(s)|^2\,ds \ge N\right\}.$$

Therefore, no matter what N is, for every $\epsilon > 0$,

$$\overline{\lim_{n\to\infty}}\, \mathbf{P}\left\{\left|\int_{t_0}^{T} f(t)\,dw(t) - \int_{t_0}^{T} f_n(t)\,dw(t)\right| > \epsilon\right\} \le 2\mathbf{P}\left\{\int_{t_0}^{T} |\varphi(s)|^2\,ds > N\right\}.$$

Passing to the limit as $N \to \infty$, we obtain the proof of Property 5. By using this property, we can show that for the integral (1.1), Property 1 is satisfied when $f(t) \in M_2(\mathbf{F}_t)$. To do this, we need to apply Property 5 to the sequence

$$\bar{f}_n(t) = \alpha f_1(t) g_n\left(\int_{t_0}^{t} |f_1(s)|^2\,ds\right) + \beta f_2(t) g_n\left(\int_{t_0}^{t} |f_2(s)|^2\,ds\right),$$

which converges to $\alpha f_1(t) + \beta f_2(t)$ and which satisfies the inequality

$$|\bar{f}_n(t)|^2 \le (\alpha^2 + \beta^2)(|f_1(t)|^2 + |f_2(t)|^2).$$

2. The stochastic integral with respect to a Brownian motion as a function of the upper limit. Let $f(t) \in M_2(\mathbf{F}_t)$, $\chi_A(t)$ be the characteristic function of some Borel set A belonging to the interval $[t_0, T]$. Then $f(t)\chi_A(t) \in M_2(\mathbf{F}_t)$, which means that $\int_{t_0}^{T} f(t)\chi_A(t)\,dw(t)$ is defined, and we shall indicate it by $\int_A f(t)\,dw(t)$. In this case, if A is an interval $[t_1, t_2]$, instead of $\int_A f(t)\,dw(t)$ we shall write $\int_{t_1}^{t_2} f(t)\,dw(t)$. In this section, the integral $\int_{t_0}^{s} f(t)\,dw(t)$ will be considered as a function of t. As we have already noted, the values of $\int_{t_0}^{s} f(t)\,dw(t)$ for different values of s agree in such a way that $\int_{t_0}^{s} f(t)\,dw(t)$ as a function of s is a separable process. Applying Property 2 of Section 1 to the integral

$$\int_{t_0}^{t} f(s)\,dw(s),$$

we have the following theorem:

THEOREM 1. *Suppose that $f(t) \in M_2(\mathbf{F}_t)$ and $\psi(t) = 0$ if $\sup_{t_0 \le s \le t} |f(s)| = 0$ and that $\psi(t) = 1$ if $\sup_{t_0 \le s \le t} |f(s)| > 0$. Then*

$$\sup_{t_0 \le t \le T} \left|\int_{t_0}^{t} f(s)\,dw(s)\right| \le \psi(T) \sup_{t_0 \le t < T} \left|\int_{t_0}^{t} f(s)\,dw(s)\right|. \tag{2.1}$$

COROLLARY. *If* $\sup_{t_0 \le t \le T} |\int_{t_0}^t f(s)\,dw(s)| < \infty$, *then*

$$\mathbf{P}\left\{\sup_{t_0 \le t \le T}\left|\int_{t_0}^t f(s)\,dw(s)\right| > 0\right\} \le \mathbf{P}\left\{\sup_{t_0 \le t \le T}|f(t)| > 0\right\}. \tag{2.2}$$

THEOREM 2. *If* $f(t) \in M_1(\mathbf{F}_t)$, *then the process*

$$\zeta(t) = \int_{t_0}^t f(s)\,dw(s)$$

is a martingale.

Proof. It follows from the way the integral was defined that $Z(t)$ is measurable with respect to $\mathbf{F}_t$. Therefore, to prove the theorem it is sufficient to show that

$$\mathbf{M}\left(\int_t^{t+h} f(s)\,dw(s)/\mathbf{F}_t\right) = 0. \tag{2.3}$$

In the case in which $f(t) \in M_0(\mathbf{F}_t) \cap M_1(\mathbf{F}_t)$, that is, in which there exist points $t_0 < t_1 < \cdots < t_n = T$ such that $f(t) = f(t_i)$ for $t \in [t_i, t_{i+1}]$, (2.3) follows from the fact that for $t \le t_k$,

$$\begin{aligned}\mathbf{M}(f(t_k)[w(t_{k+1}) - w(t_k)](\mathbf{F}_t)) &= \mathbf{M}(f(t_k))\mathbf{M}(w(t_{k+1}) - w(t_k)/\mathbf{F}_{t_k})(\mathbf{F}_t)\\ &= 0,\end{aligned}$$

because

$$\mathbf{M}(w(t_{k+1}) - w(t_k)/\mathbf{F}_{t_k}) = 0.$$

If we now choose $f_n(t) \in M_0(\mathbf{F}_t) \cap M_1(\mathbf{F}_t)$ such that

$$\int_t^{t+h} f_n(s)\,dw(s) \to \int_t^{t+h} f(s)\,dw(s)$$

in probability, then since

$$\mathbf{M}\left(\int_t^{t+h} f_n(s)\,dw(s)/\mathbf{F}_t\right) \to \mathbf{M}\left(\int_t^{t+h} f(s)\,dw(s)/\mathbf{F}_t\right)$$

in probability, we obtain (2.3) in the general case. This proves the theorem.

THEOREM 3. *For every* $f(t) \in M_2(\mathbf{F}_t)$, *the process*

$$\zeta(t) = \int_{t_0}^t f(s)\,dw(s)$$

is a continuous process with probability 1.

Proof. If $f(t) \in M_0(\mathbf{F}_t) \cap M_1(\mathbf{F}_t)$ in view of the continuity with probability 1 of the process $w(t)$, the continuity of $\zeta(t)$ follows immediately from the formula

$$\zeta(t) = \sum_{t_{k+1} \le t} f(t_k)[w(t_{k+1}) - w(t_k)] + f(t)[w(t) - w(\sup_{t_k \le t} t_k)],$$

where $f(t) = f(t_i)$ for $t \in [t_i, t_{i+1}]$.

Now assume that $f(t) \in M_1(\mathbf{F}_t)$. In this case, it is possible to construct a sequence $f_n(t) \in M_0(\mathbf{F}_t) \cap M_1(\mathbf{F}_t)$ for which (1.9) is satisfied. We set

$$\zeta_n(t) = \int_{t_0}^{t} f_n(s)\, dw(s).$$

Then the $\zeta_n(t)$ are continuous with probability 1. Further,

$$\zeta_n(t) - \zeta(t) = \int_{t_0}^{t} [f_n(s) - f(s)]\, dw(s)$$

is a martingale. Therefore, on the basis of Property 5 of Section 5, Chapter 1, we obtain

$$\mathbf{M}\ (\sup_{t_0 \le t \le T} |\zeta_n(t) - \zeta(t)|^2) \le 4\mathbf{M}|\zeta_n(T) - \zeta(T)|^2 = 4\int_{t_0}^{T} \mathbf{M}|f_n(t) - f(t)|^2\, dt$$

(we set $\alpha = 2$).

We choose a sequence n_k such that the series

$$\sum_{k=1}^{\infty} k^2 \int_{t_0}^{T} \mathbf{M}|f_{n_k}(t) - f(t)|^2\, dt$$

converges. Then the series

$$\sum_{k=1}^{\infty} \mathbf{P}\left\{\sup_{t_0 \le t \le T} |\zeta_{n_k}(t) - \zeta(t)| > \frac{1}{k}\right\} \le \sum_{k=1}^{\infty} k^2 \mathbf{M}\ (\sup_{t_0 \le t \le T} |\zeta_{n_k}(t) - \zeta(t)|^2)$$

will converge; therefore, on the basis of the Borel-Cantelli lemma, we know that with probability 1 there exists an integer beyond which the inequalities

$$\sup_{t_0 \le t \le T} |\zeta_{n_k}(t) - \zeta(t)| \le \frac{1}{k}$$

hold; that is, with probability 1 the $\zeta_{n_k}(t)$, as functions of t, converge uniformly to $\zeta(t)$. Since $\zeta_{n_k}(t)$ are continuous with probability 1, $\zeta(t)$ is continuous with probability 1.

Finally, let us examine $f(t)$ in $M_2(\mathbf{F}_t)$. Suppose that $g_N(x) = 1$ for $|x| \le N$, and $g_N(x) = 0$ for $|x| > N$. Then, on the basis of the corollary to Theorem 1,

$$\mathbf{P}\left\{\sup_{t_0 \le t \le T} \left|\int_{t_0}^{t} f(s)\, dw(s) - \int_{t_0}^{t} f(s) g_N\left(\int_{t_0}^{s} |f(u)|^2\, du\right) dw(s)\right| > 0\right\} \le \mathbf{P}\left\{\int_{t_0}^{T} |f(t)|^2\, dt > N\right\}. \tag{2.4}$$

Since
$$f(t)g_N\left(\int_{t_0}^t |f(s)|^2\,ds\right) \in M_1(\mathbf{F}_t),$$
it follows that
$$\int_{t_0}^t f(s)g_N\left(\int_{t_0}^s |f(u)|^2\,du\right) dw(s)$$

is continuous with probability 1. This means that for every $\epsilon > 0$, it is possible to find a process which is continuous with probability 1 such that $\int_{t_0}^t f(s)\,dw(s)$ will differ from this process as a function of t only on a set of measure not greater than ϵ in the variable ω. This proves the theorem.

THEOREM 4. *Suppose that $f(t)$ and $|f(t)|^2$ belong to $M_1(\mathbf{F}_t)$. Then the relation*
$$\mathbf{M}\left|\int_{t_0}^T f(t)\,dw(t)\right|^4 \leq 36(T - t_0)\int_{t_0}^T \mathbf{M}|f(t)|^4\,dt$$
is satisfied.

Proof. Suppose that $f(t) = f(t_k)$ for $t \in [t_k, t_{k+1}]$, where $t_0 < t_1 < \cdots < t_n = T$ and $f(t)$ satisfies the conditions of the theorem. Then

$$\mathbf{M}\left|\int_{t_0}^T f(t)\,dw(t)\right|^4 = \sum_{k=0}^{n-1} \mathbf{M}|f(t_k)|^4 \mathbf{M}(w(t_{k+1}) - w(t_k))^4$$
$$+ 6\sum_{k=0}^{n-1} \mathbf{M}\left|\sum_{j=1}^{k-1} f(t_j)(w(t_{j+1}) - w(t_j))\right|^2 |f(t_k)|^2 \mathbf{M}(w(t_{k+1}) - w(t_k))^2$$
$$= 3\sum_{k=0}^{n-1} \mathbf{M}|f(t_k)|^4 (t_{k+1} - t_k)^2$$
$$+ 6\sum_{k=0}^{n-1} \mathbf{M}\left|\int_{t_0}^{t_k} f(t)\,dw(t)\right|^2 |f(t_k)|^2 (t_{k+1} - t_k).$$

Since for $f(t) \in M_0(\mathbf{F}_t)$ we may take $\max_k (t_{k+1} - t_k)$ as small as we choose, then by going to the limit as $\max_k (t_{k+1} - t_k) \to 0$, we obtain
$$\mathbf{M}\left|\int_{t_0}^T f(t)\,dw(t)\right|^4 = 6\int_{t_0}^T \mathbf{M}\left|\int_{t_0}^t f(s)\,dw(s)\right|^2 |f(t)|^2\,dt. \qquad (2.5)$$
Therefore
$$\mathbf{M}\left|\int_{t_0}^T f(t)\,dw(t)\right|^4 \leq 6\sqrt{\int_{t_0}^T \mathbf{M}\left|\int_{t_0}^t f(s)\,dw(s)\right|^4 dt} \times \sqrt{\int_{t_0}^T \mathbf{M}|f(t)|^4\,dt}$$
$$\leq 6\sqrt{(T - t_0)\mathbf{M}\left|\int_{t_0}^T f(t)\,dw(t)\right|^4} \times \sqrt{\int_{t_0}^T \mathbf{M}|f(t)|^4\,dt},$$
because, on the basis of (2.5), $\mathbf{M}|\int_{t_0}^t f(s)\,dw(s)|^4$ is a monotonically in-

creasing function of t. The proof of the theorem for the case in which $f(t) \in M_0(\mathbf{F}_t)$ follows from the last inequality. The proof in the general case is easily obtained by passing to the limit for functions belonging to $M_0(\mathbf{F}_t)$. Therefore the proof in the general case may be left to the reader.

Definition. We shall say that a process $\zeta(t)$ defined on $[t_0, T]$ has a stochastic differential $d\zeta(t)$ equal to $a(t)\,dt + b(t)\,dw(t)$ if for all t_1 and t_2 in $[t_0, T]$, the relation

$$\zeta(t_2) - \zeta(t_1) = \int_{t_1}^{t_2} a(t)\,dt + \int_{t_1}^{t_2} b(t)\,dw(t) \tag{2.6}$$

is fulfilled with probability 1; here, $a(t)$ and $b(t)$ belong to $M_2(\mathbf{F}_t)$. In the future we shall use stochastic differentials since they allow us to write relations of the type (2.6) in a more compact form.

THEOREM 5. *Let $\zeta(t)$ be a numerical process satisfying the relation*

$$d\zeta(t) = a(t)\,dt + b(t)\,dw(t)$$

for $t \in [t_0, T]$, with $a(t)$, $b(t)$, and $b^2(t)$ belonging to $M_2(\mathbf{F}_t)$. If $\Phi(t, x)$ is defined and continuous and has continuous derivatives $\Phi'_t(t, x)$, $\Phi'_x(t, x)$ and $\Phi''_{xx}(t, x)$ for $t \in [t_0, T]$, $x \in (-\infty, \infty)$, then the process $\eta(t) = \Phi(t, \zeta(t))$ satisfies the relation

$$d\eta(t) = [\Phi'_t(t, \zeta(t)) + \Phi'_x(t, \zeta(t))a(t) + \tfrac{1}{2}\Phi''_{xx}(t, \zeta(t))b^2(t)]\,dt + \Phi'_x(t, \zeta(t))b(t)\,dw(t).$$

Proof. It is necessary to show that for $t' < t''$,

$$\eta(t'') - \eta(t') = \int_{t'}^{t''} [\Phi'_t(t, \zeta(t)) + \Phi'_x(t, \zeta(t))a(t) + \tfrac{1}{2}\Phi''_{xx}(t, \zeta(t))b^2(t)]\,dt + \int_{t'}^{t''} \Phi'_x(t, \zeta(t))b(t)\,dw(t). \tag{2.7}$$

Let us make a partition of the interval $[t', t'']$: $t' = t_1 < t_2 < \cdots < t_{m+1} = t''$ and write

$$\begin{aligned}\eta(t'') - \eta(t') &= \sum_{k=1}^{m} (\eta(t_{k+1}) - \eta(t_k)) \\ &= \sum_{k=1}^{m} [\Phi(t_{k+1}, \zeta(t_{k+1})) - \Phi(t_k, \zeta(t_k))] \\ &= \sum_{k=1}^{m} [\Phi(t_{k+1}, \zeta(t_{k+1})) - \Phi(t_k, \zeta(t_{k+1}))] \\ &\quad + \sum_{k=1}^{m} [\Phi(t_k, \zeta(t_{k+1})) - \Phi(t_k, \zeta(t_k))].\end{aligned}$$

Applying Taylor's formula to each of the terms, we obtain

$$\eta(t'') - \eta(t') = \sum_{k=1}^{m} \Phi_t'(t_k + \Theta_k(t_{k+1} - t_k), \zeta(t_{k+1}))(t_{k+1} - t_k)$$

$$+ \sum_{k=1}^{m} \Phi_x'(t_k, \zeta(t_k))[\zeta(t_{k+1}) - \zeta(t_k)]$$

$$+ \tfrac{1}{2} \sum_{k=1}^{m} \Phi_{xx}''[t_k, \xi(t_k) + \overline{\Theta}_k(\zeta(t_{k+1}) - \xi(t_k))]$$

$$\times [\zeta(t_{k+1}) - \zeta(t_k)]^2,$$

where Θ_k and $\overline{\Theta}_k$ are in (0, 1). It follows from the continuity of $\Phi_t'(t, x)$ and $\zeta(t)$ that the first sum on the right-hand side of the last equation will converge to $\int_{t'}^{t''} \Phi_t'(t, \zeta(t))\, dt$ as $\max_k (t_{k+1} - t_k) \to 0$. Also,

$$\sum_{k=1}^{m} \Phi_x'(t_k, \zeta(t_k))[\zeta(t_{k+1}) - \zeta(t_k)]$$

$$= \sum_{k=1}^{m} \Phi_x'(t_k, \zeta(t_k)) \int_{t_k}^{t_{k+1}} a(t)\, dt + \sum_{k=1}^{m} \int_{t_k}^{t_{k+1}} \Phi_x'(t_k, \zeta(t_k)) b(t)\, dw(t).$$

The first sum converges to $\int_{t'}^{t''} \Phi_x'(t, \zeta(t))a(t)\, dt$ as $\max_k (t_{k+1} - t_k) \to 0$. We define $f_m(t) = \Phi_x'(t_k, \zeta(t_k))b(t)$ for $t \in [t_k, t_{k+1}]$. Then

$$f_m(t) \to \Phi_x'(t, \zeta(t)b(t))$$

with probability 1 for $\max_k (t_{k+1} - t_k) \to 0$, $m \to \infty$; further,

$$|f_m(t)|^2 \leq \sup_{t_0 \leq t \leq T} |\Phi_x'(t, \zeta(t))|^2 b^2(t).$$

[The finiteness of $\sup_{t_0 \leq t \leq T} |\Phi_x'(t, \zeta(t))|$ is assured by the continuity with probability 1 of the process $\Phi_x'(t, \zeta(t))$.] Therefore, on the basis of Property 5 of Section 1, we conclude that

$$\int_{t'}^{t''} f_m(t)\, dw(t) \to \int_{t'}^{t''} \Phi_x'(t, \zeta(t))b(t)\, dw(t)$$

in probability; that is,

$$\sum_{k=1}^{m} \int_{t_k}^{t_{k+1}} \Phi_x'(t_k, \zeta(t_k))b(t)\, dw(t) \to \int_{t'}^{t''} \Phi_x'(t, \zeta(t))b(t)\, dw(t)$$

in probability.

In order to prove the theorem, we must still show that

$$\sum_{k=1}^{m} \Phi''_{xx}(t_k, \zeta(t_k) + \overline{\Theta}_k[\zeta(t_{k+1}) - \zeta(t_k)])(\zeta(t_{k+1}) - \zeta(t_k))^2 \to \int_{t'}^{t''} \Phi''_{xx}(t, \zeta(t))b^2(t)\,dt \tag{2.8}$$

in probability. To prove (2.8), we shall establish several auxiliary propositions.

LEMMA 1. *The quantity*

$$\textstyle\sum_{k=1}^{m} \left(\int_{t_k}^{t_{k+1}} b(t)\,dw(t)\right)^2$$

is uniformly bounded in probability over all subdivisions of the interval $[t', t'']$.

Proof. For every $N > 0$ and $C > 0$, we have

$$\begin{aligned}\mathbf{P}\left\{\sum_{k=1}^{m}\left(\int_{t_k}^{t_{k+1}} b(t)\,dw(t)\right)^2 > C\right\} &\le \mathbf{P}\left\{\int_{t_0}^{T} b^2(t)\,dt > N\right\} \\ &\quad + \mathbf{P}\left\{\sum_{k=1}^{m}\left(\int_{t_k}^{t_{k+1}} b(t)g_N\left(\int_{t_0}^{t} b^2(s)\,ds\right)dw(t)\right)^2 > C\right\} \\ &\le \frac{1}{C}\mathbf{M}\sum_{k=1}^{m}\left(\int_{t_k}^{t_{k+1}} b(t)g_N\left(\int_{t_0}^{t} b^2(s)\,ds\right)dw(t)\right)^2 \\ &\quad + \mathbf{P}\left\{\int_{t_0}^{T} b^2(t)\,dt > N\right\} \\ &\le \frac{N}{C} + \mathbf{P}\left\{\int_{t_0}^{T} b^2(t)\,dt > N\right\}.\end{aligned}$$

LEMMA 2. *If* $\max_k (t_{k+1} - t_k) \to 0$, *then* $\sum_{k=1}^{m} \left(\int_{t_k}^{t_{k+1}} a(t)\,dt\right)^2 \to 0$ *in probability. In fact,*

$$\begin{aligned}\sum_{k=1}^{m}\left(\int_{t_k}^{t_{k+1}} a(t)\,dt\right)^2 &\le \sum_{k=1}^{m} (t_{k+1} - t_k)\int_{t_k}^{t_{k+1}} a(t)^2\,dt \\ &\le \max_k (t_{k+1} - t_k)\int_{t'}^{t''} a^2(t)\,dt.\end{aligned}$$

From Lemmas 1 and 2, we have

$$\sum_{k=1}^{m} \Phi''_{xx}(t_k, \zeta(t_k) + \overline{\Theta}_k[\zeta(t_{k+1}) - \zeta(t_k)])\left(\int_{t_k}^{t_{k+1}} a(t)\,dt\right)^2 \to 0$$

in probability. And

$$\left|\sum_{k=1}^{m} \Phi''_{xx}(t_k, \zeta(t_k) + \overline{\Theta}_k[\zeta(t_{k+1}) - \zeta(t_k)]) \int_{t_k}^{t_{k+1}} a(t)\, dt \int_{t_k}^{t_{k+1}} b(t)\, dw(t)\right|^2$$

$$\leq \sum_{k=1}^{m} \left|\Phi''_{xx}(t_k, \zeta(t_k) + \overline{\Theta}_k[\zeta(t_{k+1}) - \zeta(t_k)])\right|^2 \left(\int_{t_k}^{t_{k+1}} a(t)\, dt\right)^2$$

$$\times \sum_{k=1}^{m} \left(\int_{t_k}^{t_{k+1}} b(t)\, dw(t)\right)^2 \to 0$$

in probability because

$$\sup_k |\Phi''_{xx}(t_k, \zeta(t_k) + \overline{\Theta}_k[\zeta(t_{k+1}) - \zeta(t_k)])| \leq \sup_t \sup_{|x| \leq \sup_s |\zeta(s)|} |\Phi''_{xx}(t, x)|.$$

Therefore, to prove the theorem it remains to show that

$$\sum_{k=1}^{m} \Phi''_{xx}(t_k, \zeta(t_k) + \overline{\Theta}_k[\zeta(t_{k+1}) - \zeta(t_k)]) \left(\int_{t_k}^{t_{k+1}} b(t)\, dw(t)\right)^2$$

$$\to \int_{t'}^{t''} \Phi''_{xx}(t, \zeta(t)) b^2(t)\, dt.$$

From the continuity of $\zeta(t)$ and $\Phi''_{xx}(t, x)$ it follows that

$$\sup_k |\Phi''_{xx}(t_k, \zeta(t_k) + \overline{\Theta}_k[\zeta(t_{k+1}) - \zeta(t_k)]) - \Phi''_{xx}(t_k, \zeta(t_k))| \to 0$$

in probability. Then

$$\lim_{\max_k (t_{k+1} - t_k) \to 0} \sum_{k=1}^{m} \Phi''_{xx}(t_k, \zeta(t_k)) \int_{t_k}^{t_{k+1}} b^2(t)\, dt = \int_{t'}^{t''} \Phi''_{xx}(t, \zeta(t)) b^2(t)\, dt.$$

Consequently the proof of the theorem will follow from the following lemma:

LEMMA 3. *If* $\max_k (t_{k+1} - t_k) \to 0$, *then*

$$\sum_{k=1}^{m} \Phi''_{xx}(t_k, \zeta(t_k)) \left[\left(\int_{t_k}^{t_{k+1}} b(t)\, dw(t)\right)^2 - \int_{t_k}^{t_{k+1}} b^2(t)\, dt\right] \to 0$$

in probability.

Proof. We define

$$b_N(t) = b(t) g_N\left(\int_{t_0}^{t} b^4(u)\, du\right).$$

Suppose further that $\chi(t) = 1$ if $\sup_{t_0 \le s \le t} |\zeta(s)| \le C$ and that $\chi(t) = 0$ if $\sup_{t_0 \le s \le t} |\zeta(s)| > C$. Then for every $\epsilon > 0$,

$$\mathbf{P}\left\{\left|\sum_{k=1}^{m} \Phi''_{xx}(t_k, \zeta(t_k)) \left[\left(\int_{t_k}^{t_{k+1}} b(t)\, dw(t)\right)^2 - \int_{t_k}^{t_{k+1}} b^2(t)\, dt\right]\right| > \epsilon\right\}$$

$$\le \mathbf{P}\left\{\left|\sum_{k=1}^{m} \chi(t_k)\Phi''_{xx}(t_k, \zeta(t_k)) \left[\left(\int_{t_k}^{t_{k+1}} b_N(t)\, dw(t)\right)^2 - \int_{t_k}^{t_{k+1}} b_N^2(t)\, dt\right]\right| > \epsilon\right\}$$

$$+ \mathbf{P}\left\{\int_{t_0}^{T} b^4(t)\, dt > N\right\} + \mathbf{P}\left\{\sup_{t_0 \le t \le T} |\zeta(t)| > C\right\}.$$

But

$$\mathbf{M}\left(\sum_{k=1}^{m} \chi(t_k)\Phi''_{xx}(t_k, \zeta(t_k)) \left[\left(\int_{t_k}^{t_{k+1}} b_N(t)\, dw(t)\right)^2 - \int_{t_k}^{t_{k+1}} b_N^2(t)\, dt\right]\right)^2$$

$$= \sum_{k=1}^{m} \mathbf{M}[\chi(t_k)\Phi''_{xx}(t_k, \zeta(t_k))]^2$$

$$\times \mathbf{M}\left(\left[\left(\int_{t_k}^{t_{k+1}} b_N(t)\, dw(t)\right)^2 - \int_{t_k}^{t_{k+1}} b_N^2(t)\, dt\right]^2 \middle| \mathbf{F}_{t_k}\right)$$

because all double products vanish after we take the mathematical expectation in view of the fact that

$$\mathbf{M}\left(\left[\int_{t_k}^{t_{k+1}} b_N(t)\, dw(t)\right]^2 - \int_{t_k}^{t_{k+1}} b_N^2(t)\, dt/\mathbf{F}_{t_k}\right) = 0.$$

Further,

$$\chi(t_k)|\Phi''_{xx}(t_k, \zeta(t_k))| \le L,$$

where $L = \sup_{\substack{t_0 \le t \le T \\ |x| \le C}} |\Phi''_{xx}(t, x)|$, and

$$\mathbf{M}\left[\left(\int_{t_k}^{t_{k+1}} b_N(t)\, dw(t)\right)^2 - \int_{t_k}^{t_{k+1}} b_N^2(t)\, dt\right]^2 \le 2\mathbf{M}\left[\int_{t_k}^{t_{k+1}} b_N(t)\, dw(t)\right]^4$$

$$+ 2\mathbf{M}\left[\int_{t_k}^{t_{k+1}} b_N^2(t)\, dt\right]^2 \le 72(t_{k+1} - t_k)\int_{t_k}^{t_{k+1}} \mathbf{M} b_N^4(t)\, dt$$

$$+ 2(t_{k+1} - t_k)\int_{t_k}^{t_{k+1}} \mathbf{M} b_N^4(t)\, dt = 74(t_{k+1} - t_k)\int_{t_k}^{t_{k+1}} \mathbf{M} b_N^4(t)\, dt.$$

(Here we used Theorem 4 for the first integral and the Cauchy inequality for the second.) Therefore

$$\mathbf{M}\left(\sum_{k=1}^{m} \chi(t_k)\Phi''_{xx}(t_k, \zeta(t_k))\left[\left(\int_{t_k}^{t_{k+1}} b_N(t)\, dw(t)\right)^2 - \int_{t_k}^{t_{k+1}} b_N^2(t)\, dt\right]\right)^2$$

$$\leq 74\lambda L^2 \int_{t'}^{t''} \mathbf{M} b_N^4(t)\, dt \leq 74L^2\lambda N,$$

where $\lambda = \max_k (t_{k+1} - t_k)$, since for $\lambda \to 0$,

$$\mathbf{P}\left\{\left|\sum_{k=1}^{m} \chi(t_k)\Phi''_{xx}(t_k, \zeta(t_k))\left[\left(\int_{t_k}^{t_{k+1}} b_N(t)\, dw(t)\right)^2\right.\right.\right.$$

$$\left.\left.\left. - \int_{t_k}^{t_{k+1}} b_N^2(t)\, dt\right]\right| > \epsilon\right\} \to 0.$$

Thus

$$\overline{\lim_{\lambda\to 0}}\, \mathbf{P}\left\{\left|\sum_{k=1}^{m} \Phi''_{xx}(t_k, \zeta(t_k))\left[\left(\int_{t_k}^{t_{k+1}} b(t)\, dw(t)\right)^2 - \int_{t_k}^{t_{k+1}} b^2(t)\, dt\right]\right| > \epsilon\right\}$$

$$\leq \mathbf{P}\left\{\int_{t_0}^{T} b^4(t)\, dt > N\right\} + \mathbf{P}\left\{\sup_{t_0 \leq t \leq T} |\zeta(t)| > C\right\}.$$

We can make the right-hand side of the last inequality arbitrarily small by a sufficiently large choice of C and N. Consequently for every $\epsilon > 0$,

$$\overline{\lim_{\lambda\to 0}}\, \mathbf{P}\left\{\left|\sum_{k=1}^{m} \Phi''_{xx}(t_k, \zeta(t_k))\left[\left(\int_{t_k}^{t_{k+1}} b(t)\, dw(t)\right)^2\right.\right.\right.$$

$$\left.\left.\left. - \int_{t_k}^{t_{k+1}} b^2(t)\, dt\right]\right| > \epsilon\right\} = 0.$$

This proves the lemma and consequently the theorem.

3. The stochastic integral with respect to a martingale. Suppose that $\eta(t)$ is a martingale with numerical values defined on $[t_0, T]$, that $\mathbf{F}_t$ is a set of σ-algebras defined for every $t \in [t_0, T]$, that $\eta(t)$ is measurable with respect to $\mathbf{F}_t$ for every t, that for $t_1 < t_2$, $\mathbf{F}_{t_1} \subset \mathbf{F}_{t_2}$, and that

$$\mathbf{M}\big(\eta(t_2) - \eta(t_1)/\mathbf{F}_{t_1}\big) = 0. \tag{3.1}$$

We assume that a nondecreasing function $\lambda(t)$ defined for $t \in [t_0, T]$ exists such that for $t_1 < t_2$,

$$\mathbf{M}\big((\eta(t_2) - \eta(t_1))^2/\mathbf{F}_{t_1}\big) = \lambda(t_2) - \lambda(t_1). \tag{3.2}$$

We denote by $M(\mathbf{F}_t)$ the set of functions $f(t)$ defined on $[t_0, T]$ such that for every $t \in [t_0, T], f(t)$ is measurable with respect to $\mathbf{F}_t$. $M_0(\mathbf{F}_t)$, as above, will denote the class of step functions belonging to $M(\mathbf{F}_t)$, and $M_1(\mathbf{F}_t; \lambda(t))$ and $M_2(\mathbf{F}_t; \lambda(t))$ will denote the classes of functions for which

$$\int_{t_0}^{T} \mathbf{M}|f(t)|^2\, d\lambda(t) < \infty \qquad \text{and} \qquad \mathbf{P}\left\{\int_{t_0}^{T} |f(t)|^2\, d\lambda(t) < \infty\right\} = 1,$$

respectively.

We now define the integral

$$\int_{t_0}^{T} f(t)\, d\eta(t) \tag{3.3}$$

for all $f(t) \in M_2(\mathbf{F}_t; \lambda(t))$. If $f(t) \in M_0(\mathbf{F}_t; \lambda(t))$, that is, if there exist points $t_0 < t_1 < \cdots < t_n = T$ such that $f(t) = f(t_k)$ for $t \in (t_k, t_{k+1})$, then

$$\int_{t_0}^{T} f(t)\, d\eta(t) = \sum_{k=0}^{n-1} f(t_k)\big(\eta(t_{k+1}) - \eta(t_k)\big). \tag{3.4}$$

The extension of the definition of the integral (3.3) to all functions in $M_2(\mathbf{F}_t, \lambda(t))$ is done in exactly the same manner as the definition of the integral (1.1) in Section 1 was extended from the class $M_0(\mathbf{F}_t)$ to the class $M_2(\mathbf{F}_t)$. Thus the definite integral (3.3) possesses the following properties:

1. If $\chi(t) = 0$ for $\sup_{t_0 \le s \le t} |f(s)| = 0$ and $\chi(t) = 1$ for $\sup_{t_0 \le s \le t} |f(s)| > 0$, then

$$\left|\int_{t_0}^{T} f(t)\, d\eta(t)\right| \le \chi(T) \left|\int_{t_0}^{T} f(t)\, d\eta(t)\right|, \tag{3.5}$$

which means that

$$\mathbf{P}\left\{\left|\int_{t_0}^{T} f(t)\, d\eta(t)\right| > 0\right\} \le \mathbf{P}\left\{\sup_{t_0 \le\ \le T} |f(t)| > 0\right\}. \tag{3.6}$$

2. If $f_1(t)$ and $f_2(t)$ belong to $M_2(\mathbf{F}_t; \lambda(t))$ and if α and β are arbitrary real numbers, then

$$\int_{t_0}^{T} (\alpha f_1(t) + \beta f_2(t))\, d\eta(t) = \alpha \int_{t_0}^{T} f_1(t)\, d\eta(t) + \beta \int_{t_0}^{T} f_2(t)\, d\eta(t). \tag{3.7}$$

3. If $f(t) \in M_1(\mathbf{F}_t; \lambda(t))$, then

$$\mathbf{M}\left(\int_{t_0}^{T} f(t)\, d\eta(t)\right) = 0, \tag{3.8}$$

and

$$\mathbf{M}\left|\int_{t_0}^{T} f(t)\, d\eta(t)\right|^2 = \int_{t_0}^{T} \mathbf{M}|f(t)|^2\, d\lambda(t). \tag{3.9}$$

4. If $f_n(t) \to f(t)$ in probability for almost all t with respect to the measure $\mu(A)$ such that $\mu(A) = \int_A d\lambda(t)$, and if there exists $\varphi(t) \in M_2(\mathbf{F}_t; \lambda(t))$ such that $|f_n(t)| \leq |\varphi(t)|$, then

$$\int_{t_0}^{T} f_n(t)\, d\eta(t) \to \int_{t_0}^{T} f(t)\, d\eta(t)$$

in probability.

Let $\chi_A(t)$ be the characteristic function of a Borel set A belonging to the interval $[t_0, T]$. Then for every function $f(t) \in M_2(\mathbf{F}_t, \lambda(t))$, there exists

$$\int f(t)\chi_A(t)\, d\eta(t),$$

which we shall denote by $\int_A f(t)\, d\eta(t)$. If A is the interval $[t_1, t_2]$, then instead of $\int_A f(t)\, d\eta(t)$ we shall write $\int_{t_1}^{t_2} f(t)\, d\eta(t)$. Let us consider $\int_{t_0}^{t} f(s)\, d\eta(s)$ as a function of the upper limit. Then, just as in Section 2, we can establish the following properties of this function:

5. If $f(t) \in M_1(\mathbf{F}_t; \lambda(t))$, then $\zeta(t) = \int_{t_0}^{t} f(s)\, d\eta(s)$ is a martingale, and for $t_1 < t$,

$$\mathbf{M}\left(\int_{t_0}^{t} f(s)\, d\eta(s)/\mathbf{F}_{t_1}\right) = \int_{t_0}^{t_1} f(s)\, d\eta(s).$$

6. If $\chi(t)$ is defined as in the statement of Property 1, then

$$\sup_{t_0 \leq t \leq T} \left| \int_{t_0}^{t} f(s)\, d\eta(s) \right| \leq \chi(T) \sup_{t_0 \leq t \leq T} \left| \int_{t_0}^{t} f(s)\, d\eta(s), \right. \tag{3.10}$$

which means that with $\sup_{t_0 \leq t \leq T} |\int_{t_0}^{t} f(s)\, d\eta(s)| < \infty$,

$$\mathbf{P}\left\{ \sup_{t_0 \leq t \leq T} \left| \int_{t_0}^{t} f(s)\, d\eta(s) \right| > 0 \right\} \leq \mathbf{P}\left\{ \sup_{t_0 \leq t \leq T} |f(t)| > 0 \right\} \cdot \tag{3.11}$$

7. $\int_{t_0}^{t} f(s)\, d\eta(s)$ as a function of t with probability 1 does not have discontinuities of the second kind. In the case where $f(t) \in M_1(\mathbf{F}_t; \lambda(t))$, this property follows from Property 3 of martingales, which was given in Section 5 of Chapter 1. The extension of this property to

$$f(t) \in M_2(\mathbf{F}_t; \lambda(t))$$

is carried out in the same manner as in Theorem 3 of the preceding section.

We shall need the following theorem on passage to the limit under the stochastic integral sign.

THEOREM. *Suppose that a sequence of martingales* $\eta_n(t)$ *converges for every* t *in probability to a Brownian motion* $w(t)$ *and that* $\lambda_n(t) = \mathbf{M}\eta_n(t)^2$ *as* $n \to \infty$ *converges to* $t - t_0$. *Suppose further that* $f_n(t)$ *is a sequence such that the integrals* $\int_{t_0}^T f_n(t)\, d\eta_n(t)$ *exist, and that for every* t, $f_n(t)$ *converges in probability to a function* $f(t)$ *for which*

$$\int_{t_0}^T f(t)\, dw(t)$$

exists. Suppose, in addition, that the following conditions are satisfied for $f_n(t)$:

(a) *for every* $\epsilon > 0$, *there exists a* $C > 0$ *such that for all* n,

$$\mathbf{P}\left\{\sup_{t_0 \le t \le T} |f_n(t)| > C\right\} \le \epsilon;$$

(b) *for every* $\epsilon > 0$,

$$\lim_{h \to 0} \lim_{n \to \infty} \sup_{|t_1 - t_2| \le h} \mathbf{P}\{|f_n(t_2) - f_n(t_1)| > \epsilon\} = 0.$$

Then

$$\int_{t_0}^T f_n(t)\, d\eta_n(t) \to \int_{t_0}^T f(t)\, dw(t)$$

in probability.

Proof. Suppose that $\varphi_C(x) = 1$ for $|x| \le C$, that $\varphi_C(x) = |x| + 1 - C$ for $|x| \in [C, C+1]$, and that $\varphi_C(x) = 0$ for $|x| > C + 1$. Then for all n,

$$\mathbf{P}\left\{\left|\int_{t_0}^T \varphi_C(f_n(t)) f_n(t)\, d\eta_n(t) - \int_{t_0}^T f_n(t)\, d\eta_n(t)\right| > 0\right\} \le \mathbf{P}\left\{\sup_{t_0 \le t \le T} |f_n(t)| > C\right\},$$

$$\mathbf{P}\left\{\left|\int_{t_0}^T \varphi_C(f(t)) f(t)\, dw(t) - \int_{t_0}^T f(t)\, dw(t)\right| > 0\right\} \le \mathbf{P}\left\{\sup_{t_0 \le t \le T} |f(t)| > C\right\} \le \overline{\lim_{n \to \infty}}\, \mathbf{P}\left\{\sup_{t_0 \le t \le T} |f_n(t)| > C\right\}.$$

Therefore, from condition (a), to prove the theorem it is sufficient to show that for all $C > 0$,

$$\int_{t_0}^T \varphi_C(f_n(t)) f_n(t)\, d\eta_n(t) \to \int_{t_0}^T \varphi_C(f(t)) f(t)\, dw(t)$$

in probability.

From condition (b), it follows that $f(t)$ is a stochastically continuous process. Let $t_0 < t_1 < \cdots < t_k = T$ be an arbitrary partition of the interval $[t_0, T]$. Then

$$\mathbf{P}\left\{\left|\int_{t_0}^{T} \varphi_C(f_n(t))f_n(t)\,d\eta_n(t) - \int_{t_0}^{T} \varphi_C(f(t))f(t)\,dw(t)\right| > \epsilon\right\}$$
$$\leq \mathbf{P}\left\{\left|\int_{t_0}^{T} \varphi_C(f_n(t))f_n(t)\,d\eta_n(t) - \sum_{r=0}^{k-1} f_n(t_r)\varphi_C(f_n(t_r))(\eta_n(t_{r+1}) - \eta_n(t_r))\right| > \frac{\epsilon}{3}\right\} + \mathbf{P}\left\{\left|\int_{t_0}^{T} \varphi_C(f(t))f(t)\,dw(t) - \sum_{r=0}^{k-1} f(t_r)\varphi_C(f(t_r))(w(t_{r+1}) - w(t_r))\right| > \frac{\epsilon}{3}\right\}$$
$$+ \mathbf{P}\left\{\left|\sum_{r=0}^{k-1} f_n(t_r)\varphi_C(f_n(t_r))[\eta_n(t_{r+1}) - \eta_n(t_r)] - \sum_{r=0}^{k-1} f(t_r)\varphi_C(f(t_r))[w(t_{r+1}) - w(t_r)]\right| > \frac{\epsilon}{3}\right\}.$$

The last probability converges to zero as $n \to \infty$ because

$$f_n(t_r),\ \varphi_C(f_n(t_r)), \qquad \text{and} \qquad \eta_n(t_{r+1}) - \eta_n(t_r)$$

converge in probability to

$$f(t),\ \varphi_C(f(t)) \qquad \text{and} \qquad w(t_{r+1}) - w(t_r),$$

respectively. Let us show that

$$\lim_{\max_r |t_{r+1}-t_r|\to 0} \overline{\lim_{n\to\infty}}\, \mathbf{P}\left\{\left|\int_{t_0}^{T} \varphi_C(f_n(t))f_n(t)\,d\eta_n(t) - \sum_{r=0}^{k-1} f_n(t_r)\varphi_C(f_n(t_r))[\eta_n(t_{r+1}) - \eta_n(t_r)]\right| > \frac{\epsilon}{3}\right\} = 0 \tag{3.12}$$

and

$$\lim_{\max_r |t_{r+1}-t_r|\to 0} \mathbf{P}\left\{\left|\int_{t_0}^{T} \varphi_C(f(t))f(t)\,dw(t) - \sum_{r=0}^{k-1} f(t_r)\varphi_C(f(t_r))[w(t_{r+1}) - w(t_r)]\right| > \frac{\epsilon}{3}\right\} = 0. \tag{3.13}$$

Both these relationships may be proved in the same way. Let us prove the first. Suppose that $\lambda = \max_r (t_{r+1} - t_r)$. Then

$$\mathbf{P}\left\{\left|\int_{t_0}^{T} \varphi_C(f_n(t))f_n(t)\, d\eta_n(t) - \sum_{r=0}^{k-1} f_n(t_r)\varphi_C(f_n(t_r)) \times (\eta_n(t_{r+1}) - \eta_n(t_r))\right| > \frac{\epsilon}{3}\right\}$$

$$\leq \frac{9}{\epsilon^2}\sum_{r=0}^{k-1}\int_{t_r}^{t_{r+1}} \mathbf{M}\,|\,\varphi_C(f_n(t))f_n(t) - \varphi_C(f_n(t_r))f_n(t_r)^2\, d\lambda_n(t)$$

$$\leq \frac{9}{\epsilon^2}[\lambda_n(T) - \lambda_n(t_0)] \max_{|t'-t''|\leq\lambda} \mathbf{M}\,|\,\varphi_C(f_n(t'))f_n(t') - \varphi_C(f_n(t''))f_n(t'')|^2.$$

From this, condition (b) implies that

$$\lim_{\lambda\to 0}\lim_{n\to\infty}\max_{|t'-t''|\leq\lambda} \mathbf{M}|\varphi_C(f_n(t'))f_n(t') - \varphi_C(f_n(t''))f_n(t'')|^2 = 0.$$

Also, $\lambda_n(T) - \lambda_n(t_0) \to T - t_0$. Thus (3.12) is proved. The proof of the theorem follows from (3.12) and (3.13).

4. Stochastic integrals with respect to certain random measures. Let X be a set, $\mathbf{B}$ a σ-algebra of subsets. We shall say that a random measure μ is given on $\mathbf{B}$ if for every $A \in \mathbf{B}$, a random variable $\mu(A)$ is defined, and if the following condition is met: For any sequence A_k of disjoint sets in $\mathbf{B}$ such that $A = \cup_k A_k$, the series $\sum_{k=1}^{\infty} \mu(A_k)$ converges in probability to $\mu(A)$.

If for the disjoint sets $A_1, A_2, \ldots, A_k$, the variables $\mu(A_1), \mu(A_2), \ldots, \mu(A_k)$ are mutually independent, then we shall call μ a measure with independent values. In what follows, we shall make extensive use of two particular measures with independent values.

Let $[t_0, T] \times R^{(m)}$ be the space of pairs of points (t, u) for which $t \in [t_0, T]$, $u \in R^{(m)}$, and let $\mathbf{B}$ be the ring of all the Borel sets A in $[t_0, T] \times R^{(m)}$ for which

$$\int_A \frac{du\, dt}{|u|^{m+1}} < \infty. \tag{4.1}$$

We shall denote by $p(A)$ a random measure with independent values that is defined on $\mathbf{B}$ and for which $p(A)$ has a Poisson distribution with parameter (4.1) for every $A \in \mathbf{B}$. We shall denote by $q(A)$ the random measure defined by the relation

$$q(A) = p(A) - \mathbf{M}p(A). \tag{4.2}$$

It is possible to construct the measure $p(A)$ in the following manner. Let $\xi(t)$ be a homogeneous process with independent increments which takes values in $R^{(m)}$ and for which

$$\mathbf{M}e^{i(z,\xi(t))} = \exp\left\{t\left[\int_{|u|\leq 1} (e^{i(z,u)} - 1 - i(z, u)) \frac{du}{|u|^{m+1}} + \int_{|u|>1} (e^{i(z,u)} - 1) \frac{du}{|u|^{m+1}}\right]\right\}. \quad (4.3)$$

The process $\xi(t)$ with probability 1 does not have discontinuities of the second kind. If we indicate by $p(A)$ the number of discontinuities of the process $\xi(t)$ for which the point $(t, \xi(t + 0) - \xi(t - 0))$ belongs to A, then $p(A)$ will have a Poisson distribution with parameter (4.1), and for the disjoint sets A_k, the variables $p(A_k)$ will be independent.

We suppose that for every $t \in [t_0, T]$, a σ-algebra of events $\mathbf{F}_t$ is defined such that $A \subset [t_0, T] \times R^{(m)}$ (that is, if only those points (s, u) for which $s \in [t_0, t]$ can belong to A). Also we suppose that $p(A)$ is measurable with respect to $\mathbf{F}_t$ and that no matter what the sets $A_1, A_2, \ldots, A_k$ in $[t, T] \times R^{(m)}$ may be, the quantities $p(A_k)$ are mutually independent of any event in $\mathbf{F}_t$. We indicate by $\overline{M}(\mathbf{F}_t)$ the collection of measurable random functions $f(t, u)$ such that $f(t, u)$ is measurable with respect to $\mathbf{F}_t$ for every t irrespective of the value of u. We shall call the function $f(t, u)$ a step function if there exist $t_0 < t_1 < \cdots < t_n = T$ and Borel sets $B_1, B_2, \ldots, B_n$ in $R^{(m)}$ such that $f(t, u)$ is constant on every set $[t_k, t_{k+1}] \times B_j$, $k = 0, 1, \ldots, n - 1$, $j = 1, 2, \ldots, n$, and $\cup B_j = R^{(m)}$. The collection of step functions in $M_0(\mathbf{F}_t)$ for which there exists $\epsilon > 0$ such that $f(t, u) = 0$ for $|u| \leq \epsilon$ will be denoted by $\overline{M}_0(\mathbf{F}_t)$. Let $\overline{M}_p^{(1)}(\mathbf{F}_t)$ be the collection of functions for which

$$\int_{t_0}^{T} \int_{R^{(m)}} \mathbf{M}|f(t, u)| \frac{dt\, du}{|u|^{m+1}} < \infty;$$

let $M_p^{(2)}(\mathbf{F}_t)$ be the collection of functions $f(t, u)$ for which

$$\mathbf{P}\left\{\int_{t_0}^{T} \int_{R^{(m)}} |f(t, u)| \frac{dt\, du}{|u|^{m+1}} < \infty\right\} = 1;$$

let $M_q^{(1)}(\mathbf{F}_t)$ be the collection of functions $f(t, u)$ for which

$$\int_{t_0}^{T} \int_{R^{(m)}} \mathbf{M}|f(t, u)|^2 \frac{dt\, du}{|u|^{m+1}} < \infty;$$

and let $\overline{M}_q^{(2)}(\mathbf{F}_t)$ be the collection of functions $f(t, u)$ for which

$$\mathbf{P}\left\{\int_{t_0}^{T} \int_{R^{(m)}} |f(t, u)|^2 \frac{dt\, du}{|u|^{m+1}} < \infty\right\} = 1.$$

We now define the integrals with respect to the measures p and q. This definition is completely consistent with the definition of the integral (1.1), and therefore the proofs will not be detailed.

We begin with the second integral. Assume that $f(t, u) \in \overline{M}_0(\mathbf{F}_t)$, that is, that there exist $t_0 < t_1, < \cdots < t_n = T$ and $B_1, B_2, \ldots B_n$ in $R^{(m)}$, $\bigcup_j B_j = R^{(m)}$, such that $f(t, u)$ is constant over the sets $[t_k, t_{k+1}) \times B_j$. Taking $u_j \in B_j$, we set

$$\int_{t_0}^{T} \int_{R^{(m)}} f(t, u) q\,(dt \times du) = \sum_{\substack{0 \le k \le n-1 \\ 1 \le j \le n}} f(t_k, u_j) q\big((t_k, t_{k+1}) \times B_j\big). \tag{4.4}$$

The integral defined by (4.4) possesses the following properties:

1. If α and β are real numbers, then

$$\int_{t_0}^{T} \int_{R^{(m)}} \big(\alpha f_1(t, u) + \beta f_2(t, u)\big) q\,(dt \times du)$$
$$= \alpha \int_{t_0}^{T} \int_{R^{(m)}} f_1(t, u) q\,(dt \times du) + \beta \int_{t_0}^{T} \int_{R^{(m)}} f_2(t, u) q\,(dt \times du).$$

2. If $\chi(t) = 0$ for $\sup_{u \in R^{(m)}} \sup_{s \in [t_0, t]} |f(s, u)| = 0$ and $\chi(t) = 1$ for $\sup_{u \in R^{(m)}} \sup_{s \in [t_0, t]} |f(s, u) > 0$, then

$$\left| \int_{t_0}^{T} \int_{R^{(m)}} f(t, u) q\,(dt \times du) \right| \le \chi(T) \left| \int_{t_0}^{T} \int_{R^{(m)}} f(t, u) q\,(dt \times du) \right| \tag{4.5}$$

and hence

$$\mathbf{P}\left\{ \left| \int_{t_0}^{T} \int_{R^{(m)}} f(t, u) q\,(dt \times du) \right| > 0 \right\} \le \mathbf{P}\left\{ \sup_{u \in R^{(m)}} \sup_{s \in [t_0, T]} |f(s, u)| > 0 \right\} \cdot \tag{4.6}$$

3. If, in addition, $f(t, u) \in \overline{M}_q^{(1)}(\mathbf{F}_t)$, then

$$\mathbf{M}\left(\int_{t_0}^{T} \int_{R^{(m)}} f(t, u) q\,(dt \times du) \right) = 0, \tag{4.7}$$

$$\mathbf{M}\left| \int_{t_0}^{T} \int_{R^{(m)}} f(t, u) q\,(dt \times du) \right|^2 = \int_{t_0}^{T} \int_{R^{(m)}} \mathbf{M}|f(t, u)|^2 \frac{dt\,du}{|u|^{m+1}} \cdot \tag{4.8}$$

For every $f(t, u) \in \overline{M}_q^{(1)}(\mathbf{F}_t)$, there exists a sequence $f_n(t, u) \in \overline{M}_0(\mathbf{F}_t) \cap \overline{M}_q^{(1)}(\mathbf{F}_t)$ such that

$$\lim_{n \to \infty} \int_{t_0}^{T} \int_{R^{(m)}} \mathbf{M}|f_n(t, u) - f(t, u)|^2 \frac{dt\,du}{|u|^{m+1}} = 0. \tag{4.9}$$

[This is shown in the same manner as (1.9).] Then

$$\lim_{\substack{n\to\infty \\ l\to\infty}} \int_{t_0}^{T} \int_{R^{(m)}} \mathbf{M}|f_l(t, u) - f_n(t, u)|^2 \frac{dt\, du}{|u|^{m+1}} = 0 \tag{4.10}$$

and hence the sequence of random variables $\int_{t_0}^{T} \int_{R^{(m)}} f(t, u)q(dt \times du)$ will converge in probability to some particular random variable which we shall denote by

$$\int_{t_0}^{T} \int_{R^{(m)}} f(t, u)q\,(dt \times du)$$

whenever $f(t, u) \in \overline{M}_q^{(1)}(\mathbf{F}_t)$. Properties 1 through 3 are satisfied for this integral.

Now let $f(t, u) \in \overline{M}_q^{(2)}(\mathbf{F}_t)$, $g_N(x) = 1$ for $|x| \le N$ and $g_N(x) = 0$ for $|x| > N$. Then for all N,

$$f_N(t, u) = f(t, u)g_N\left(\int_{t_0}^{t} \int_{R^{(m)}} |f(s, u)|^2 \frac{ds\, du}{|u|^{m+1}}\right)$$

belongs to $\overline{M}_q^{(1)}(\mathbf{F}_t)$, and consequently the expression

$$\int_{t_0}^{T} \int_{R^{(m)}} f_N(t, u)q\,(dt \times du)$$

is meaningful. But by using (4.6) we obtain for $N' > N$,

$$\mathbf{P}\left\{\left|\int_{t_0}^{T} \int_{R^{(m)}} f_{N'}(t, u)q\,(dt \times du) - \int_{t_0}^{T} \int_{R^{(m)}} f_N(t, u)q\,(dt \times du)\right| > 0\right\}$$
$$\le \mathbf{P}\left\{\int_{t_0}^{T} \int_{R^{(m)}} |f(t, u)|^2 \frac{dt\, du}{|u|^{m+1}} > N\right\}. \tag{4.11}$$

Since the probability on the right-hand side of (4.11) approaches zero as $N \to \infty$, the limit of the variables $\int_{t_0}^{T} \int_{R^{(m)}} f_N(t, u)q\,(dt \times du)$ exists as $N \to \infty$ in the sense of convergence in probability, and we shall indicate it by $\int_{t_0}^{T} \int_{R^{(m)}} f(t, u)q\,(dt \times du)$. In addition to Properties 1 and 2, this integral has the following property:

4. If $f_n(t, u) \to f(t, u)$ in probability for almost all t and u, and if there exists $\varphi(t, u) \in \overline{M}_q^{(2)}(\mathbf{F}_t)$ such that $|f_n(t, u)| \le |\varphi(t, u)|$, then

$$\int_{t_0}^{T} \int_{R^{(m)}} f_n(t, u)q\,(dt \times du) \to \int_{t_0}^{T} \int_{R^{(m)}} f(t, u)q\,(dt \times du)$$

in probability as $n \to \infty$.

Let us turn to the definition of the integral with respect to the measure p. If $f(t, u) \in \overline{M}_q^{(2)}(\mathbf{F}_t) \cap \overline{M}_p^{(2)}(\mathbf{F}_t)$, we set

$$\int_{t_0}^{T} \int_{R^{(m)}} f(t, u)p\,(dt \times du) = \int_{t_0}^{T} \int_{R^{(m)}} f(t, u)q\,(dt \times du) + \int_{t_0}^{T} \int_{R^{(m)}} f(t, u) \frac{dt\,du}{|u|^{m+1}}. \tag{4.12}$$

The integral defined will possess these properties:

5. For real α and β,

$$\int_{t_0}^{T} \int_{R^{(m)}} [\alpha f_1(t, u) + \beta f_2(t, u)]p\,(dt \times du) = \alpha \int_{t_0}^{T} \int_{R^{(m)}} f_1(t, u)p\,(dt \times du) + \beta \int_{t_0}^{T} \int_{R^{(m)}} f_2(t, u)p\,(dt \times du).$$

6. If $\chi(t)$ is defined as in Property 2, then

$$\mathbf{P}\left\{\left|\int_{t_0}^{T} \int_{R^{(m)}} f(t, u)p\,(dt \times du)\right| > 0\right\} \leq \mathbf{P}\left\{\sup_{\substack{t_0 \leq t \leq T \\ u \in R^{(m)}}} |f(t, u)| > 0\right\} \tag{4.13}$$

and

$$\left|\int_{t_0}^{T} \int_{R^{(m)}} f(t, u)p\,(dt \times du)\right| \leq \chi(T)\left|\int_{t_0}^{T} \int_{R^{(m)}} f(t, u)p\,(dt \times du)\right|. \tag{4.14}$$

7. If $f(t, u) \in \overline{M}_p^{(1)}(\mathbf{F}_t)$, then

$$\mathbf{M}\left|\int_{t_0}^{T} \int_{R^{(m)}} f(t, u)p\,(dt \times du)\right| \leq \int_{t_0}^{T} \int_{R^{(m)}} \mathbf{M}|f(t, u)| \frac{du\,dt}{|u|^{m+1}}. \tag{4.15}$$

For every function $f(t, u) \in \overline{M}_p^{(1)}(\mathbf{F}_t)$, it is possible to construct a sequence $f(t, u) \in \overline{M}_p^{(1)} \cap \overline{M}_p^{(2)}(\mathbf{F}_t)$ such that

$$\lim_{n \to \infty} \int_{t_0}^{T} \int_{R^{(m)}} \mathbf{M}|f_n(t, u) - f(t, u)| \frac{dt\,du}{|u|^{m+1}} = 0. \tag{4.16}$$

Then

$$\lim_{\substack{n \to \infty \\ l \to \infty}} \int_{t_0}^{T} \int_{R^{(m)}} \mathbf{M}|f_l(t, u) - f_n(t, u)| \frac{dt\,du}{|u|^{m+1}} = 0.$$

Therefore the sequence of the variables $\int_{t_0}^{T} \int_{R^{(m)}} f_n(t, u)p\,(dt \times du)$ will converge in probability to some limit, which we shall denote by

$$\int_{t_0}^{T} \int_{R^{(m)}} f(t, u)p\,(dt \times du).$$

Assume that $f(t, u) \in \overline{M}_p^{(2)}(\mathbf{F}_t)$. Then for every N,

$$f_N(t, u) = f(t, u) g_N\left(\int_{t_0}^{t} |f(t, u)| \frac{dt\, du}{|u|^{m+1}}\right)$$

will belong to $\overline{M}_p^{(1)}(\mathbf{F}_t)$, and for $N' > N$,

$$\mathbf{P}\left\{\left|\int_{t_0}^{T}\int_{R^{(m)}} f_N(t, u) p\,(dt \times du) - \int_{t_0}^{T}\int_{R^{(m)}} f_{N'}(t, u) p\,(dt \times du)\right| > 0\right\}$$
$$\leq \mathbf{P}\left\{\int_{t_0}^{T}\int_{R^{(m)}} |f(t, u)| \frac{dt\, du}{|u|^{m+1}} > N\right\}.$$

Consequently $\lim_{N\to\infty} \int_{t_0}^{T} \int_{R^{(m)}} f_N(t, u) p\,(dt \times du)$ exists in the sense of convergence in probability. We shall denote this limit by

$$\int_{t_0}^{T}\int_{R^{(m)}} f(t, u) p\,(dt \times du) \qquad \text{for} \qquad f(t, u) \in \overline{M}_p^{(2)}(\mathbf{F}_t).$$

This integral will possess Properties 5 and 6 and also the following properties:

8. If $f_n(t, u) \to f(t, u)$ in probability for almost all t and u and if there exists $\varphi(t, u) \in \overline{M}_p^{(2)}(\mathbf{F}_t)$ such that $|f_n(t, u)| \leq |\varphi(t, u)|$, then

$$\int_{t_0}^{T}\int_{R^{(m)}} f_n(t, u) p\,(dt \times du) \to \int_{t_0}^{T}\int_{R^{(m)}} f(t, u) p\,(dt \times du)$$

in probability as $n \to \infty$.

9. If $f(t, u)$ is different from zero only on a set A in the algebra $\mathbf{B}$, then

$$\mathbf{P}\left\{\left|\int_{t_0}^{T}\int_{R^{(m)}} f(t, u) p\,(dt \times du)\right| > 0\right\} \leq \mathbf{P}\{p(A) > 0\}$$
$$= 1 - \exp\left(-\int_A \frac{du\, dt}{|u|^{m+1}}\right). \qquad (4.17)$$

We denote by $\overline{M}_p(\mathbf{F}_t)$ the collection of functions $f(t, u)$ belonging to $\overline{M}(\mathbf{F}_t)$ and having the property that for every $N > 0$,

$$\mathbf{P}\left\{\int_{t_0}^{T}\int_{|u|\leq N} |f(t, u)| \frac{dt\, du}{|u|^{m+1}} < \infty\right\} = 1.$$

If $f(t, u) \in \overline{M}_p(\mathbf{F}_t)$, then $f_N(t, u) = f(t, u) g_N(u)$ belongs to $\overline{M}_p^{(2)}(\mathbf{F}_t)$ and the expression

$$\int_{t_0}^{T}\int_{R^{(m)}} f_N(t, u) p\,(dt \times du)$$

is meaningful. Since for $N < N'$, we know from (4.17) that

$$\mathbf{P}\left\{\left|\int_{t_0}^{T}\int_{R^{(m)}} f_N(t, u)p\,(dt \times du) - \int_{t_0}^{T}\int_{R^{(m)}} f_N\,(t, u)p\,(dt \times du)\right| > 0\right\} \leq 1 - \exp\left\{-\int_{t_0}^{T}\int_{|u|\geq N} \frac{du\,dt}{|u|^{m+1}}\right\},$$

then as $N \to \infty$ the sequence of random variables

$$\int_{t_0}^{T}\int_{R^{(m)}} f_N(t, u)p\,(dt \times du)$$

converges in probability to some random variable. Thus we extend the definition of the integral with respect to the measure $\mathbf{P}$ to all functions in $\overline{M}_p(\mathbf{F}_t)$.

This integral satisfies Properties 5, 6, 9, and 8 even in the case where $\varphi(t, u) \in \overline{M}_p(\mathbf{F}_t)$.

The integrals

$$\int_{t_0}^{t}\int_{R^{(m)}} f(s, u)p\,(ds \times du) \qquad \text{and} \qquad \int_{t_0}^{t}\int_{R^{(m)}} f(s, u)q\,(ds \times du)$$

can be considered as functions of the upper limit. These functions possess the following properties:

10. If $f(t, u) \in \overline{M}_q^{(1)}(\mathbf{F}_t)$, then

$$\int_{t_0}^{t}\int_{R^{(m)}} f(s, u)q\,(ds \times du)$$

is a martingale with respect to t, and

$$\mathbf{M}\left(\int_{t}^{t+h}\int_{R^{(m)}} f(s, u)q\,(ds \times du)/\mathbf{F}_t\right) = 0,$$

$$\mathbf{M}\left(\left|\int_{t}^{t+h}\int_{R^{(m)}} f(s, u)q\,(ds \times du)\right|^2/\mathbf{F}_t\right) = \mathbf{M}\left(\int_{t}^{t+h}\int_{R^{(m)}} |f(s, u)|^2 \frac{ds\,du}{|u|^{m+1}}\Big/ \mathbf{F}_t\right).$$

11. The integrals

$$\int_{t_0}^{t}\int_{R^{(m)}} f(s, u)p\,(ds \times du) \qquad \text{and} \qquad \int_{t_0}^{t}\int_{R^{(m)}} f(s, u)q\,(ds \times du)$$

as functions of t with probability 1 do not have discontinuities of the second kind. If for some $\epsilon > 0$, $f(t, u) = 0$ when $|u| \leq \epsilon$, then

$$\int_{t_0}^{t} \int_{R^{(m)}} f(s, u) p \,(ds \times du)$$

with probability 1 is a piecewise constant function; that is, for some subdivision of $[t_0, T]$: $t_0 < t_1 < t_2 < \cdots < t_n = T$, this function is constant on each of the intervals $[t_i, t_{i+1}]$ (the points $t_1, \ldots, t_{n-1}$ are random).

12. If $\chi(t)$ satisfies Condition 6, then

$$\sup_{t_0 \leq t \leq T} \left| \int_{t_0}^{t} \int_{R^{(m)}} f(s, u) q \,(ds \times du) \right| \leq \chi(T) \sup_{t_0 \leq t \leq T} \left| \int_{t_0}^{t} \int_{R^{(m)}} f(s, u) q \,(ds \times du) \right|,$$

$$\sup_{t_0 \leq t \leq T} \left| \int_{t_0}^{T} \int_{R^{(m)}} f(s, u) p \,(ds \times du) \right| \leq \chi(T) \sup_{t_0 \leq t \leq T} \left| \int_{t_0}^{t} \int_{R^{(m)}} f(s, u) p \,(ds \times du) \right| \cdot$$

13. If $f(t, u)$ is different from zero only for $(t, u) \in A$, then

$$\mathbf{P}\left\{ \sup_{t} \left| \int_{t_0}^{t} \int_{R^{(m)}} f(s, u) p \,(ds \times du) \right| > 0 \right\} \leq 1 - \exp\left\{ - \int_{A} \frac{dt\, du}{|u|^{m+1}} \right\}.$$

CHAPTER 3

MARKOV PROCESSES AND STOCHASTIC EQUATIONS

1. The form of stochastic equations for Markov processes. The first general method used for studying Markov processes was Kolmogorov's method of differential equations for the transition probability functions of a process. This analytic method makes it possible to compute the probabilities of certain events that are connected with the behavior of a process during a finite interval of time if we know the infinitesimal characteristics of the process which determine the probabilities of a transition for an infinitesimal interval of time. Thus with the analytic approach we deal not with the process itself as a random function of time but only with its transition probability function.

A more direct method of studying Markov processes is based on a preliminary construction of the process itself on the basis of simple processes that easily lend themselves to study. The theory of stochastic equations for the processes themselves can serve as an apparatus for such a construction.

In this chapter we shall examine stochastic equations for Markov processes; we shall establish existence and uniqueness theorems for these equations; we shall study the dependence of the solutions of the equations on initial conditions; and, on the basis of this study, we shall derive the usual integro-differential equations for certain conditional mathematical expectations which will make it possible to determine a transition probability function.

Let us examine the construction of the stochastic equations for different classes of Markov processes.

Consider a Markov process $\xi(t)$ that is continuous with probability 1 and that takes values in $R^{(m)}$. Suppose that for a transition function $\mathbf{P}(t, x, t_1, A)$ of this process the following conditions are satisfied:

1. For every $\delta > 0$,

$$\lim_{\Delta t \to 0} \frac{1}{\Delta t} \int_{|x-y|>\delta} \mathbf{P}(t, x, t + \Delta t, dy) = 0.$$

2. For some $\delta > 0$ and every x and t, the limit

$$a(t, x) = \lim_{\Delta t \to 0} \frac{1}{\Delta t} \int_{|x-y| \leq \delta} (y - x)\mathbf{P}(t, x, t + \Delta t, dy)$$

exists.

3. For every t and x, there exists a linear symmetric nonnegative transformation $A(t, x)$ to $R^{(m)}$ such that for every $\delta > 0$, $z \in R^{(m)}$,

$$\lim_{\Delta t \to 0} \frac{1}{\Delta t} \int_{|x-y| \leq \delta} (y - x, z)^2 \mathbf{P}(t, x, t + \Delta t, dy) = (A(t, x)z, z).$$

Here the process $\xi(t)$ is called a *diffusion process*, the vector $a(t, x)$ is called a *transition vector*, and the operator $A(t, x)$ is called a *diffusion operator*. (In the one-dimensional case, $a(t, x)$ and $A(t, x)$ will be numerical functions, so that $A(t, x) \geq 0$; in this case $A(t, x)$ is called the *coefficient of diffusion* and $a(t, x)$ the *transition coefficient*.)

If $\xi(t)$ has the properties just listed, then the increment of $\xi(t)$, that is, $\xi(t + \Delta t) - \xi(t)$, no matter what $\xi(t)$ may be, will have the same conditional truncated moments of the first two orders as the quantity

$$a(t, \xi(t))\, \Delta t + \sum_{k=1}^{l} b_k(t, \xi(t))[w_k(t + \Delta t) - w(t)], \tag{1.1}$$

where $b_k(t, x) = \sqrt{\lambda_k(t, x)} e_k(t, x)$, $e_k(t, x)$ are the eigenvectors of the operator $A(t, x)$ corresponding to the nonzero eigenvalues $\lambda_k(t, x)$, and the $w_k(t)$ are Brownian processes that are mutually independent. Since the truncated moments of higher order of $\xi(t + \Delta t) - \xi(t)$ and the quantity (1.1) will be of the order of $o(\Delta t)$, and since for every $\delta > 0$,

$$\mathbf{P}\left\{\left|a(t, \xi(t))\, \Delta t + \sum_{k=1}^{l} b_k(t, \xi(t))[w_k(t + \Delta t) - w_k(t)]\right| > \delta / \xi(t)\right\} = o(\Delta t),$$

the distributions of $\xi(t + \Delta t) - \xi(t)$ and (1.1) coincide with accuracy up to $o(\Delta t)$. Therefore it is natural to expect that when we pass to differentials we shall obtain exact equality of the distributions. This consideration leads us to a stochastic differential equation for the diffusion process

$$d\xi(t) = a(t, \xi(t))\, dt + \sum_{k=1}^{l} b_k(t, \xi(t))\, dw_k(t), \tag{1.2}$$

which, as with an ordinary differential equation, must be solved with certain given initial conditions. Equation (1.2) can be written in the following integral form: If the initial condition is given at a point t_0 and the equation is solved for $t \geq t_0$, then

$$\xi(t) = \xi(t_0) + \int_{t_0}^{t} a(s, \xi(s))\, ds + \sum_{k=1}^{l} \int_{t_0}^{t} b_k(s, \xi(s))\, dw_k(s). \tag{1.3}$$

Let us now examine simple discontinuous Markov processes. Let us first assume that $\xi(t)$ can have jumps only of height h and that the transition probability function satisfies the following condition: There exists a function $\lambda(t, x)$ such that

$$\lim_{\Delta t \to 0} \frac{1}{\Delta t} (\mathbf{P}(t, x, t + \Delta t, \{x\}) - 1 = -\lambda(t, x),$$

$$\lim_{\Delta t \to 0} \frac{1}{\Delta t} \mathbf{P}(t, x, t + \Delta t, \{x + h\}) = \lambda(t, x).$$

(Here $\{y\}$ denotes the set consisting only of the point y.) Let $p(A)$ be a random measure as defined in Section 4 of Chapter 2. We shall assume that the space $R^{(m)}$ is the real line ($m = 1$). Under these conditions, if $\Delta_{x,t}$ is an arbitrary interval on the straight line such that

$$\int_{\Delta_{x,t}} \frac{du}{u^2} = \lambda(t, x),$$

then the quantities $\xi(t + \Delta t) - \xi(t)$ and $hp([t, t + \Delta t] \times \Delta_{\xi(t),t})$ have coincident conditional distributions for fixed $\xi(t)$ with accuracy up to $o\,(\Delta t)$. We set $f(t, x, u) = h$ for $u \in \Delta_{x,t}$ and $f(t, x, u) = 0$ for $u \in \Delta_{x,t}$; then

$$hp([t, t + \Delta t] \times \Delta_{\xi(t),t}) = \int_t^{t+\Delta t} \int_{R^{(1)}} f(t, \xi(t), u) p\,(dt \times du).$$

It is natural to expect exact coincidence of the distributions of $d\xi(t)$ and of the variable

$$\int_{u \in R^{(1)}} f(t, \xi(t), u) p\,(dt \times du).$$

Therefore, for a given process we shall examine the stochastic equations of the form

$$d\xi(t) = \int_{u \in R^{(1)}} f(t, \xi(t), u) p\,(dt \times du). \tag{1.4}$$

If the initial condition is given at a point t_0 and the equation is solved for $t_0 \leq t$, then (1.4) can be written in integral form

$$\xi(t) = \xi(t_0) + \int_{t_0}^{t} \int_{R^{(1)}} f(s, \xi(s), u) p\,(ds \times du). \tag{1.5}$$

If the function $f(t, x, u)$ takes the values $h_1, h_2, \ldots, h_k, 0$, then Eq. (1.5) can determine simple discontinuous processes with jumps of height h_1, $h_2, \ldots, h_k$. When we examine Eq. (1.5) with the function $f(t, x, u)$ assuming a continuous set of values, we obtain a stochastic equation for Markov processes with jumps of arbitrary height. Henceforth we shall consider equations of the form (1.5) where the functions $f(t, x, u)$ are bounded in every bounded region of variation of u. Then for the integral on the right-hand side of (1.5) to exist, it is necessary that

$$\int_{|u|\leq C} |f(t, x, u)| \frac{du}{|u|^2} < \infty.$$

We rewrite Eq. (1.5) in the form

$$\begin{aligned}\xi(t) = \xi(t_0) &+ \int_{t_0}^t \int_{|u|\leq 1} f(s, \xi(s), u) \frac{ds\,du}{u^2} \\ &+ \int_{t_0}^t \int_{|u|\leq 1} f(s, \xi(s), u) q\,(ds \times du) \\ &+ \int_{t_0}^t \int_{|u|>1} f(s, \xi(s), u) p\,(ds \times du)\end{aligned}$$

and define

$$\int_{|u|\leq 1} f(s, x, u) \frac{du}{u^2} = a(s, x).$$

Then

$$\begin{aligned}\xi(t) = \xi(t_0) + \int_{t_0}^t a(s, \xi(s)\,ds) &+ \int_{t_0}^t \int_{|u|\leq 1} f(s, \xi(s), u) q\,(ds \times du) \\ &+ \int_{t_0}^t \int_{|u|>1} f(s, \xi(s), u) p\,(ds \times du). \qquad (1.6)\end{aligned}$$

In this form, the equation is meaningful for those $f(t, x, u)$ for which

$$\int_{|u|\leq 1} |f(t, x, u)|^2 \frac{du}{|u|^2} < \infty,$$

that is, for a larger class of functions $f(t, x, u)$. Equation (1.6) can also be used for defining Markov processes in $R^{(m)}$, in which case $a(t, x)$ and $f(t, x, u)$ must also take values in $R^{(m)}$, and it is natural to examine the measure p on $[t_0, T] \times R^{(m)}$.

By combining Eqs. (1.3) and (1.6), we shall obtain the most general stochastic equation for Markov processes. In this case, it is possible to obtain processes whose jumps are superposed on the continuous diffusion component. This equation has the form:

$$\begin{aligned}\xi(t) = \xi(t_0) &+ \int_{t_0}^t a(s, \xi(s))\,ds + \sum_{k=1}^{l} \int_{t_0} b_k(s, \xi(s))\,dw_k(s) \\ &+ \int_{t_0}^t \int_{|u|\leq 1} f(s, \xi(s), u) q\,(ds \times du) \\ &+ \int_{t_0}^t \int_{|u|>1} f(s, \xi(s), u) p\,(ds \times du). \qquad (1.7)\end{aligned}$$

In addition to these equations, we shall examine the narrower class of equations for which

$$\int |f(t, x, u)|^2 \frac{du}{|u|^{m+1}} < \infty.$$

Then Eq. (1.7) can be more conveniently written in the form

$$\xi(t) = \xi(t_0) + \int_{t_0}^{t} a(s, \xi(s))\, ds + \sum_{k=1}^{l} \int_{t_0}^{t} b_k(s, \xi(s))\, dw_k(s)$$

$$+ \int_{t_0}^{t} sf(s, \xi(s), u) q\,(ds \times du). \tag{1.8}$$

These equations will be studied in this chapter.

2. Existence and uniqueness of the solution to stochastic equations. In this section, we shall examine stochastic equations of a more general type than the equations in the preceding section. Such equations will be used for studying the dependence of the solutions of stochastic equations on the initial conditions.

Suppose that for every $t \in [t_0, T]$, a σ-algebra $\mathbf{F}_t$ is defined, that for $t_1 < t_2$, $\mathbf{F}_{t_1} \subset \mathbf{F}_{t_2}$, l mutually independent Brownian processes $w_1(t), w_2(t), \ldots, w_l(t)$ are defined on the interval $[t_0, T]$, and finally, that the measures p and q, independent of $w_k(t)$ and possessing the properties shown in Section 4 of Chapter 2, are defined on $[t_0, T] \times R^{(m)}$. Suppose, further, that no matter what $t_1 \in [t_0, T]$ is, the random variables $w_1(t_1), \ldots, w_l(t_1)$, $p(\Delta \times A)$, where $\Delta[t_0, t_1]$, $A \subset R^{(m)}$, are measurable with respect to $\mathbf{F}_{t_1}$, and that for $t_1 < t_2 < \cdots < t_r$, the variables $w_1(t_2) - w_1(t_1), \ldots, w_1(t_r) - w_1(t_{r-1}), \ldots, w_l(t_2) - w_l(t_1), \ldots, w_l(t_r) - w_l(t_{r-1})$, $p([t_1, t_2] \times A_j), \ldots, p([t_{r-1}, t_r] \times A_j)$, $j = 1, 2, \ldots, k$, where $A_1, A_2, \ldots, A_k$ are arbitrary Borel sets in $R^{(m)}$ for which the corresponding variables are meaningful, do not depend mutually on each of the events of the σ-algebra $\mathbf{F}_{t_1}$.

Suppose that the functions $\varphi(t)$, $A(t, x)$, $B_1(t, x), \ldots, B_l(t, x)$, $F(t, x, u)$, for every $t \in [t_0, T]$, $x \in R^{(m)}$, $u \in R^{(m)}$, are measurable with respect to $\mathbf{F}_t$ and that they are measurable with respect to the totality of all variables (including the "random" variable ω).

Let us consider the equation

$$\xi(t) = \varphi(t) + \int_{t_0}^{t} A(s, \xi(s))\, ds + \sum_{k=1}^{l} \int_{t_0}^{t} B_k(s, \xi(s))\, dw_k(s)$$

$$+ \int_{t_0}^{t} \int_{R^{(m)}} F(s, \xi(s), u) q\,(ds \times du). \tag{2.1}$$

THEOREM 1. *Suppose that the following conditions are fulfilled:*

1. $\int_{t_0}^{T} \mathbf{M}|\varphi(t)|^2\,dt < \infty.$

2. *There exists an* $L > 0$ *such that, for every* x *and* y *in* $R^{(m)}$ *and* $t \in [t_0, T]$,

$$(T - t_0)|A(t, x) - A(t, y)|^2 + \sum_{k=1}^{l} |B_k(t, x) - B_k(t, y)|^2 + \int_{R^{(m)}} |F(t, x, u) - F(t, y, u)|^2 \frac{du}{|u|^{m+1}} \leq L^2|x - y|^2.$$

Under these conditions, (2.1) *has a solution* $\xi(t)$ *satisfying the following conditions:*

(a) $\xi(t)$ *is measurable with respect to* $\mathbf{F}_t$, *and*

(b)
$$\int_{t_0}^{t} \mathbf{M}|\xi(t)|^2\,dt < \infty;$$

also, this solution is unique up to stochastic equivalence [*that is, any two solutions of* (2.1) *satisfying conditions* (a) *and* (b) *are stochastically equivalent*].

Proof. Let us consider the linear normed space Ξ of processes $\xi(t)$ that satisfy conditions (a) and (b) of the theorem. Assume that

$$\|\xi(t)\|^2 = \int_{t_0}^{T} \mathbf{M}|\xi(t)|^2\,dt.$$

In Ξ, let us define the mapping S:

$$S\xi(t) = \varphi(t) + \int_{t_0}^{t} A(s, \xi(s))\,ds + \sum_{k=1}^{l} \int^{t} B_k(s, \xi(s))\,dw_k(s) + \int_{t_0}^{t} \int_{R^{(m)}} F(s, \xi(s), u)q\,(ds \times du).$$

We shall show that the operator S^n is a contraction operator for sufficiently large values of n:

$$S(\xi_1(t) - S\xi_2(t)) = \int_{t_0}^{t} [A(s, \xi_1(s)) - A(s, \xi_2(s))]\,ds + \sum_{k=1}^{l} \int_{t_0}^{t} [B_k(s, \xi_1(s)) - B_k(s, \xi_2(s))]\,dw_k(s) + \int_{t_0}^{t} \int_{R^{(m)}} [F(s, \xi_1(s), u) - F(s, \xi_2(s), u)]q\,(ds \times du).$$

By using the inequality $(\sum_1^n Q_k)^2 \leq n\sum_{k=1}^n Q_k^2$ and formulas (1.6) and (4.8) in Chapter 2, we obtain

$$\mathbf{M}|S\xi_1(t) - S\xi_2(t)|^2 \leq (l+2)\Big[(t-t_0)\int_{t_0}^t \mathbf{M}|A(s, \xi_1(s)) - A(s, \xi_2(s))|^2\, ds + \sum_{k=1}^{l}\int_{t_0}^t \mathbf{M}|B_k(s, \xi_1(s)) - B_k(s, \xi_2(s))|^2\, ds + \int_{t_0}^t \int_{R^{(m)}} \mathbf{M}|F(s, \xi_1(s), u) - F(s, \xi_2(s), u)|^2 \frac{ds\, du}{|u|^{m+1}}\Big].$$

By considering Condition 2 of the theorem, we can write

$$\mathbf{M}|S\xi_1(t) - S\xi_2(t)|^2 \leq C\int_{t_0}^t \mathbf{M}|\xi_1(s) - \xi_2(s)|^2\, ds, \tag{2.2}$$

where $C = (l+2)L^2$. From (2.2), it is easy to obtain the inequality

$$\mathbf{M}|S^n\xi_1(t) - S^n\xi_2(t)|^2 \leq C^n \int_{t_0<s_1<s_2<\cdots<s_n=T}\!\!\cdots\!\!\int \mathbf{M}|\xi_1(s_1) - \xi_2(s_1)|^2\, ds_1 \cdots ds_n \leq C^n \frac{(t-t_0)^{n-1}}{(n-1)!}\|\xi_1(s) - \xi_2(s)\|,$$

and hence

$$\|S^n\xi_1(t) - S^n\xi_2(t)\| \leq \frac{C^n(T-t_0)^n}{n!}\|\xi_1(t) - \xi_2(t)\|. \tag{2.3}$$

Therefore for sufficiently large values of n_0, S^{n_0} is in fact a contraction operator. Let $\bar{\xi}(t)$ be the fixed point of S^{n_0}. Since such a point is unique, S also has no more than one fixed point. We shall show that $\bar{\xi}(t)$ is also a fixed point of S. From (2.3) and the relation $S^{kn_0}\bar{\xi}(t) = \bar{\xi}(t)$, we have the inequality

$$\|\bar{\xi}(t) - S\bar{\xi}(t)\| = \|S^{kn_0}\bar{\xi}(t) - S^{kn_0+1}\bar{\xi}(t)\| \leq \frac{C^{kn_0}(T-t_0)^{kn_0}}{(kn_0)!}\|\bar{\xi}(t) - S\bar{\xi}(t)\|$$

for all k. Taking the limit in this inequality as $k \to \infty$, we obtain

$$\|\bar{\xi}(t) - S\bar{\xi}(t)\| = 0.$$

The unique fixed point of the mapping S will also be unique up to stochastic continuity. [This follows from (2.2) by solving Eq. (2.1).] This proves the theorem.

Remark. Since the integrals on the right-hand side of (2.1) do not have discontinuities of the second kind (see Theorem 3 of Section 2 and Property 10 of Section 4, Chapter 2), if $\varphi(t)$ with probability 1 has no discontinuities of the second kind, then $\xi(t)$ also has no discontinuities of the second kind. Therefore any two solutions of Eq. (2.1) will in this case coincide with probability 1 at all points of continuity because they coincide with probability 1 on an arbitrary countable set.

The following theorem establishes a local dependence of the solution of Eq. (2.1) on the coefficients $A(t, x)$, $B_k(t, x)$, $F(t, x, u)$ as functions of x.

THEOREM 2. *Let $\xi_1(t)$ and $\xi_2(t)$ be solutions of the equation*

$$\xi_i(t) = \varphi_i(t) + \int_{t_0}^{t} A_i(s, \xi_i(s))\, ds + \sum_{k=1}^{l} B_k^{(i)}(s, \xi_i(s))\, dw_k(s) + \int_{t_0}^{t} \int_{R^{(m)}} F_i(s, \xi_i(s), u) q\,(ds \times du), \tag{2.4}$$

where the $\varphi_i(t)$ do not have discontinuities of the second kind and all the coefficients satisfy the conditions of Theorem 1. Suppose further that $\psi(t)$ is a nonincreasing function that takes only the values 1 and 0 and is measurable with respect to $\mathbf{F}_t$ for every value of t, so that with probability 1 the following relations are satisfied:

$$\psi(t)(\varphi_1(t) - \varphi_2(t)) = 0,$$
$$\psi(t)[A_1(t, \xi_1(t)) - A_2(t, \xi_1(t))] = 0,$$
$$\psi(t)[B_k^{(1)}(t, \xi_1(t)) - B_k^{(2)}(t, \xi_1(t))] = 0,$$
$$\psi(t)[F_1(t, \xi_1(t), u) - F_2(t, \xi_1(t), u)] = 0.$$

Then with probability 1, $\psi(t)\xi_1(t) = \psi(t)\xi_2(t)$ for all t that are points of continuity $\xi_1(t)$ and $\xi_2(t)$.

Proof. It is easy to show that

$$\left|\psi(t)\int_{t_0}^{t} [A_1(s, \xi_1(s)) - A_2(s, \xi_1(s))]\, ds\right| \leq \left|\int_{t_0}^{t} \psi(s)[A_1(s, \xi_1(s)) - A_2(s, \xi_1(s))]\, ds\right| = 0.$$

In the same way, by using Property 2 of Section 1 and Property 2 of Section 4, Chapter 2, we can show that

$$\psi(t)\int_{t_0}^{t} [B_k^{(1)}(s, \xi_1(s)) - B_k^{(2)}(s, \xi_2(s))]\, dw_k(s) = 0,$$
$$\psi(s)\int_{t_0}^{t} \int_{R^{(m)}} [F_1(s, \xi_1(s), u) - F_2(s, \xi_1(s), u)]q\,(ds \times du) = 0.$$

Therefore

$$\psi(t)[\xi_1(t) - \xi_2(t)] = \psi(t)\int_{t_0}^{t} [A_2(s, \xi_1(s)) - A_2(s, \xi_2(s))]\, ds$$
$$+ \sum_{k=1}^{l} \psi(t)\int_{t_0}^{t} [B_k^{(2)}(s, \xi_1(s)) - B_k^{(2)}(s, \xi_2(s))]\, dw_k(s)$$
$$+ \psi(t)\int_{t_0}^{t}\int_{R^{(m)}} [F_2(s, \xi_1(s), u) - F_2(s, \xi_2(s), u)]q\,(ds \times du). \quad (2.5)$$

From this relation we obtain, by the same method as in the proof of Theorem 1, the inequality

$$\mathbf{M}\psi(t)|\xi_1(t) - \xi_2(t)|^2 \le C\int_{t_0}^{t} \mathbf{M}\psi(s)|\xi_1(s) - \xi_2(s)|^2\, ds.$$

By substituting the left-hand side of the last inequality into the right-hand side and repeating such an operation n times, we obtain the inequality

$$\mathbf{M}\psi(t)|\xi_1(t) - \xi_2(t)|^2 \le \frac{C^n}{n!}\int_{t_0}^{T} \mathbf{M}\psi(s)|\xi_1(s) - \xi_2(s)|^2\, ds,$$

from which, by passing to the limit as $n \to \infty$ and remembering the remark following the preceding theorem, we obtain the proof of the theorem.

COROLLARY 1. *Suppose that for $x \in G$, where G is some open set in $R^{(n)}$, $A_1(t, x) = A_2(t, x)$, $B_k^{(1)}(t, x) = B_k^{(2)}(t, x)$, $k = 1, \ldots, l$, $F_1(t, x, u) = F_2(t, x, u)$. We denote by τ an instant of time such that for $t < \tau$, $\xi_1(t) \in G$, $\xi_2(t) \in G$, $\varphi_1(t) = \varphi_2(t)$. Then with probability 1, $\xi_1(t) = \xi_2(t)$ for $t < \tau$, for all t that are points of continuity of $\xi_1(t)$ and $\xi_2(t)$.*

COROLLARY 2. *Let $\xi(t)$ be a solution of Eq. (2.1) with coefficients satisfying the conditions of Theorem* 1. *Let us set $g_N(x) = 1$* for *$|x| \le N$, $g_N(x) = N + 1 - |x|$* for *$N < |x| \le N + 1$, and $g_N(x) = 0$* for *$|x| > N + 1$;*

$$A_N(t, x) = g_N(x)A(t, x);$$
$$B_k^{(N)}(t, x) = g_N(x)B_k(t, x);$$
$$F_N(t, x, u) = g_N(x)F(t, x, u). \quad (2.6)$$

If $\xi_N(t)$ is a solution of (2.1) in which, instead of $A(t, x)$, $B_k(t, x)$, $F(t, x, u)$, we substitute $A_N(t, x)$, $B_k^{(N)}(t, x)$, $F_N(t, x, u)$, then with probability 1, for all $\xi(t)$ for which $\sup_{t_0 \le t \le T} |\xi(t)| \le N$, $\xi(t)$ and $\xi_N(t)$ will coincide with probability 1 at all points of continuity of both functions. Thus in the case in which $\varphi(t)$ is bounded with probability 1, if we neglect the ω-set with arbitrarily small probability, we may assume that $\xi(t)$ is a solution of the equation of the form (2.1), whose coefficients are different from zero only in some bounded interval where x varies.

COROLLARY 3. *We denote by* $\psi(t)$ *the function that is equal to* 0 *if* $p([t_0, T] \times \{|u| > N\}) > 0$ *and equal to* 1 *if* $p([t_0, T)] \times \{|u| > N\}) = 0$. *This function* $\psi(t)$ *satisfies the conditions of Theorem* 2. *Consequently if* $\overline{F}_N(t, x, u) = g_N(u)F(t, x, u)$ *and* $\bar{\xi}_N(t)$ *is a solution of Eq.* (2.1) *in which, instead of* $F(t, x, u)$, *we substitute* $\overline{F}_N(t, x, u)$, *then for all* $t \in [t_0, T]$, $\mathbf{P}\{\psi(t)(\xi(t) - \xi_N(t)) = 0\} = 1$. *Since by choosing N sufficiently large it is possible to make* $\mathbf{P}\{\psi(T) = 1\}$ *arbitrarily close to unity, we can, by neglecting the* ω-*set of arbitrarily small probability, assume that* $\xi(t)$ *is a solution of an equation of the form* (2.1) *such that* $F(t, x, u)$ *is different from zero only in some bounded region where u varies.*

THEOREM 3. *Assume that the coefficients* $A(s, x), B_1(s, x), \ldots, B_l(s, x), F(s, x, u)$ *satisfy the following conditions:*

1. *For every C there exists an* L_C *such that*

$$(T - t_0)|A(s, x) - A(s, y)|^2 + \sum_1^l |B_k(s, x) - B_k(s, y)|^2 + \int_{u\in R^{(m)}} |F(s, x, u) - F(s, y, u)|^2 \frac{du}{|u|^{m+1}} \leq L_C^2|x - y|^2$$

whenever $|x| \leq C$, $|y| \leq C$.

2. *There exists a K such that*

$$(T - t_0)|A(s, x)|^2 + \sum_1^l |B_k(s, x)|^2 + \int_{u\in R^{(m)}} |F(s, x, u)|^2 \frac{du}{|u|^{m-1}} \leq K(|x|^2 + 1).$$

3. $\varphi(t)$ *with probability 1 does not have discontinuities of the second kind. Then Eq.* (2.1) *has with probability 1 a bounded solution that is unique up to stochastic equivalence.*

Proof of the uniqueness. Suppose that $\xi(t_1)$ and $\xi(t_2)$ are solutions of (2.1) and that with probability 1 they are bounded. Set $\psi(t) = 1$ if $\sup_{t_0\leq s\leq t} |\xi_1(s)| \leq N$ and $\sup_{t_0\leq s\leq t} |\xi_2(s)| \leq N$, and otherwise $\psi(t) = 0$. Since $\psi(t)$ satisfies the conditions of Theorem 2, we may write formula (2.5) for $\psi(t)[\xi_1(t) - \xi_2(t)]$. From Condition 1,

$$[T - t_0]\psi(t)|A(t, \xi_1(t)) - A(t, \xi_2(t))|^2 + \sum_{k=1}^l \psi(t)|B_k(t, \xi_1(t)) - B_k(t, \xi_2(t))|^2 + \psi(t)\int_{u\in R^{(m)}} |F(t, \xi_1(t), u) - F(t, \xi_2(t)u)|^2 \frac{du}{|u|^{m+1}} \leq L_N^2|\xi_1(t) - \xi_2(t)|^2\psi(t),$$

and from (2.5) we obtain the inequality

$$\mathbf{M}\psi(t)|\xi_1(t) - \xi_2(t)|^2 \leq C\int_{t_0}^{t} \mathbf{M}\psi(s)|\xi_1(s) - \xi_2(s)|^2\, ds,$$

from which, as in Theorem 2, we obtain

$$\mathbf{M}\psi(t)|\xi_1(t) - \xi_2(t)|^2 = 0.$$

From the last inequality, it follows that

$$\mathbf{P}\{\psi(t)|\xi_1(t) - \xi_2(t)| = 0\} = 1.$$

Since $\psi(t)$ as a result of the boundedness of $\xi_1(t)$ and $\xi_2(t)$ with probability 1 approaches unity as $N \to \infty$, we obtain the proof of uniqueness by passing to the limit as $N \to \infty$ in the last equation.

Proof of existence. Let $A_N(s, x)$, $B_1^{(N)}(s, x), \ldots, B_l^{(N)}(s, x)$, $F_N(s, x, u)$ be defined by (2.6), $\varphi_N(t) = g_N(\varphi(t))\varphi(t)$. We denote by $\xi_N(t)$ the solution of the equation

$$\xi_N(t) = \varphi_N(t) + \int_{t_0}^{t} A_N(s, \xi_N(s))\, ds + \sum_{k=1}^{l} \int_{t_0}^{t} B_k^{(N)}(s, \xi_N(s))\, dw_k(s)$$
$$+ \int_{t_0}^{t} \int_{u\in R^{(m)}} F_N(s, \xi_N(s), u) q\,(ds \times du). \qquad (2.7)$$

From Corollary 1 of the preceding theorem, it follows that for $N' > N$, $\xi_{N'}(t) = \xi_N(t)$ for all points t at which $\xi_{N'}(t)$ and $\xi_N(t)$ are continuous provided that $\sup_t |\xi_{N'}(t)| < N$ and $\sup_t |\xi_N(t)| < N$. Let us assume that the $\varphi(t)$ are continuous from the right and that the solutions $\xi_{N'}(t)$ are also continuous from the right. From the fact that $\xi_N(t) = \xi_{N'}(t)$ for almost all t, if $\sup_t |\xi_N(t)| < N$, $\sup_t |\xi_{N'}(t)| < N$, we have the inequality

$$\mathbf{P}\left\{\sup_{t_0\leq t\leq T} |\xi_{N'}(t) - \xi_N(t)| > 0\right\}$$
$$\leq \mathbf{P}\left\{\sup_{t_0\leq t\leq T} |\xi_{N'}(t)| \geq N\right\} + \mathbf{P}\left\{\sup_{t_0\leq t\leq T} |\xi_N(t)| \geq N\right\}. \qquad (2.8)$$

Let us show that $\mathbf{P}\{\sup_{t_0\leq t\leq T} |\xi_N(t)| > C\} \to 0$ uniformly with respect to N as $C \to \infty$. For any N_0, we note that if $\bar{\xi}_N(t)$ is a solution to equation (2.7) with $\varphi_{N_0}(t)$ substituted for $\varphi_N(t)$, then

$$\mathbf{P}\left\{\sup_{t_0\leq t\leq T} |\xi_N(t) - \bar{\xi}_N(t)| > 0\right\} \leq \mathbf{P}\left\{\sup_{t_0\leq t\leq T} |\varphi(t)| > N_0\right\}.$$

Therefore

$$\mathbf{P}\left\{\sup_{t_0\leq t\leq T} |\xi_N(t)| > C\right\} \leq \mathbf{P}\left\{\sup_{t_0\leq t\leq T} |\bar{\xi}_N(t)| > \epsilon\right\}$$
$$+ \mathbf{P}\left\{\sup_{t_0\leq t\leq T} |\varphi(t)| > N_0\right\}. \qquad (2.9)$$

Since we have

$$|\bar{\xi}_N(t)|^2 \leq (l+3)\Big[|\varphi_{N_0}(t)|^2 + (t-t_0)\int_{t_0}^{t} |A_N(s, \bar{\xi}_N(s))|^2\,ds + \sum_{k=1}^{l}\Big(\int_{t_0}^{t} B_k^{(N)}(s, \bar{\xi}_N(s))\,dw_k(s)\Big)^2 + \Big(\int_{t_0}^{t}\int_{u\in R^{(m)}} F_N(s, \bar{\xi}_N(s), u)q\,(ds\times du)\Big)^2\Big],$$

from Condition 2 we obtain

$$\mathbf{M}|\bar{\xi}_N(t)|^2 \leq \Big(\mathbf{M}|\varphi_{N_0}(t)|^2 + K\int_{t_0}^{t}\mathbf{M}|\bar{\xi}_N(s)|^2\,ds\Big)(l+3).$$

From this inequality, it follows that

$$\mathbf{M}|\bar{\xi}_N(t)|^2 \leq (l+3)\mathbf{M}\,(\sup_t |\varphi_{N_0}(t)|^2)e^{(l+3)K(t+t_0)}.$$

Also,

$$\sup_{t_0\leq t\leq T} |\bar{\xi}_N(t)| \leq \sup_{t_0\leq t\leq T} |\varphi_{N_0}(t)| + \int_{t_0}^{T} |A_N(s, \bar{\xi}_N(s))|\,ds + \sum_{k=1}^{l}\sup_{t_0\leq t\leq T}\Big|\int_{t_0}^{t} B_k^{(N)}(s, \bar{\xi}_N(s))\,dw_k(s)\Big| + \sup_{t_0\leq t\leq T}\Big|\int_{t_0}^{t}\int_{u\in R^{(m)}} F(s, \bar{\xi}_N(s), u)q\,(ds\times du)\Big|.$$

Since stochastic integrals when considered as functions of the upper limit are martingales, and for martingales we have Property 5 of Section 5, Chapter 1 (here, $\alpha = 2$), we obtain

$$\mathbf{M}\sup_{t_0\leq t\leq T} |\bar{\xi}_N(t)|^2 \leq (l+3)\Big[\mathbf{M}\sup_{t_0\leq t\leq T}|\varphi_{N_0}(t)|^2 + (T-t_0)\int_{t_0}^{T}\mathbf{M}|A_N(s, \bar{\xi}_N(s))|^2\,ds + 4\sum_{k=1}^{l}\int_{t_0}^{T}|B_k^{(N)}(s, \bar{\xi}_N(s))|^2\,ds + 4\int_{t_0}^{T}\int_{u\in R^{(m)}}|F(s, \bar{\xi}_N(s), u)|^2\frac{ds\,du}{|u|^{m+1}}\Big]$$

$$\leq (l+3)(N_0+1)^2 + 4K\int_{t_0}^{T}\mathbf{M}|\bar{\xi}_N(s)|^2\,ds$$

$$\leq (l+3)(N_0+1)^2[1 + 4K(T-t_0)e^{(l+3)K(T-t_0)}].$$

Thus there exists a constant H depending only on K, l, $(T - t_0)$ such that

$$\mathbf{M} \sup_{t_0 \leq t \leq T} |\bar{\xi}_N(t)|^2 \leq H\mathbf{M} \sup_{t_0 \leq t \leq T} |\varphi_{N_0}(t)|^2. \tag{2.10}$$

From (2.9), it follows that

$$\mathbf{P}\left\{\sup_{t_0 \leq t \leq T} |\xi_N(t)| > C\right\} < \frac{H}{C^2} \mathbf{M}\left(\sup_{t_0 \leq t \leq T} |\varphi_{N_0}(t)|^2\right) + \mathbf{P}\left\{\sup_{t_0 \leq t \leq T} |\varphi(t)| > N_0\right\}.$$

Taking $\sup_N$ on both sides of the last inequality and passing to the limit as $C \to \infty$ and $N_0 \to \infty$, we obtain

$$\lim_{C \to \infty} \sup_N \mathbf{P}\left\{\sup_{t_0 \leq t \leq T} |\xi_N(t)| > C\right\} = 0.$$

We choose an increasing sequence N_k such that

$$\mathbf{P}\left\{\sup_{t_0 \leq t \leq T} |\xi_N(t)| > N_k\right\} \leq \frac{1}{k^2}$$

for $N > N_k$. Then from (2.8), it follows that

$$\mathbf{P}\left\{\sup_{t_0 \leq t \leq T} |\xi_{N_k}(t) - \xi_{N_{k+1}}(t)| > 0\right\} \leq \frac{2}{k^2}.$$

From the Borel-Cantelli lemma, in the sequence of events

$$\left\{\sup_{t_0 \leq t \leq T} |\xi_{N_k}(t) - \xi_{N_{k+1}}(t)| > 0\right\},$$

only a finite number of events takes place with probability 1, and therefore, from some number on, all the $\xi_{N_k}(t)$ are equal to one another. Consequently the $\xi_{N_k}(t)$ with probability 1 converge uniformly to some process $\xi(t)$. The process $\xi(t)$ also will have no discontinuities of the second kind with probability 1. We note further that in the case in which

$$\sup_{t_0 \leq t \leq T} |\xi_{N_k}(t)| < N_k,$$

we have $\xi(t) = \xi_{N_k}(t)$, and when

$$\sup_{t_0 \leq t \leq T} |\xi_{N_k}(t)| \geq N_{N_k},$$

we have

$$\sup_{t_0 \leq t \leq T} |\xi(t)| \geq N_k.$$

Since $|\xi_{N_k}(t)| \le N_k + 1$, with probability 1 the inequality

$$\sup_{t_0 \le t \le T} |\xi_{N_k}(t)| \le \sup_{t_0 \le t \le T} |\xi(t)| + 1 \tag{2.11}$$

holds. If we substitute N_k for N in (2.7) and pass to the limit as $N_k \to \infty$ [we can make this passage to the limit under the stochastic integral signs in connection with (2.11) because of Condition 2 of the theorem and Properties 5 of Section 1 and 4 of Section 4, Chapter 2], we see that $\xi(t)$ is a solution of Eq. (2.1). Thus the existence of a solution is also shown.

Remark 1. If $\mathbf{M} \sup_{t_0 \le t \le T} |\varphi(t)|^2 < \infty$, then, going to the limit as $N_0 \to \infty$ and $N \to \infty$ in (2.10), we see that there exists an H depending only on K, l, $T - t_0$ such that

$$\mathbf{M}\left(\sup_{t_0 \le t \le T} |\xi(t)|^2\right) \le \mathbf{HM}\left(\sup_{t_0 \le t \le T} |\varphi(t)|^2\right). \tag{2.12}$$

Therefore

$$\sup_{t_0 \le t \le T} \mathbf{M}|\xi(t)|^2 \le H\mathbf{M}\left(\sup_{t_0 \le t \le T} |\varphi(t)|^2\right). \tag{2.13}$$

Remark 2. Under the conditions of the preceding remark, there exists an H_1 such that

$$\mathbf{M} \sup_{t_1 \le t_2 \le t_1 + h} |\xi(t_2) - \varphi(t_2) - \xi(t_1) + \varphi(t_1)|^2 \le H_1 \mathbf{M} \sup_{t_0 \le t \le T} |\varphi(t)|^2 h. \tag{2.14}$$

In fact,

$$\begin{aligned}\mathbf{M} \sup_{t_1 \le t_2 \le t_1 + h} &|\xi(t_2) - \varphi(t_2) - \xi(t_1) + \varphi(t_1)|^2 \\ &\le (l+2)\Big[h \int_{t_1}^{t_1+h} \mathbf{M}|A(s, \xi(s))|^2 \, ds \\ &\quad + \sum_{k=1}^{l} \mathbf{M} \sup_{t_1 \le t_2 \le t_1 + h} \left| \int_{t_1}^{t_2} B_k(s, \xi(s)) \, dw_k(s) \right|^2 \\ &\quad + \mathbf{M} \sup_{t_1 \le t_2 \le t_1 + h} \left| \int_{t_1}^{t_1+h} \int_{u \in R^{(m)}} F(s, \xi_1(s), u) q\,(ds \times du) \right|^2 \Big].\end{aligned}$$

Using Property 5 of martingales in Section 5 of Chapter 1, we obtain

$$\begin{aligned}\mathbf{M} \sup_{t_1 \le t_2 \le t_1 + h} &|\xi(t_2) - \varphi(t_2) - \xi(t_1) + \varphi(t_1)|^2 \\ &\le (l+2)K \int_{t_1}^{t_1+h} \mathbf{M}|\xi(t)|^2 \, dt \le HK(l+2)\mathbf{M} \sup_{t_0 \le t \le T} |\varphi(t)|^2 \cdot h,\end{aligned}$$

in view of (2.13). Setting $H_1 = HK(l+2)$, we obtain (2.14). From (2.14), it follows that, in particular, $\xi(t)$ will be continuous in probability provided $\varphi(t)$ is continuous in probability.

Theorem 4 follows from Theorem 3:

THEOREM 4. *Suppose that $\xi(t_0)$ is independent of the measure q and of the processes $w_k(t)$ and that $a(t, x)$, $b_k(t, x)$, $f(t, x, u)$, defined for $t \in [t_0, T]$, $x \in R^{(m)}$, $u \in R^{(m)}$, are measurable with respect to the set of variables, and that they satisfy the following conditions:*

1. *For every $C > 0$, there exists an L_C such that*

$$(T - t_0)|a(t, x) - a(t, y)|^2 + \sum_{k=1}^{l} |b_k(t, x) - b_k(t, y)|^2 + \int_{u \in R^{(m)}} |f(t, x, u) - f(t, y, u)|^2 \frac{du}{|u|^{m+1}} \leq L_C^2 |x - y|^2$$

if $|x| \leq C$ and $|y| \leq C$.

2. *There exists a K at which*

$$|a(t, x)|^2 + \sum_{k=1}^{l} |b_k(t, x)|^2 + \int_{u \in R^{(m)}} |f(t, x, u)|^2 \frac{du}{|u|^{m+1}} \leq K(|x|^2 + 1).$$

In such a case, Eq. (1.8) *has a solution without discontinuities of the second kind with probability 1; also, both solutions of* (1.8) *with probability 1 coincide at all the points t that are points of continuity of both processes.*

To prove the theorem, we note that the existence of a solution follows from Theorem 3; the fact that this solution will not have discontinuities of the second kind with probability 1 follows from Formula (1.8), because the stochastic integrals of the right-hand side of (1.8) do not have discontinuities of the second kind with probability 1. Finally, it is easy to see that two stochastically equivalent processes having no discontinuities of the second kind and coinciding on an arbitrary countable set with probability 1 will also coincide with probability 1 at all points of continuity of both processes.

3. Existence and uniqueness of the solution to stochastic equations (continued). In this section, we shall examine the questions of existence and uniqueness of Eq. (1.7). To study Eq. (1.7), we need a certain property of integrals with respect to the measure p.

LEMMA 1. *There exist a random number ν and random points $\tau_1, \tau_2, \ldots, \tau_\nu$ on $[t_0, T]$ with $u_1, u_2, \ldots, u_\nu \in R^{(m)}$, satisfying the condition $|u_i| > \epsilon$, such that for every function $\varphi(t, u)$ in $\mathbf{M}_p(\mathbf{F}_t)$, the following equation is* *satisfied with probability 1:*

$$\int_{t_0}^{T}\int_{|u|>\epsilon} \varphi(t, u)p\,(dt \times du) = \sum_{k=1}^{\nu} \varphi(\tau_k, u_k). \tag{3.1}$$

Proof. Let

$$\xi(t) = \int_{t_0}^{t}\int_{|u|>\epsilon} up\,(ds \times du).$$

It is easy to see that $\xi(t)$ is a process with independent increments whose characteristic function is

$$\mathbf{M}e^{i(z,\xi(t))} = \exp\left\{(t - t_0)\int_{|u|>\epsilon} (e^{i(u,z)} - 1)\frac{du}{|u|^{m+1}}\right\}.$$

Since the process $\xi(t)$ does not have discontinuities of the second kind and all its discontinuities exceed ϵ in absolute value, it must have only a finite number of discontinuities (with probability 1).

We denote the number of discontinuities by ν, the instants of time at which discontinuities appear by $\tau_1, \tau_2, \ldots, \tau_\nu$, and the values of the discontinuities $(u_k = \xi(\tau_k + 0) - \xi(\tau_k - 0))$ by $u_1, u_2, \ldots, u_\nu$. Then for every set $A \subset [t_0, T] \times \{|u| > \epsilon\}$, $p(A) = k$, where k is the number of points (τ_i, u_i) for which $(\tau_i, u_i) \in A$. If $\varphi(t, u)$ is a step function, that is, if there exist disjoint sets $A_j \subset [t_0, T] \times \{|u| > \epsilon\}$, $j = 1, \ldots, N$ for which $U_jA_j = [t_0, T] \times \{|u| > \epsilon\}$ and $\varphi(t, u)$ is constant on each of the sets A_j, then with probability 1,

$$\int_{t_0}^{T}\int_{|u|>\epsilon} \varphi(t, u)p\,(dt \times du) = \sum_{j=1}^{N} p(A_j)\varphi(\bar{t}_j, \bar{u}_j),$$

where $(\bar{t}_j, \bar{u}_j)$ is an arbitrary point in A_j. If $P(A_j) = k$, then in the preceding formula we may substitute $(1/k)\sum_{(\tau_i,u_i)\in A_j} \varphi(\tau_i, u_i)$ for $\varphi(\bar{t}_j, \bar{u}_j)$. With such a substitution, we can see that Formula (3.1) holds for an arbitrary step function $\varphi(t, u)$. We note now that the variables τ_k and u_k have a bounded common distribution function, that τ_k and u_k are independent, that the distribution of u_k is given by the formula

$$\mathbf{P}\,\{u_k \in B\} = \int_{B\cap\{|u|>\epsilon\}} \frac{du}{|u|^{m+1}} \cdot \left(\int_{|u|>\epsilon} \frac{du}{|u|^{m+1}}\right)^{-1},$$

and that the distribution of the quantity τ_k with the condition that $\nu \geq k$ is given by the formula

$$\mathbf{P}\{t_0 < \tau_k < t\} = \left[\frac{1 - \exp\left(-\,(t - t_0)\displaystyle\int_{|u|>\epsilon} \frac{du}{|u|^{m+1}}\right)}{1 - \exp\left(-\,(T - t_0)\displaystyle\int_{|u|>\epsilon} \frac{du}{|u|^{m+1}}\right)}\right]^k.$$

For an arbitrary function $\varphi(t, u)$ in $\overline{M}_p(\mathbf{F}_t)$, it is possible to construct a sequence of step functions $\varphi_n(t, u)$ such that $\varphi_n(t, u) \to \varphi(t, u)$ in probability for almost all (t, u) and $|\varphi_n(t, u)| \le |\varphi(t, u)|$. From the boundedness of the density of the variables (τ_k, u_k), it follows that $\varphi_n(\tau_k, u_k) \to \varphi(\tau_k, u_k)$ for almost all values of (τ_k, u_k); this means that $\varphi_n(\tau_k, u_k) \to \varphi(\tau_k, u_k)$ in probability. Writing (3.1) for $\varphi_n(t, u)$ and passing to the limit as $n \to \infty$ [on the left-hand side, it is possible to pass to the limit under the integral sign because of Condition 8 for functions in $\overline{M}_p(\mathbf{F}_t)$, Section 24, Chapter 2], we obtain proof of the lemma.

Remark. In exactly the same manner, we can show that with probability 1

$$\int_{t_0}^{t} \int_{|u|>\epsilon} \varphi(s, u) p\,(ds \times du) = \sum_{\tau_i < t} \varphi(\tau_i, u_i),$$

no matter what the function $\varphi(t, u)$ in $\overline{M}_p(\mathbf{F}_i)$, $t \in [t_0, T]$ may be.

A THEOREM ON THE UNIQUENESS OF THE SOLUTION OF (1.7). *Suppose that the following conditions are fulfilled:*

1. *$\xi(t_0)$ does not depend on the processes $w_1(t), \ldots, w_l(t)$ or on the measures p and q.*

2. *For every $C > 0$, there exists an L_C such that for $|x| \le C$, $y \le C$, the inequality*

$$(T - t_0)|a(t, x) - a(t, y)|^2 + \sum_{k=1}^{l} |b_k(t, x) - b_k(t, y)|^2$$
$$+ \int_{|u|\le 1} \frac{|f(t, x, u) - f(t, y, u)|^2}{|u|^{m+1}}\, du \le L_C|x - y|^2$$

is fulfilled.

3. *There exists a K such that*

$$|a(t, x)|^2 + \sum_{k=1}^{l} |b_k(t, x)|^2 + \int_{|u|\le 1} |f(t, x, u)|^2 \frac{du}{|u|^{m+1}} \le K(1 + |x|^2).$$

4. For every $N > 0$, $\displaystyle\sup_{\substack{t_0 \le t \le T \\ |x|\le N,\, |u|\le N}} |f(t, x, u)| < \infty.$

Then the left-continuous solution of Eq. (1.7) is unique in the sense that any two such solutions coincide with probability 1 at all points.

Proof. Let $\xi_1(t)$ and $\xi_2(t)$ be two left-continuous solutions of (1.7). On the basis of the preceding lemma, we can find points $(\tau_1, u_1), (\tau_2, u_2), \ldots,$

(τ_ν, u_ν) that satisfy the relations

$$\xi_i(t) = \xi(t_0) + \int_{t_0}^t a(s, \xi_i(s))\, ds + \sum_{k=1}^{l} \int_{t_0}^t b_k(s, \xi_i(s))\, dw_k(s)$$

$$+ \int_{t_0}^t \int_{|u|\le 1} f(s, \xi_i(s), u) q\, (ds \times du) + \sum_{\tau_k < t} f(\tau_k, \xi_i(\tau_k), u_k).$$

Assume that $\varphi_i(t) = \sum_{\tau_k<t} f(\tau_k, \xi_i(\tau_k), u_k) + \xi(t_0)$, that $\psi(t) = 1$ for $t \le \tau_1$, and that $\psi(t) = 0$ for $t > \tau_1$. Since $[\varphi_1(t) = \varphi_2(t)]\psi(t) = 0$, on the basis of Theorem 2, Section 2, we may state that $[\xi_1(t) - \xi_2(t)]\psi(t) = 0$. Since $\xi_1(t) = \xi_2(t)$ for $t \le \tau_1$, this implies in particular that $\xi_1(\tau_1) = \xi_2(\tau_1)$ and therefore $f(\tau_1, \xi_1(\tau_1), u_1) = f(\tau_1, \xi_2(\tau_1), u_1)$. If $\psi_1(t) = 1$ for $t \le \tau_2$, and $\psi_1(t) = 0$ for $t > \tau_2$, we know on the basis of what was just said that $\psi_1(t)[\varphi_1(t) - \varphi_2(t)] = 0$ and therefore $\psi_1(t)[\xi_1(t) - \xi_2(t)] = 0$; that is, $\xi_1(t) = \xi_2(t)$ for $t \le \tau_2$. By continuing this line of reasoning, we see that $\xi_1(t) = \xi_2(t)$ for all $t \in [t_0, T]$ with probability 1. This proves the theorem.

A THEOREM ON THE EXISTENCE OF A SOLUTION TO EQ. (1.7). This theorem is characterized by the fact that there is no Lipschitz condition imposed on the coefficients of the equation. In fact, in the absence of such conditions, we cannot guarantee the uniqueness of the solution.

Suppose that the following conditions are fulfilled:

1. *$a(t, x), b_1(t, x), \ldots, b_l(t, x)$ are continuous with respect to the set of variables with $x \in R^{(m)}$, $t \in [t_0, T]$.*

2. *No matter what $t_1 \in [t_0, T]$ and $x_1 \in R^{(m)}$ may be,*

$$\lim_{\substack{t\to t_1 \\ x\to x_1}} \int_{|u|\le 1} |f(t, x, u) - f(t_1, x_1, u)|^2 \frac{du}{|u|^{m+1}} = 0.$$

3. *There exists a K such that*

$$|a(t, x)|^2 + \sum_{k=1}^{l} |b_k(t, x)|^2 + \int_{|u|\le 1} |f(t, x, u)|^2 \frac{du}{|u|^{m+1}} \le K(1 + |x|^2).$$

4. *$f(t, x, u)$ is bounded in every bounded region of variation of x and u and is continuous for almost all u with respect to t and x. Then Eq.* (1.7) *has a bounded solution with probability 1.*

The proof of the existence of a solution will be given by the method of finite differences. This method is also useful for investigating other questions to be considered in this book.

Let us consider a sequence of subdivisions of the interval $[t_0, T]$:

$$t_0 = t_0^{(n)} < t_1^{(n)} < \cdots < t_n^{(n)} = T,$$

such that

$$\lim_{k\to\infty} \max_k (t_{k+1}^{(n)} - t_k^{(n)}) = 0.$$

We define the variables $\xi_k^{(n)}$ by the relations: $\xi_0^{(n)} = \xi(t_0)$,

$$\xi_{k+1}^{(n)} = \xi_k^{(n)} + a(t_k^{(n)}, \xi_k^{(n)})\, \Delta t_k^{(n)} + \sum_{j=1}^{l} b_i(t_k^{(n)}, \xi_k^{(n)})[w_j(t_{k+1}^{(n)}) - w_j(t_k^{(n)})]$$
$$+ \int_{t_k^{(n)}}^{t_{k+1}^{(n)}} \int_{|u|\le 1} f(t_k^{(n)}, \xi_k^{(n)}, u) q\,(ds \times du)$$
$$+ \int_{t_k^{(n)}}^{t_{k+1}^{(n)}} \int_{|u|>1} f(t_k^{(n)}, \xi_k^{(n)}, u) p\,(ds \times du). \tag{3.2}$$

Here and elsewhere in this section, $\Delta t_k^{(n)} = t_{k+1}^{(n)} - t_k^{(n)}$. Let us show that under these assumptions, the variables $\xi_k^{(n)}$ converge in a certain sense to a solution of Eq. (1.7) as $t_k^{(n)} \to t$, $n \to \infty$. To prove this, we need certain auxiliary propositions.

LEMMA 2. *Suppose that there exists a constant K such that*

$$|a(t, x)|^2 + \sum_{j=1}^{l} |b_j(t, x)|^2 + \int_{|u|\le 1} |f(t, x, u)|^2 \frac{du}{|u|^{m+1}} \le K(1 + |x|^2),$$

and let $f(t, x, u)$ be bounded in every bounded region of variation of x and u. Then the values of $\sup_k |\xi_k^{(n)}|$ are bounded in probability uniformly with respect to n.

Proof. Define $g_N(x) = 1$ for $|x| \le N$, $g_N(x) = 0$ for $|x| > N$, and $\eta_0^{(n)} = g_N(\xi_0^{(n)})\,\xi_0^{(n)}$, and

$$\eta_{k+1}^{(n)} = \eta_k^{(n)} + a_N(t_k^{(n)}, \eta_k^{(n)})\, \Delta t_k^{(n)} + \sum_{j=1}^{l} b_j^{(N)}(t_k^{(n)}, \eta_k^{(n)})[w_j(t_{k+1}^{(n)}) - w_j(t_k^{(n)})]$$
$$+ \int_{t_k^{(n)}}^{t_{k+1}^{(n)}} \int_{|u|\le 1} f_N(t_k^{(n)}, \eta_k^{(n)}, u) q\,(ds \times du)$$
$$+ \int_{t_k^{(n)}}^{t_{k+1}^{(n)}} \int_{|u|>1} f_N(t_k^{(n)}, \eta_k^{(n)}, u) p\,(ds \times du), \tag{3.3}$$

where $a_N(t, x) = g_N(x)a(t, x)$, $b_j^{(N)}(t, x) = g_N(x)b_j(t, x)$, $f_N(t, x, u) = g_N(x)g_N(u)f(t, x, u)$. Obviously, $|\eta_k^{(n)}| = \eta_{k+1}^{(n)}$ for $|\eta_k^{(n)}| > N$, and for $|\eta_k^{(n)}| \leq N$,

$$
\begin{aligned}
|\eta_{k+1}^{(n)}| \leq N &+ |a_N(t_k^{(n)}, \eta_k^{(n)})| \, \Delta t_k^{(n)} \\
&+ \sum_{j+1}^{l} |b_j^{(N)}(t_k^{(n)}, \eta_k^{(n)})| \, |w_j(t_{k+1}^{(n)}) - w_j(t_k^{(n)})| \\
&+ \left| \int_{t_k^{(n)}}^{t_{k+1}^{(n)}} \int_{|u| \leq 1} f_N(t_k^{(n)}, \eta_k^{(n)}, u) q \, (ds \times du) \right| \\
&+ \sup_{\substack{t_0 \leq t \leq T \\ |u| \leq N, |x| \leq N}} |f(t, x, u)| p([t_0, T] \times \{|u| > 1\}).
\end{aligned}
$$

Therefore the inequality

$$
\begin{aligned}
|\eta_k^{(n)}| \leq N &+ \sum_{r=0}^{n-1} |a_N(t_r^{(n)}, \eta_r^{(n)})| \, \Delta t_r^{(n)} \\
&+ \sum_{r=0}^{n-1} \sum_{j=1}^{l} |b_j^{(N)}(t_r^{(n)}, \eta_r^{(n)})| w_j(t_{r+1}^{(n)}) - w_j(t_r^{(n)})| \\
&+ \sum_{r=0}^{n-1} \left| \int_{t_r^{(n)}}^{t_{r+1}^{(n)}} \int_{|u| \leq 1} f_N(t_r^{(n)}, \eta_r^{(n)}, u) q \, (ds \times du) \right| \\
&+ \sup_{t_0 \leq t \leq T} \sup_{|u| \leq N, |x| \leq N} |f(t, x, u)| p([t_0, T] \times \{|u| > 1\})
\end{aligned}
$$

holds for all k. Since the mathematical expectation of the square exists for every term on the right-hand side, $\mathbf{M}|\eta_k^{(n)}|^2 < \infty$.

It follows from Lemma 1 that the measure p, taken over $[t_0, T] \times \{|u| > 1\}$, is concentrated in a finite number of points (τ_1, u_1), (τ_2, u_2), $\ldots$, (τ_ν, u_ν) and that the number of these points is random and coincides with $p([t_0, T] \times \{|u| > 1\})$; also,

$$
\mathbf{P}\left\{ \sup_{1 \leq i \leq \nu} |u_i| > C \right\} = 1 - \exp\left\{ - (T - t_0) \int_{|u| < C} \frac{du}{|u|^{m+1}} \right\}
$$

can be made arbitrarily small by a large enough choice of C. We denote by $[t_{k_i}^{(n)}, t_{k_i+1}^{(n)}]$, $i = 1, 2, \ldots, \nu' \leq \nu$, those intervals that contain at least one of the points $\tau_1, \ldots, \tau_\nu$. Since for $k_i + 1 \leq k < k_{i+1}$,

$$
\int_{t_k^{(n)}}^{t_{k+1}^{(n)}} \int_{|u| > 1} f_N(t_k^{(n)}, \eta_k^{(n)}, u) p \, (ds \times du) = 0,
$$

it follows that

$$\mathbf{M}|\eta_{k+1}^{(n)}|^2 = \mathbf{M}|\eta_k^{(n)}|^2 + 2\mathbf{M}\left(a_N(t_k^{(n)}, \eta_k^{(n)}), \eta_k^{(n)}\right)\Delta t_k^{(n)} + \mathbf{M}|a_N(t_k^{(n)}, \eta_k^{(n)})|^2 (\Delta t_k^{(n)})^2 + \mathbf{M}\left|\sum_{j=1}^{l} b_j^{(N)}(t_k^{(n)}, \eta_k^{(n)})[w_j(t_{k+1}^{(n)}) - w_j(t_k^{(n)})] + \int_{t_k^{(n)}}^{t_{k+1}^{(n)}}\int_{|u|\le 1} f_N(t_k^{(n)}, \eta_k^{(n)}, u)q\,(ds\times du)\right|^2.$$

Consequently there exists a constant H dependent only upon K and $(T - t_0)$ such that

$$\mathbf{M}|\eta_{k+1}^{(n)}|^2 \le \mathbf{M}|\eta_k^{(n)}|^2(1 + H\,\Delta t_k^{(n)}) + H\,\Delta t_k^{(n)} \le \mathbf{M}|\eta_k^{(n)}|^2 e^{H\,\Delta t_k^{(n)}} + H\,\Delta t_k^{(n)}.$$

This means that

$$\mathbf{M}|\eta_{k+1}^{(n)}|^2 + 1 \le (\mathbf{M}|\eta_k^{(n)}|^2 + 1)e^{H\,\Delta t_k^{(n)}};$$

therefore

$$\mathbf{M}|\eta_k^{(n)}|^2 \le \mathbf{M}(|\eta_{k_i+1}^{(n)}|^2 + 1)e^{H(T-t_0)} - 1 \tag{3.4}$$

if $k_i + 1 \le k < k_{i+1}$. Further,

$$\sup_{k_i+1\le k<k_{i+1}} |\eta_k^{(n)}| \le |\eta_{k_i+1}|^2 + \sum_{k=k_i+1}^{k_{i+1}-1} |a_N|(t_k^{(n)}, \eta_k^{(n)})|\,\Delta t_k^{(n)} + \sum_{j=1}^{l}\sup_{k_i+1<r<k_{i+1}}\left|\sum_{k=k_i+1}^{r} b_j^{(N)}(t_k^{(n)}, \eta_k^{(n)})\left(w_j(t_{k+1}^{(n)}) - w_j(t_k^{(n)})\right)\right| + \sup_{k_i+1<r<k_{i+1}}\left|\sum_{k=k_i+1}^{r}\int_{t_k^{(n)}}^{t_{k+1}^{(n)}}\int_{|u|\le 1} f_N(t_k^{(n)}, \eta_k^{(n)}, u)q\,(ds\times du)\right|.$$

The sums

$$\sum_{k=k_i+1}^{r} b_j^{(N)}(t_k^{(n)}, \eta_k^{(n)})\left(w_j(t_{k+1}^{(n)}) - w_j(t_k^{(n)})\right)$$

and

$$\sum_{k=k_i+1}^{r}\int_{t_k^{(n)}}^{t_{k+1}^{(n)}}\int_{|u|\le 1} f_N(t_k^{(n)}, \eta_k^{(n)}, u)q\,(ds\times du)$$

constitute martingales with respect to r. Therefore, on the basis of Property 2 of Section 5, Chapter 1,

$$\mathbf{M} \sup_{k_i+1<r\le k_{i+1}} \left| \sum_{k=k_i+1}^{r} b_j^{(N)}(t_k^{(n)}, \eta_k^{(n)})\left(w_j(t_{k+1}^{(n)}) - w_j(t_k^{(n)})\right)\right|^2$$

$$\le 4 \sum_{k=k_i+1}^{k_{i+1}-1} \mathbf{M}\, |b_j^{(N)}(t_k^{(n)}, \eta_k^{(n)})|^2 \,\Delta t_k^{(n)} \le 4K \sum_{k=k_i+1}^{k_{i+1}-1} (\mathbf{M}|\eta_k^{(n)}|^2 + 1)\,\Delta t_k^{(n)}.$$

Analogously,

$$\mathbf{M} \sup_{k_i+1\le r<k_{i+1}} \left| \sum_{k=k_{i+1}}^{r} \int_{t_k^{(n)}}^{t_{k+1}^{(n)}} \int_{|u|\le 1} f_N(t_k^{(n)}, \eta_k^{(n)}, u) q\,(ds \times du)\right|^2$$

$$\le 4K \sum_{k=k_{i+1}}^{k_{i+1}-1} (\mathbf{M}|\eta_k^{(n)}|^2 + 1)\,\Delta t_k^{(n)},$$

$$\mathbf{M}\left(\sum_{k=k_i+1}^{k_{i+1}-1} |a_N(t_k^{(n)}, \eta_k^{(n)})|\,(\Delta t_k^{(n)})^2\right)$$

$$\le 4K(T - t_0) \sum_{k=k_i+1}^{k_{i+1}-1} (\mathbf{M}|\eta_k^{(n)}|^2 + 1)\,\Delta t_k^{(n)}.$$

From the inequality (3.4), we see that there exist constants A and B depending only on K and $T - t_0$ such that

$$\mathbf{M} \sup_{k_i+1\le k\le k_{i+1}} |\eta_k^{(n)}|^2 \le A + B\mathbf{M}|\eta_{k_i+1}^{(n)}|^2. \tag{3.5}$$

When we examine the conditional mathematical expectation for fixed $\eta_{k_i+1}^{(n)}$, we can, in the same way as for (3.5), show that

$$\mathbf{M}\,(\sup_{k_i+1\le k\le k_{i+1}} |\eta_k^{(n)}|^2/\eta_{k_i+1}^{(n)}) \le A + B|\eta_{k_i+1}^{(n)}|^2.$$

We note further that from the hypothesis of the lemma,

$$\mathbf{M}(|\eta_{k_i+1}^{(n)} - \eta_{k_i}^{(n)}|^2/\eta_{k_i}^{(n)}) \le C(1 + |\eta_{k_i}^{(n)}|^2) + \sup_{\substack{t\in[t_0,1]\\ j\le\nu}} |f(t, \eta_{k_i}^{(n)}, u_j|^2\nu^2$$

for some C. Thus $\sup |\eta_k^{(n)}|$ is uniformly bounded with respect to N and n in probability for $k_i + 1 < k < k_{i+1}$ provided $|\eta_{k_i+1}^{(n)}|$ possesses this property, and $|\eta_{k_i+1}^{(n)}|$ is uniformly bounded with respect to n and N in probability if $|\eta_{k_i}^{(n)}|$ possesses this property. Since $|\eta_0^{(n)}|$ is bounded in probability uniformly with respect to n and N, we see by induction on i that $\sup_k |\eta_k^{(n)}|$ is bounded in probability uniformly with respect to n

and N. But for $\sup_k |\eta_k^{(n)}| < N$,

$$\sup_k |\eta_k^{(n)}| = \sup_k |\xi_k^{(n)}|;$$

consequently $\sup_k |\xi_k^{(n)}|$ is also bounded in probability uniformly with respect to n. This proves the lemma.

Remark. It follows from this lemma that for every $\epsilon > 0$, it is possible to find an N and $a_N(t, x)$, $b_k^{(N)}(t, x)$, $f_N(t, x, u)$ different from zero only for $|x| < N$, $|u| < N$ such that if $\xi_k^{(n)}$ is defined by the relation (3.2) and $\eta_k^{(n)}$ by the relations (3.3), then

$$\mathbf{P}\left\{\sup_k |\xi_k^{(n)} - \eta_k^{(n)}| > 0\right\} < \epsilon.$$

LEMMA 3. *We define $\xi_n(t) = \xi_k^{(n)}$ if $t \in [t_k^{(n)}, t_{k+1}^{(n)}]$. Then for every $\delta > 0$,*

$$\lim_{h\to 0} \overline{\lim_{n\to\infty}} \sup_{|t_1-t_2|\le h} \mathbf{P}\{|\xi_n(t_1) - \xi_n(t_2)| > \delta\} = 0.$$

Proof. Let $\eta_k^{(n)}$ be defined by Formula (3.3) and let $a_N(t, x)$, $b_j^{(n)}(t, x)$ and $f_N(t, x, u)$ be such that the conditions of the preceding remark are fulfilled. We set $\eta_n(t) = \eta_k^{(n)}$ for $t \in [t_k^{(n)}, t_{k+1}^{(n)}]$. Then

$$\sup_{|t_1-t_2|\le h} \mathbf{P}\{|\xi_n(t_2) - \xi_n(t_1)| > \delta\} \le \sup_{|t_1-t_2|\le h} \mathbf{P}\{|\eta_n(t_2) - \eta_n(t_1)| > \delta\}$$
$$+ \mathbf{P}\{\sup_k |\xi_k^{(n)} - \eta_k^{(n)}| > 0\}.$$

From (3.3), in view of the boundedness of $a_N(t, x)$, $b_k^{(N)}(t, x)$ and $f_N(t, x, u)$ it is easy to see that

$$\lim_{h\to 0} \overline{\lim_{n\to\infty}} \sup_{|t_2-t_1|\le h} \mathbf{P}\{|\eta_n(t_2) - \eta_n(t_1)| > \delta\} = 0.$$

For example,

$$\mathbf{P}\left\{\left|\sum_{t_1<t_{k+1}^{(n)}\le t_2} \int_{t_k^{(n)}}^{t_{k+1}^{(n)}} \int_{|u|\le 1} f(t_k^{(n)}, \eta_k^{(n)}, u) q\,(ds \times du)\right| > \delta\right\}$$
$$\le \frac{1}{\delta^2} \sum_{t_1\le t_{k+1}^{(n)}\le t_2} \int_{t_k^{(n)}}^{t_{k+1}^{(n)}} \int_{|u|\le 1} M|f(t_k^{(n)}, \eta_k^{(n)}, u)|^2 \frac{ds\,du}{|u|^{m+1}}$$
$$\le \frac{1}{\delta^2} K\big(t_2 - t_1 + \max_k \Delta t_k^{(n)}(1 + \sup_k \mathbf{M}|\eta_k^{(n)}|^2)\big).$$

Since $\sup_k \mathbf{M}|\eta_k^{(n)}|^2$ is uniformly bounded with respect to n,

$$\lim_{n\to\infty} \sup_{|t_1-t_2|\le h} \mathbf{P}\left\{\left| \sum_{t_1<t_{k+1}^{(n)}\le t_2} \int_{t_k^{(n)}}^{t_{k+1}^{(n)}} \int_{|u|\le 1} f(t_k^{(n)}, \eta_k^{(n)}, u) q\,(ds\times du)\right| > \delta\right\}$$

$$\le \frac{1}{\delta^2} Kh(1 + \overline{\lim_{n\to\infty}} \sup_k \mathbf{M}|\eta_k^{(n)}|^2).$$

Therefore

$$\lim_{h\to 0} \lim_{n\to\infty} \sup_{|t_1-t_2|\le h} \mathbf{P}\{|\xi_n(t_2) - \xi_n(t_1)| > \delta\}$$

$$\le \overline{\lim_{n\to\infty}} \mathbf{P}\,\{\sup_k |\xi_k^{(n)} - \eta_k^{(n)}| > 0\}.$$

Since we can make the right-hand side of the last inequality arbitrarily small, the proof of the lemma follows.

LEMMA 4. *Let p_n be a sequence of Poisson measures with independent values defined on $[t_0, T] \times R^{(m)}$ for which*

$$\mathbf{M}p_n(A) = \int_A \frac{dt\,du}{|u|^{m+1}}, \qquad q_n(A) = p_n(A) - \mathbf{M}p_n(A).$$

(That is, the measures p_n and q_n have the same properties as p and q, as defined in Section 4, Chapter 2.) *Consider the sequence of processes*

$$\zeta_n(t) = \int_{t_0}^t \int_{|u|\le 1} uq_n\,(ds\times du) + \int_{t_0}^t \int_{|u|>1} up_n\,(ds\times du).$$

If $\zeta_n(t)$ converges in probability for some t to a particular process $\zeta_0(t)$ as $n\to\infty$, then there exists a Poisson measure with independent values p_0 in the space $[t_0, T] \times R^{(m)}$ such that

$$\mathbf{M}p_0(A) = \int_A \frac{dt\,du}{|u|^{m+1}},$$

and if $q_0(A) = p_0(A) - \mathbf{M}p_0(A)$, then

1.

$$\zeta_0(t) = \int_{t_0}^t \int_{|u|\le 1} uq_0\,(ds\times du) + \int_{t_0}^t \int_{|u|>1} up_0\,(ds\times du); \tag{3.6}$$

2. *for an arbitrary measurable finite function $\varphi(u)$,*

$$\int_{t_0}^T \int_{|u|>\epsilon} \varphi(u)p_n\,(ds\times du) \to \int_{t_0}^T \int_{|u|>\epsilon} \varphi(u)p_0\,(ds\times du)$$

in probability.

Proof. Since the processes $\zeta_n(t)$ all have identical finite-dimensional distributions and are processes with independent increments, the process $\zeta_0(t)$ must be a process with independent increments and the finite-dimensional distributions of the process $\zeta_0(t)$ will coincide with the finite-dimensional distributions of the processes $\zeta_n(t)$.

Processes with independent increments, with probability 1, do not have discontinuities of the second kind. If $A \subset [t_0, T] \times R^{(m)}$, we denote by $p_0(A)$ the number of jumps of the process $\zeta_0(t)$ for which $(t, \zeta_0(t+0) - \zeta_0(t-0), p) \in A$. As we can see, $p_0(A)$ has the same joint distributions (for different A) as do the measures $p_n(A)$; that is, $p_0(A)$ is a Poisson measure with independent values for which

$$\mathbf{M}p_0(A) = \int_A \frac{dt\,du}{|u|^{m+1}}.$$

We denote by $\nu_n, \tau_1^{(n)}, \ldots, \tau_{\nu_n}^{(n)}, u_1^{(n)}, \ldots, u_{\nu_n}^{(n)}$ the number, the instants, and the values of the jumps of $\zeta_n(t)$ that exceed ϵ in absolute value; we denote by $\nu_1, \tau_1, \ldots, \tau_\nu, u_1, \ldots, u_\nu$ the number, the instants, and the values of the jumps of $\zeta_0(t)$ that exceed ϵ in absolute value. We can show that as $n \to \infty$, $\nu_n \to \nu$, $\tau_k^{(n)} \to \tau_k$, $u_k^{(n)} \to u_k$ in probability. Since

$$\int_{t_0}^T \int_{|u|>\epsilon} \varphi(u)p_n\,(ds \times du) = \sum_{k=1}^{\nu_n} \varphi(u_k^{(n)}),$$

$$\int_{t_0}^T \int_{|u|>\epsilon} \varphi(u)p_0\,(ds \times du) = \sum_{k=1}^{\nu} \varphi(u_k),$$

the variables u_k and $u_k^{(n)}$ have identical distributions,

$$\mathbf{P}\{u_k \in A\} = \left(\int_{\substack{|u|>\epsilon \\ u\in A}} \frac{du}{|u|^{m+1}}\right)\left(\int_{|u|>\epsilon} \frac{du}{|u|^{m+1}}\right)^{-1},$$

and since for a continuous function $g(u)$, $g(u_k^{(n)}) \to g(u)$ in probability, it follows that

$$\int_{t_0}^T \int_{|u|>\epsilon} \varphi(u)p_n\,(ds \times du) \to \int_{t_0}^T \int_{|u|>\epsilon} \varphi(u)p_0\,(ds \times du)$$

in probability, because it is possible to choose a continuous function g for which $\mathbf{M}|g(u_k) - \varphi(u_k)|$ will be arbitrarily small. This proves assertion (2) of the Lemma.

To prove assertion (1), we note on the basis of the second assumption that

$$\int_{t_0}^t \int_{|u|>\epsilon} up_n\,(ds \times du) \to \int_{t_0}^t \int_{|u|>\epsilon} up_0\,(ds \times du)$$

in probability. From this, we see that

$$\int_{t_0}^{t}\int_{\epsilon<|u|\le 1} uq_n\,(ds\times du) + \int_{t_0}^{t}\int_{|u|>1} up_n\,(ds\times du)$$
$$\to \int_{t_0}^{t}\int_{\epsilon<|u|\le 1} uq_0\,(ds\times du) + \int_{t_0}^{t}\int_{|u|>1} up_0\,(ds\times du)$$

in probability. Therefore for every $\rho > 0$,

$$\mathbf{P}\left\{\left|\zeta_0(t) - \int_{t_0}^{t}\int_{|u|\le 1} uq_0\,(ds\times du) - \int_{t_0}^{t}\int_{|u|>1} up_0\,(ds\times du)\right| > \rho\right\}$$
$$\le \mathbf{P}\left\{\left|\int_{t_0}^{t}\int_{|u|\le\epsilon} uq_0\,(ds\times du)\right| > \frac{\rho}{3}\right\}$$
$$+ \varlimsup_{n\to\infty}\mathbf{P}\left\{\left|\int_{t_0}^{t}\int_{|u|\le\epsilon} uq_n\,(ds\times du)\right| > \frac{\rho}{3}\right\}$$
$$\le \frac{18}{\rho^2}(t-t_0)\int_{|u|\le\epsilon}\frac{du}{|u|^{m+1}} \to 0$$

as $\epsilon \to \infty$. This proves the lemma.

Proof of the existence theorem. Let $\xi_n(t)$ be defined as in Lemma 3 and $\zeta_n(t)$ as in Lemma 4. Then for each of the processes $\xi_n(t)$, $\zeta_n(t)$, $w_1(t), \ldots,$ $w_l(t)$, the conditions of Remark 2 of Section 6, Chapter 1, will be fulfilled. Therefore, on the basis of Corollary 2 of Section 6, Chapter 1, it is possible to choose a sequence n' and to construct processes $\tilde{\xi}_{n'}(t)$, $\tilde{\zeta}_{n'}(t)$, $\tilde{w}_1^{(n')}(t), \ldots,$ $\tilde{w}_l^{(n')}(t)$ such that their joint finite-dimensional distributions coincide for every n' with the joint finite-dimensional distributions of the processes $\xi_{n'}(t)$, $\zeta_{n'}(t)$, $w_1(t), \ldots, w_l(t)$, and such that $\tilde{\xi}_{n'}(t) \to \tilde{\xi}(t)$ as $n' \to \infty$, and $\tilde{w}_j^{(n')}(t) \to \tilde{w}_j(t)$ as $\tilde{\zeta}_{n'}(t) \to \tilde{\zeta}(t)$ in probability, where $\tilde{\xi}(t)$, $\tilde{\xi}(t)$, $\tilde{w}_1(t),$ $\ldots, \tilde{w}_l(t)$ are certain random processes.

Suppose that $\tilde{p}_n$ and $\tilde{p}$ are measures defined for the processes $\tilde{\zeta}_{n'}(t)$ and $\tilde{\zeta}(t)$ in the same way as the measure p_0 was defined in the proof of Lemma 4 for the process $\zeta_0(t)$. The measures $\tilde{p}_{n'}$ and $\tilde{p}$ are Poisson measures with independent values for which

$$\mathbf{M}\tilde{p}_n(A) = \mathbf{M}\tilde{p}(A) = \int_A \frac{dt\,du}{|u|^{m+1}}.$$

We define

$$\tilde{q}(A) = \tilde{p}(A) - \mathbf{M}\tilde{p}(A), \quad \tilde{q}_n(A) = \tilde{p}_n(A) - \mathbf{M}\tilde{p}_n(A).$$

The independence of the processes $\tilde{\zeta}_{n'}(t)$, $\tilde{w}_1^{(n')}(t), \ldots, \tilde{w}_l^{(n')}(t)$ implies the independence of the processes $\tilde{w}_1(t), \ldots, \tilde{w}_l(t)$ and the measure $\tilde{p}$. We see that the joint finite-dimensional distributions of the processes $\tilde{w}_1(t)$, $\ldots$, $\tilde{w}_l(t)$ and of the measures $\tilde{p}$, $\tilde{q}$ are exactly the same as the joint finite-dimensional distributions of the processes $w_1(t)$, $\ldots$, $w_l(t)$ and of the measures p and q. If we establish the existence of a solution to (1.7), when $w_1(t)$, $\ldots$, $w_l(t)$, p, q are replaced with $\tilde{w}_1(t)$, $\ldots$, $\tilde{w}_l(t)$, $\tilde{p}$, $\tilde{q}$, then this very fact establishes the existence of a solution to (1.7). An existing solution defined in this manner will be expressed by $\tilde{\xi}(t_0)$, $w_1(t)$, $\ldots$, $\tilde{w}_l(t)$, $\tilde{p}$, $\tilde{q}$; therefore, when we examine just such a function of $\xi(t_0)$, $w_1(t)$, $\ldots$, $w_l(t)$, p, q, we obtain a solution of Eq. (1.7). Let us show that $\tilde{\xi}(t)$ will be a solution of Eq. (1.7) if in this equation we replace $w_1(t)$, $\ldots$, $w_l(t)$, p, q with $\tilde{w}_1(t)$, $\ldots$, $\tilde{w}_1(t)$, $\tilde{p}$, $\tilde{q}$. $\tilde{\xi}_{n'}(t)$ satisfies the equation

$$
\begin{aligned}
\tilde{\xi}_{n'}(t) = \tilde{\xi}_{n'}(t_0) &+ \sum_{t_{k+1}^{(n')} \leq t} a\big(t_k^{(n')}, \tilde{\xi}_{n'}(t_k^{(n')})\big)\, \Delta t_k^{(n')} \\
&+ \sum_{t_{k+1}^{(n')} \leq t} \sum_{j=1}^{l} b_j\big(t_k^{(n')}, \tilde{\xi}_{n'}(t_k^{(n')})\big)[\tilde{w}_j^{(n')}(t_{k+1}^{(n')}) - \tilde{w}_j^{(n')}(t_k^{(n')})] \\
&+ \sum_{t_{k+1}^{(n')} \leq t} \bigg(\int_{t_k^{(n')}}^{t_{k+1}^{(n')}} \int_{|u| \leq 1} f(t_k^{(n')}, \tilde{\xi}_{n'}(t_k^{(n')}), u)\tilde{q}_{n'}\,(ds \times du) \\
&+ \int_{t_k^{(n')}}^{t_{k+1}^{(n')}} \int_{|u|>1} f(t_k^{(n')}, \tilde{\xi}_{n'}(t_k^{(n')}), u)\tilde{p}_{n'}\,(ds \times du)\bigg). \qquad (3.7)
\end{aligned}
$$

By making use of the uniform convergence of

$$
\mathbf{P}\Big\{\sup_{t_0 \leq t \leq T} |\tilde{\xi}_{n'}(t)| > N\Big\} = \mathbf{P}\Big\{\sup_{t_0 \leq t \leq T} |\xi_{n'}(t)| > N\Big\}
$$

to zero with respect to n' as $N \to \infty$, of the continuity of $a(t, x)$ and $b_j(t, x)$, and of the convergence of $\tilde{\xi}_{n'}(t)$ and $\tilde{w}_j^{(n')}(t)$ to $\tilde{\xi}(t)$ and $\tilde{u}_j(t)$, respectively, we can, by using the theorem of Section 3, Chapter 2, show that

$$
\sum_{t_{k+1}^{(n')} \leq t} b_j\big(t_k^{(n')}, \tilde{\xi}_{n'}(t_k^{(n')})\big)[w_j^{(n')}(t_{k+1}^{(n')}) - w_j^{(n')}(t_k^{(n')})] \to \int_{t_0}^{t} b_j\big(s, \tilde{\xi}(s)\big)\, d\tilde{w}_j(s)
$$

in probability as $n' \to \infty$, that

$$
\sum_{t_{k+1}^{(n')}} a\big(t_k^{(n')}, \tilde{\xi}_{n'}(t_k^{(n')})\big)\, \Delta t_k^{(n')} \to \int_{t_0}^{t} a\big(s, \tilde{\xi}(s)\big)\, ds
$$

in probability as $n' \to \infty$, and that for every $\epsilon > 0$,

$$\sum_{t_{k+1}^{(n')} \le t} \int_{t_k^{(n')}}^{t_{k+1}^{(n')}} \int_{\epsilon < |u| \le 1} f(t_k^{(n')}, \xi_{n'}(t_k^{(n')}), u) \frac{dt\, du}{|u|^{m+1}} \to \int_{t_0}^{t} \int_{\epsilon < |u| \le 1} f(s, \xi(s), u) \frac{ds\, du}{|u|^{m+1}}$$

in probability as $n' \to \infty$.

Let us consider the variables

$$\sum_{t_{k+1}^{(n')} \le t} \int_{t_k^{(n')}}^{t_{k+1}^{(n')}} \int_{|u| \le \epsilon} f(t_k^{(n')}, \xi_{n'}(t_k^{(n')}), u) \tilde{q}_{n'} (ds \times du).$$

If $g_N(x) = 1$ as $|x| \le N$ and $g_N(x) = 0$ as $|x| > N$, then for every $\rho > 0$,

$$\mathbf{P}\left\{\left| \sum_{t_{k+1}^{(n')} \le t} \int_{t_k^{(n')}}^{t_{k+1}^{(n')}} \int_{|u| \le \epsilon} f(t_k^{(n')}, \xi_{n'}(t_k^{(n')}), u) \tilde{q}_{n'} (ds \times du) \right| > \rho\right\}$$
$$\le \mathbf{P}\left\{\sup_t |\xi_{n'}(t)| > N\right\} + \frac{1}{\rho^2} \sum_{t_{k+1}^{(n')} \le t} \int_{t_k^{(n')}}^{t_{k+1}^{(n')}} \int_{|u| \le \epsilon} \mathbf{M} |f(t_k^{(n')}, \xi_{n'}(t_k^{(n')}), u)|^2 g_N(\xi_{n'}(t_k^{(n')}))$$
$$\times \frac{dt\, du}{|u|^{m+1}} \le \frac{T - t_0}{\rho^2} \sup_{\substack{|x| \le N \\ t \in [t_0, T]}} \int_{|u| \le \epsilon} |f(t, x, u)|^2 \frac{dt\, du}{|u|^{m+1}} + \mathbf{P}\left\{\sup_t |\xi_{n'}(t)| > N\right\}.$$

From Condition 2 of the theorem, it follows that for every N,

$$\lim_{\epsilon \to 0} \sup_{\substack{|x| \le N \\ t \in [t_0, T]}} \int_{|u| \le \epsilon} |f(t, x, u)|^2 \frac{du}{|u|^{m+1}} = 0.$$

Then by a sufficiently small choice of $\epsilon > 0$, it is possible to make

$$\sum_{t_{k+1}^{(n')} \le t} \int_{t_k^{(n')}}^{t_{k+1}^{(n')}} \int_{|u| \le \epsilon} f(t_k^{(n')}, \xi_{n'}(t_k^{(n')}), u) \tilde{q}_{n'} (ds \times du)$$

arbitrarily small in probability uniformly with respect to n'. An analogous calculation shows that

$$\int_{t_0}^{t} \int_{|u| \leq \epsilon} f(s, \xi(s), u) \tilde{q}\ (ds \times du)$$

can be made arbitrarily small in probability by a sufficiently small choice of ϵ. If we can show that for every $\epsilon > 0$,

$$\sum_{t_{k+1}^{(n')} \leq t} \int_{t_k^{(n')}}^{t_{k+1}^{(n')}} \int_{|u| > |\epsilon} f(t_k^{(n')}, \xi_{n'}(t_k^{(n')}), u) \tilde{p}_{n'}\ (ds \times du) \rightarrow \int_{t_0}^{t} \int_{|u| > \epsilon} f(s, \xi(s), u) \tilde{p}\ (ds \times du) \tag{3.8}$$

in probability as $n' \rightarrow \infty$, then by going to the limit as $n' \rightarrow \infty$ in (3.7), we see that $\xi(t)$ satisfies Eq. (1.7), in which $w_k(t)$, p, q are replaced with $\tilde{w}_k(t)$, $\tilde{p}$, $\tilde{q}$; that is, we obtain the proof of the theorem.

To prove relation (3.8), we first prove the following lemma.

LEMMA 5. *If $t_0 \leq t < t + h \leq T$, then*

$$\int_{t}^{t+h} \int_{|u| > \epsilon} f(t, \xi_{n'}(t), u) \tilde{p}_{n'}\ (ds \times du) \rightarrow \int_{t}^{t+h} \int_{|u| > \epsilon} f(t, \xi(t), u) \tilde{p}\ (ds \times du)$$

in probability as $n' \rightarrow \infty$.

Proof. We define

$$\varphi_n(x) = \int_{t}^{t+h} \int_{|u| > \epsilon} f(t, x, u) \tilde{p}_n\ (ds \times du)$$

and

$$\varphi_0(x) = \int_{t}^{t+h} \int_{|u| > \epsilon} f(t, x, u) \tilde{p}\ (ds \times du).$$

The random functions $\varphi_n(x)$ and $\varphi_0(x)$ have identical finite-dimensional distributions. Also, $\varphi_n(x)$ and $\varphi_0(x)$ are continuous with probability 1 because, on the basis of Lemma 1, $\varphi_n(x) = \sum_{k+1}^{\nu_n} f(t, x, u_k^n)$; also $f(t, x, u)$ is continuous with respect to x and t for almost all u [the same applies to $\varphi_0(x)$]. From the continuity of $\varphi_0(x)$ with probability 1, it follows that for every $\epsilon > 0$ and $C > 0$,

$$\lim_{\delta \rightarrow 0} \mathbf{P} \left\{ \sup_{\substack{|x-y| \leq \delta \\ |x| \leq C}} |\varphi_0(x) - \varphi_0(y)| > \epsilon \right\} = 0.$$

Therefore no matter what $\epsilon > 0$, $C > 0$, $\eta > 0$ may be, it is possible to find $\delta > 0$ for which

$$\mathbf{P}\left\{\sup_{\substack{|x-y|\leq\delta\\|x|\leq C}} |\varphi_0(x) - \varphi_0(y)| > \epsilon\right\} > \eta.$$

Since $\varphi_n(x)$ and $\varphi_0(x)$ have identical distributions, it follows that

$$\mathbf{P}\left\{\sup_{\substack{|x-y|\leq\delta\\|x|\leq C}} |\varphi_n(x) - \varphi_n(y)| > \epsilon\right\} < \eta.$$

From Lemma 4, it follows that $\varphi_n(x) \to \varphi(x)$ in probability for all x. Suppose that x_k, $k = 1, 2, \ldots, N$, forms a δ-net on the set $\{|x| \leq C\}$. Then

$$\begin{aligned}\mathbf{P}\left\{\sup_{|x|\leq C} |\varphi_{n'}(x) - \varphi_0(x)| > 3\epsilon\right\} &\leq \mathbf{P}\left\{\sup_k |\varphi_{n'}(x_k) - \varphi_0(x_k)| > \epsilon\right\}\\ &\quad + \mathbf{P}\left\{\sup_{\substack{|x-y|\leq\delta\\|x|\leq C}} |\varphi_0(x) - \varphi_0(y)| > \epsilon\right\}\\ &\quad + \mathbf{P}\left\{\sup_{\substack{|x-y|\leq\delta\\|x|\leq C}} |\varphi_{n'}(x) - \varphi_{n'}(y)| > \epsilon\right\}\\ &\leq \sum_{k=1}^{N} \mathbf{P}\{|\varphi_{n'}(x_k) - \varphi_0(x_k)| > \epsilon\} + 2\eta.\end{aligned}$$

Therefore for every $\epsilon > 0$ and C > 0,

$$\lim_{n'\to\infty} \mathbf{P}\left\{\sup_{|x|\leq C} |\varphi_{n'}(x) - \varphi_0(x)| > \epsilon\right\} = 0. \tag{3.9}$$

To prove the lemma, we need to show that for every $\epsilon > 0$,

$$\lim_{n'\to\infty} \mathbf{P}\{|\varphi_{n'}(\xi_{n'}(t)) - \varphi_0(\xi(t))| > \epsilon\} = 0. \tag{3.10}$$

But

$$\begin{aligned}\mathbf{P}\{|\varphi_{n'}(\xi_{n'}(t)) - \varphi_0(\xi(t))| > \epsilon\} &\leq \mathbf{P}\{|\xi(t)| > C\}\\ &\quad + \mathbf{P}\left\{\sup_{|x|\leq C} |\varphi_n(x) - \varphi_0(x)| > \frac{\epsilon}{2}\right\}\\ &\quad + \mathbf{P}\left\{|\varphi_0(\xi_{n'}(t)) - \varphi_0(\xi(t))| > \frac{\epsilon}{2}\right\};\end{aligned}$$

then from (3.9) it follows that

$$\begin{aligned}\overline{\lim_{n'\to\infty}} \mathbf{P}\{|\varphi_{n'}(\xi_{n'}(t)) - \varphi_0(\xi(t))| > \epsilon\} &\leq \mathbf{P}\{|\xi(t)| > C\}\\ &\quad + \overline{\lim_{n'\to\infty}} \mathbf{P}\left\{|\varphi_0(\xi_{n'}(t)) - \varphi_0(\xi(t))| < \frac{\epsilon}{2}\right\}.\end{aligned}$$

The second term on the right-hand side of the last equation is equal to zero because $\varphi_0(x)$ is continuous with probability 1 and $\xi_{n'}(t) \to \xi(t)$ in probability. By a suitable choice of C, we can make $\mathbf{P}\{|\xi(t)| > C\}$ arbitrarily small. Consequently (3.10) is satisfied. This proves the lemma.

To derive Formula (3.8), we take an arbitrary subdivision of the interval $[t_0, t] : t_0 < \bar{t}_1 < \cdots < \bar{t}_r = t$. Then

$$\Delta_{n'} = \sum_{t_{k+1}^{(n')} \le t} \int_{t_k^{(n')}}^{t_{k+1}^{(n')}} \int_{|u|<\epsilon} f(t_k^{(n')}, \xi_{n'}(t_k^{(n')}), u)\tilde{p}_{n'}\,(ds \times du)$$
$$- \int_{t_0}^{t} \int_{|u|>\epsilon} f(s, \xi(s), u)\tilde{p}\,(ds \times du)$$
$$= \sum_{i=0}^{r-1} \left(\sum_{\bar{t}_i < t_{k+1}^{(n')} \le \bar{t}_{i+1}} \int_{t_k^{(n')}}^{t_{k+1}^{(n')}} \int_{|u|>\epsilon} f(\bar{t}_i, \xi_{n'}(\bar{t}_i), u)\tilde{p}_{n'}\,(ds \times du) - \int_{\bar{t}_i}^{\bar{t}_{i+1}} \int_{|u|>\epsilon} f(\bar{t}_i, \xi(\bar{t}_i), u)\tilde{p}\,(ds \times du) \right)$$
$$+ \sum_{i=0}^{r-1} \left(\sum_{\bar{t}_i < t_{k+1}^{(n')} \le \bar{t}_i} \int_{t_k^{(n')}}^{t_{k+1}^{(n')}} \int_{|u|>\epsilon} [f(t_k^{(n')}, \xi_{n'}(t_k^{(n')}), u) - f(\bar{t}_i, \xi_{n'}(\bar{t}_i), u)]\tilde{p}_{n'}\,(ds \times du) \right)$$
$$+ \sum_{i=0}^{r-1} \left(\int_{\bar{t}_i}^{\bar{t}_{i+1}} \int_{|u|>\epsilon} [f(\bar{t}_i, \xi(\bar{t}_i), u) - f(s, \xi(s), u)]p\,(ds \times du) \right). \tag{3.11}$$

The first sum $\sum_{i=0}^{r-1}$ on the right-hand side of the last equation approaches zero in probability according to Lemma 5. For the second sum, the inequality

$$\overline{\lim_{n'\to\infty}}\, \mathbf{P}\left\{\left| \sum_{i=0}^{r-1} \sum_{\bar{t}_i < t_{k+1}^{(n')} \le \bar{t}_{i+1}} \int_{t_k^{(n')}}^{t_{k+1}^{(n')}} \int_{|u|>\epsilon} [f(t_k^{(n')}, \xi_{n'}(t_k^{(n')}), u) - f(\bar{t}_i, \xi_{n'}(\bar{t}_i), u)]\tilde{p}_n\,(ds \times du)\right| > \rho\right\}$$
$$\le \overline{\lim_{n'\to\infty}}\, \mathbf{P}\,\{\sup_t |\xi_{n'}(t)| > C\} + \mathbf{P}\left\{\int_{t_0}^{T} \int_{|u|>C} \tilde{p}_n\,(ds \times du) > 0\right\}$$
$$+ \overline{\lim_{n'\to\infty}}\, \frac{1}{\rho} \sum_{i=0}^{r-1} \sum_{\bar{t}_i < t_{k+1}^{(n')} \le \bar{t}_{i+1}} \int_{t_k^{(n')}}^{t_{k+1}^{(n')}} \int_{|u|>\epsilon} \mathbf{M}|f(t_k^{(n)}, \xi_{n'}(t_k^{(n')}), u)$$
$$- f(\bar{t}_i, \xi_{n'}(\bar{t}_i), u) g_C(u) g_C(\xi_{n'}(\bar{t}_i)) g_C(\xi_{n'}(t_k^{(n')}))\, \frac{dt\,du}{|u|^{m+1}}$$

holds, where $g_C(x) = 1$ for $|x| \leq C_u$ and $g_C(x) = 0$ for $|x| > C$. By Lemma 3 and Condition 4 of the theorem, the third term can be made arbitrarily small by a sufficiently small choice of $\max_i (\bar{t}_{i+1} - \bar{t}_i)$. Since, by a sufficiently large choice of C, the first two terms can be made arbitrarily small, the whole estimate can be made arbitrarily small by a sufficiently small choice of $\max_i (\bar{t}_{i+1} - \bar{t}_i)$. In exactly the same manner, it can be shown that the third sum in (3.11) can, by a sufficiently small choice of $\max_i (\bar{t}_{i+1} - \bar{t}_i)$, be made arbitrarily small. Thus (3.8), and hence the theorem, is proved.

COROLLARY. *If the $\xi_n(t)$ are defined as in Lemma 3, and if the coefficients of the equation satisfy the conditions of both theorems of this section, the finite-dimensional distributions of $\xi_n(t)$ will converge to finite-dimensional distributions of the process $\xi(t)$ which is the solution of Eq. (1.7).*

In fact, it follows from the second theorem that no matter what the sequence n' may be, it is possible to choose from it a subsequence n'_k such that the finite-dimensional distributions of the processes $\xi_{n'_k}(t)$ will converge to finite-dimensional distributions of the unique solution of Eq. (1.7). Therefore the finite-dimensional distributions of all the sequences $\xi_n(t)$ will converge to this same limit.

4. The solutions of stochastic equations as Markov processes. Let us examine the solution of Eq. (1.7) when the coefficients of the equation satisfy the conditions of the existence and uniqueness theorems of the preceding section.

Let $\mathbf{F}_t$ be the minimal σ-algebra of events relative to which the variables $w_1(s), \ldots, w_l(s)$, $\xi(t_0)$ and $p([t_0, s] \times A)$ are measurable for $s \leq t$ and for an arbitrary Borel set A in $R^{(m)}$; let $\mathbf{F}_{[t,t+h]}$ be the minimal σ-algebra relative to which the variables $w_1(s) - w_1(t), \ldots, w_l(s) - w_l(t)$ and $p([t, s] \times A)$ are measurable for $s \in [t, t + h]$ and for an arbitrary Borel set A in $R^{(m)}$; let $\mathbf{F}_{[t]}$ be the minimal σ-algebra relative to which the variable $\xi(t)$ is measurable. We define

$$\begin{aligned} g(h, x, \omega) = x &+ \int_0^h a\big(t + s, g(s, x, \omega)\big)\, ds \\ &+ \sum_{j=1}^{l} \int_0^h b_j\big(t + s, g(s, x, \omega)\big)\, dw_j(s + t) \\ &+ \int_0^h \int_{|u| \leq 1} f\big(t + s, g(s, x, \omega), u\big) q\, \big(d(t + s) \times du\big) \\ &+ \int_0^h \int_{|u| > 1} f\big(t + s, g(s, x, \omega), u\big) p\, \big(d(t + s) \times du\big). \end{aligned} \tag{4.1}$$

(Here, it is convenient to point out the dependence on the random variable ω.) The variable $g(h, x, \omega)$ is measurable with respect to $\mathbf{F}_{[t,t+h]}$ for all x and $h > 0$. Clearly, $g(h, \xi(t), \omega) = \xi(t+h)$ if $\xi(t)$ is a solution of Eq. (1.7).

LEMMA 1. *For every bounded Borel function* $\lambda(x)$,

$$\mathbf{M}(\lambda(\xi(t+h)) \mid \mathbf{F}_t) = \mathbf{M}(\lambda\xi(t+h)) \mid \mathbf{F}_{[t]}).$$

To prove the lemma, it is sufficient to show that for every bounded measurable function $\psi(x, \omega)$ for which $\psi(x, \omega)$ is measurable with respect to $\mathbf{F}_{[t,t+h]}$ at every x, the relation

$$\mathbf{M}(\psi(\xi(t+h), \omega)/\mathbf{F}t)$$
$$= \mathbf{M}(\psi(\xi(t+h), \omega)/\xi(t)) = C(\xi(t)), \tag{4.2}$$

where $C(x) = \mathbf{M}\psi(x, \omega)$, is fulfilled. First we show that (4.2) is fulfilled for step functions $\psi(x, \omega)$; that is, for those functions for which there exist disjoint Borel sets $A_1, A_2, \ldots, A_k$, $\cup_i A_i = R^{(m)}$ such that on each of the sets $\psi(x, \omega)$ is constant; that is, there exist random variables ψ_i, $i = 1, 2, \ldots, k$, that are measurable with respect to $\mathbf{F}_{[t,t+h]}$ such that $\psi(x, \omega) = \sum \chi_{A_i}(x)\psi_i$ if $\chi_A(x)$ is the characteristic function of the set A_i. To establish (4.2), as derived from Section 2, Chapter 1, we need to show that for every bounded variable η that is measurable with respect to F_t, the relation

$$\mathbf{M}\eta\psi(\xi(t+h), \omega) = \mathbf{M}\eta C(\xi(t))$$

holds. But

$$\mathbf{M}\eta\psi(\xi(t), \omega) = \mathbf{M}\eta\sum\chi_{A_i}(\xi(t))\psi_i.$$

Since the variables ψ_i do not depend on $\eta \cdot \chi_{A_i}(\xi(t))$ (any variable that is measurable with respect to $\mathbf{F}_{[t,t+h]}$ is independent of any variable that is measurable with respect to $\mathbf{F}_t$), it follows that

$$\mathbf{M}\eta\psi(\xi(t), \omega) = \mathbf{M}\eta\sum\chi_{A_i}(\xi(t))\cdot\mathbf{M}\psi_i.$$

Since $\mathbf{M}\psi(x, \omega) = \sum\chi_{A_i}(x)\mathbf{M}\psi_i$, we may write the preceding formula in the following form: $\mathbf{M}\eta\psi(\xi(t), \omega) = \mathbf{M}\eta C(\xi(t))$. Formula (4.2) is established for finite-valued functions. The proof of this formula in the general case is obtained by passing to the limit from finite-valued functions. From the equations $\xi(t+h) = g(h, \xi(t), \omega)$ and (4.2), we obtain the proof of the lemma.

COROLLARY. *If* $\lambda(x)$ *is the characteristic function of some Borel set* A, *then*

$$\mathbf{P}\{\xi(t+h) \in A/\mathbf{F}_t\} = \mathbf{P}\{\xi(t+h) \in A/\xi(t)\}. \tag{4.3}$$

From this formula, when we remember that $\xi(s)$ is measurable with respect to $\mathbf{F}_t$ for $s \leq t$, we obtain

$$\mathbf{P}\{\xi(t+h) \in A/\xi(s),\, s \in [t_0, t]\} = \mathbf{P}\{\xi(t+h) \in A/\xi(t)\}. \tag{4.4}$$

We then have the following theorem:

THEOREM 1. *If the conditions of the existence and uniqueness theorems of the preceding sections are satisfied, then the solution to Eq. (1.7) will be a Markov process.*

How wide is the class of Markov processes that are solutions to the stochastic equations examined in the preceding sections? To investigate this question, let us show that among the solutions to equations of the form (1.7) there are homogeneous processes with independent increments with all possible characteristic functions given by Formula (3.2) of Chapter 1. For this, we shall prove the following assertion:

THEOREM 2. *There exist vectors $\overline{a}, b_1, b_2, \ldots, b_m$ and a Borel function $f(u)$ that is defined for $u \in R^{(m)}$, that takes values in $R^{(m)}$, that is bounded in every bounded region of variation of u, that is continuous for $u = 0$, and that satisfies the relation*

$$\int_{|u|\leq 1} |f(u)|^2 \frac{du}{|u|^{m+1}} < \infty,$$

such that the process

$$\xi(t) = \int_0^t \overline{a}\, ds + \sum_{j=1}^{m} \int_0^t b_j\, dw(s)$$
$$+ \int_0^t \int_{|u|\leq 1} f(u) q\, (ds \times du) + \int_0^t \int_{|u|>1} f(u) p\, (ds \times du) \tag{4.5}$$

[where $w_1(t), \ldots, w_m(t)$, p and q are defined as in the preceding sections] *will be a homogeneous process with independent increments with the characteristic function defined by Formula* (3.2) *of Chapter* 1.

Proof. That the process $\xi(t)$ will be homogeneous with independent increments follows immediately from the properties of the measures p and q and the processes $w_j(t)$. To calculate the characteristic function $\xi(t)$, we note that all terms on the right-hand side of (4.5) are mutually independent. Therefore it is sufficient to find the characteristic function of each term and then multiply the expressions found. We have

$$\mathbf{M} \exp\left[i\left(z, \int_0^t b_j\, dw(s)\right)\right] = \mathbf{M} \exp\left[i(z, b_j)\big(w(t) - w(0)\big)\right] = e^{-(z,b_j)^2 t/2}.$$

For an arbitrary step function $f(u)$ such that $f(u) = f(u_k)$ for $u \in G_k$, $\bigcup_k G_k = R^{(m)}$, we have

$$\mathbf{M} \exp\left[i\left(z, \int_0^t f(u)p\,(ds \times du)\right)\right]$$
$$= \mathbf{M} \exp\left[i \sum_k (z, f(u_k))p([0, t] \times G_k)\right]$$
$$= \prod_k \mathbf{M} \exp\,[i(z, f(u_k))p([0, t] \times G_k)]$$
$$= \prod_k \exp\left((e^{i(z,f(u_k))} - 1)\int_{G_k} \frac{t\,du}{|u|^{m+1}}\right) = \exp\left[t\int (e^{i(z,f(u))} - 1)\frac{du}{|u|^{m+1}}\right].$$

[We used here the fact that the $p([0, t]) \times G_k)$ are mutually independent and are distributed in accordance with Poisson's law with parameter $t\int_{u\in G_k} (du/|u|^{m+1})$.] Therefore for every function $f(u)$ for which

$$\int_0^t \int_{R^{(m)}} f(u)p\,(ds \times du)$$

is meaningful, the relation

$$\mathbf{M} \exp\left\{i\left(z, \int_0^t \int_{R^{(m)}} f(u)p\,(ds \times du)\right)\right\}$$
$$= \exp\left\{t\int_{u\in R^{(m)}} (e^{i(z,f(u))} - 1)\frac{du}{|u|^{m+1}}\right\} \tag{4.6}$$

will be fulfilled. From Formula (4.6) it follows that for every function for which

$$\int |f(u)|^2 \frac{du}{|u|^{m+1}} < \infty,$$

the equation

$$\mathbf{M} \exp\left\{i\left(z, \int_0^t \int_{R^{(m)}} f(u)q\,(ds \times du)\right)\right\}$$
$$= \exp\left\{t\int_{R^{(m)}} (e^{i(z,f(u))} - 1 - i(z, f(u)))\frac{du}{|u|^{m+1}}\right\} \tag{4.7}$$

will be satisfied. Thus the characteristic function of the process $\xi(t)$

defined by Formula (4.5) will be equal to

$$\mathbf{M} \exp \{i(z, \xi(t))\} = \exp \Bigg\{ t \Bigg[i(\bar{a}, z) - \frac{1}{2} \sum_{j=1}^{m} (b_j, z)^2 + \int_{|u| \leq 1} (e^{i(z, f(u))} - 1 - i(z, f(u))) \frac{du}{|u|^{m+1}} + \int_{|u|>1} (e^{i(z, f(u))} - 1) \frac{du}{|u|^{m+1}} \Bigg] \Bigg\}. \tag{4.8}$$

In order for $\sum_{j=1}^{m} (\bar{b}_j, z)^2 = (Az, z)$ to hold, it is sufficient to take $\bar{b}_j = \sqrt{\lambda_j}\, e_j$, where e_j is an orthonormal system of eigenvectors of the operator A and the λ_j are the corresponding eigenvalues. If we make the substitution $f(u) = x$ in (4.8), then to make the formula obtained coincide with Formula (3.2), Chapter 1, it is sufficient that

$$\bar{a} = a + \int_{\substack{|u| \leq 1 \\ |f(u)| > 1}} f(u) \frac{du}{|u|^{m+1}} - \int_{\substack{|u| > 1 \\ |f(u)| \leq 1}} f(u) \frac{du}{|u|^{m+1}}$$

and that for every Borel set for which $\Pi(A) < \infty$, the following is true:

$$\Pi(A) = \int_{f(u) \in A} \frac{du}{|u|^{m+1}}.$$

The proof of the theorem, therefore, follows from our next lemma.

LEMMA 2. *For every measure $\Pi(A)$ defined on the Borel sets $R^{(m)}$ for which*

$$\int \frac{|x|^2}{1 + |x|^2} \Pi\, (dx) < \infty,$$

there exists a measurable function $f(u)$ defined for $u \in R^{(m)}$ that takes values in $R^{(m)}$ and is bounded in every finite region of variation of u such that:

1. $\lim_{u \to 0} f(u) = 0;$
2. *For every Borel set A for which $\Pi(A) < \infty$, the relation*

$$\int_{f(u) \in A} \frac{du}{|u|^{m+1}} = \Pi(A)$$

holds;

3. $\displaystyle\int_{|u| \leq 1} |f(u)|^2 \frac{du}{|u|^{m+1}} < \infty.$

Proof. Let us partition the space $R^{(m)}$ into spherical layers C_n, $n = 0, 1, -1, 2, -2, \ldots$; C_n is the set of points u for which $2^n \leq |u| < 2^{n+1}$. Let r_n be a number such that

$$\int_{|u|\geq 2^n} \Pi\,(du) = \int_{|u|\geq r_n} \frac{du}{|u|^{m+1}},$$

and let C'_n be the set of points u for which $r_n \leq |u| < r_{n+1}$. We denote by $T_n(u) = 2^n u$ the mapping of C_0 to C_n.

$$T'_n(u) = (u/|u|)[r_n + (r_{n+1} - r_n)(|u| - 1)]$$

is the mapping of C_0 to C'_n. We also define for every natural number k and every choice of indices $i_1, i_2, \ldots, i_k$, with values 0 and 1, the sets $S_{i_1,i_2,\ldots,i_k}$ such that the following conditions are fulfilled:

(a) $S_{i_1,\ldots,i_{k-1},0} \cup S_{i_1,\ldots,i_{k-1},1} = S_{i_1,\ldots,i_{k-1}}$, $S_0 \cup S_1 = C_0$;

(b) $S_{i_1,i_2,\ldots,i_k}$ does not intersect $S_{j_1,j_2,\ldots,j_k}$ if at least one i_l is different from j_l;

(c) the diameter of $S_{i_1,i_2,\ldots,i_k}$ approaches zero as $k \to \infty$.

We set $T_n S_{i_1,i_2,\ldots,i_k} = S^{(n)}_{i_1,\ldots,i_k}$, $T'_n(S_{i_1,i_2,\ldots,i_k}) = S'^{(n)}_{i_1,\ldots,i_k}$.

The sets $S^{(n)}_{i_1,\ldots,i_k}$ and $S'^{(n)}_{i_1,\ldots,i_k}$ satisfy the same conditions except that

$$S_0^{(n)} \cup S_1^{(n)} = C_n, \qquad S_0'^{(n)} \cup S_1'^{(n)} = C'_n.$$

Suppose that

$$\pi^{(n)}_{i_1,\ldots,i_k} = \Pi(S^{(n)}_{i_1,\ldots,i_k}), \qquad \pi'^{(n)}_{i_1,\ldots,i_k} = \int_{u\in S'^{(n)}_{i_1,\ldots,i_k}} \frac{du}{|u|^{m+1}}.$$

We now define a function $g_n(\alpha)$ by the formula

$$g_n(\alpha) = \lim_{k\to\infty} \sum_{\frac{i_1}{2}+\frac{i_2}{4}+\cdots+\frac{i_k}{2^k}\leq\alpha} \pi'^{(n)}_{i_1,i_2,\ldots,i_k}, \qquad 0 \leq \alpha \leq 1.$$

The given function will be nondecreasing and continuous. [The continuity follows from the fact that $\lim_{k\to\infty} \pi'^{(n)}_{i_1,i_2,\ldots,i_k} = 0$ because the diameter of $S'^{(n)}_{i_1,\ldots,i_k}$ approaches zero as $k \to \infty$.] Let $\alpha^{(n)}_{i_1,\ldots,i_k}$ be such that

$$g_n(\alpha^{(n)}_{i_1,i_2,\ldots,i_k}) = \sum_{\frac{j_1}{2}+\frac{j_2}{4}+\cdots+\frac{j_k}{2^k}\leq\frac{i_1}{2}+\frac{i_2}{4}+\cdots+\frac{i_k}{2^k}} \pi^{(n)}_{j_1,j_2,\ldots,j_k}.$$

[The existence of the $\alpha^{(n)}_{i_1,\ldots,i_k}$ follows from the fact that

$$g_n(0) = 0, \qquad g_n(1) = \int_{u\in C'_n} \frac{du}{|u|^{m+1}} = \Pi(C_n) = \sum_{i_1,\ldots,i_k} \pi^{(n)}_{i_1,i_2,\ldots,i_k}.\Big]$$

We also define

$$U_n(\alpha) = \bigcup_{k=1}^{\infty} \left(\bigcup_{\frac{i_1}{2}+\frac{i_2}{4}+\cdots+\frac{i_k}{2^k}\leq\alpha} S'^{(n)}_{i_1,i_2,\ldots,i_k} \right),$$

$$V^{(n)}_{i_1,\ldots,i_k} = U_n(\alpha^{(n)}_{i_1,\ldots,i_k}) - U_n(\alpha^{(n)}_{j_1,\ldots,j_k}),$$

where

$$\frac{j_1}{2} + \frac{j_2}{4} + \cdots + \frac{j_k+1}{2^k} = \frac{i_1}{2} + \frac{i_2}{4} + \cdots + \frac{i_k}{2^k}.$$

It is obvious that $V^{(n)}_{i_1,\ldots,i_k}$ and $V^{(n)}_{j_1,\ldots,j_k}$ are disjoint if at least one i_l is not equal to j_l, and that

$$\bigcup_{\frac{j_1}{2}+\cdots+\frac{j_k}{2^k}\leq\frac{i_1}{2}+\cdots+\frac{i_k}{2^k}} V^{(n)}_{j_1,\ldots,j_k}$$

$$= U_n(\alpha^{(n)}_{i_1,\ldots,i_k}),\ V^{(n)}_{i_1,\ldots,i_{k-1},0} \cup V^{(n)}_{i_1,\ldots,i_{k-1},1} = V^{(n)}_{i_1,\ldots,i_{k-1}},$$

$$\int_{u\in V^{(n)}_{i_1,i_2,\ldots,i_k}} \frac{du}{|u|^{m+1}} = \pi^{(n)}_{i_1,\ldots,i_k}, \qquad \text{because} \qquad \int_{u\in U_n(\alpha)} \frac{du}{|u|^{m+1}} = g_n(\alpha).$$

We choose in each of the sets $S^{(n)}_{i_1,i_2,\ldots,i_k}$ a point $x^{(n)}_{i_1,\ldots,i_k}$ and we let $f_k(x) = x^{(n)}_{i_1,\ldots,i_k}$ if $x \in V^{(n)}_{i_1,\ldots,i_k}$. Since the diameters of the sets $S^{(n)}_{i_1,i_2,\ldots,i_k}$ approach zero as $k \to \infty$, it follows that $f_k(x)$ has a limit as $k \to \infty$ for every x; also, this limit exists uniformly on every bounded set. Define

$$\lim_{k\to\infty} f_k(x) = f(x).$$

Let us show that the conditions of the lemma are satisfied for the function $f(x)$. We note that for $|u| \leq r_n$, $|f(u)| \leq 2^n$. Therefore $f(u)$ is bounded in every bounded interval and it satisfies Condition 1 of the lemma. To prove Condition 2, we consider the measure

$$\bar{\Pi}(A) = \int_{f(u)\in A} \frac{du}{|u|^{m+1}}$$

and show that it coincides with the measure $\Pi(A)$. For this it is sufficient to show the coincidence of $\Pi(A)$ and $\bar{\Pi}(A)$ on some algebra of sets whose Borel closure coincides with the σ-algebra of Borel sets. The algebra of bounded sets A for which $\Pi(\bar{A} \cap \overline{R^{(m)}\setminus A}) = 0$, $\bar{A}$ being the closure of the set A, can serve as such an algebra. Let A^ϵ be the set of points whose distance from A is no greater than ϵ and let A_ϵ be the set of interior points of A whose distance from the boundary of the set A is no less than ϵ. Let us choose a k such that the diameter of $S^{(n)}_{i_1,\ldots,i_k}$ is less than $\epsilon/2$. Under these conditions, if $f(x) \in A$, then $f_k(x) \in A_{\epsilon/2}$; and if $f_k(x) \in A_{\epsilon/2}$, then $f(x) \in A$. Therefore

$$\bar{\Pi}(A) = \int_{f(u)\in A} \frac{du}{|u|^{m+1}} \le \int_{f_k(u)\in A^{\epsilon/2}} \frac{du}{|u|^{m+1}}$$

$$\le \sum_{S^{(n)}_{i_1,\ldots,i_k}\cap A^{\epsilon/2}\neq 0} \Pi(S^{(n)}_{i_1,\ldots,i_k}) \le \Pi(A^{\epsilon}).$$

In an analogous manner, we can show that $\bar{\Pi}(A) \ge \Pi(A_\epsilon)$.

By taking the limit as $\epsilon \to 0$ and remembering that

$$\lim_{\epsilon\to 0} \left(\Pi(A^\epsilon) - \Pi(A_\epsilon)\right) = \lim_{\epsilon\to 0} \Pi(A^\epsilon - A_\epsilon) = 0,$$

we obtain $\bar{\Pi}(A) = \Pi(A)$. Thus Property 2 of the lemma is fulfilled for our function. Property 3 follows from the relation

$$\int_{2^{-n}\le|x|\le 2^0} |x|^2\, \Pi\,(dx) = \int_{r_{-n}\le|u|\le r_0} |f(u)|^2 \frac{du}{|u|^{m+1}},$$

which is a consequence of Property 2 and of the fact that $r_{-n} \le |u| < r_0$ holds for $2^{-n} \le |f(u)| < 2^0$. This proves the lemma.

We have introduced above the concept of the Itō differential of a Markov process. We shall say that a Markov process $\xi(t)$ is *differentiable in the sense of Itō* at a point t if for every $z \in R^{(m)}$, the expression

$$\frac{1}{\Delta} \log \mathbf{M}\left(e^{i(z,\xi(t+\Delta)-\xi(t))}/\xi(t)\right) \tag{4.9}$$

has a limit in the sense of convergence in probability as $\Delta \to +0$. The value of this limit is called the *Itō differential* of the process $\xi(t)$ at the point t. The value of the Itō differential is a function of z and of $\xi(t)$; also, since it is a function of z, it has for almost all ω the form of the characteristic function of an infinitely divisible law, that is, of the characteristic function of the distribution of the process with independent increments. We denote the limit of the expression (4.9) by $\varphi(t, \xi(t), z)$. It follows from Theorem 2 that there exist vectors $a(t, x)$, $b_1(t, x)$, $\ldots$, $b_l(t, x)$ and a function $f(t, x, u)$ such that with probability 1 the relation

$$\varphi(t, \xi(t), z) = i(a(t, \xi(t)), z) - \frac{1}{2}\sum_1^l (b_k(t, \xi(t)), z)^2$$

$$+ \int_{|u|\le 1} [e^{i(z,f(t,\xi(t),u))} - 1 - i(z, f(t, \xi(t), u))] \frac{du}{|u|^{m+1}}$$

$$+ \int_{|u|>1} (e^{i(z,f(t,\xi(t),u))} - 1) \frac{du}{|u|^{m+1}} \tag{4.10}$$

is fulfilled. If a process has an Itō differential at every point of an interval $[t_0, T]$, we shall say that it is *differentiable in the Itō sense on* $[t_0, T]$. The solutions of stochastic equations have Itō differentials under broad assumptions; also, these differentials can be of a rather arbitrary form. Let us establish a very simple theorem on the existence of a process with a given Itō differential.

THEOREM 3. *Suppose that $\xi(t)$ is the solution of Eq. (1.8) the coefficients of which satisfy the conditions of Theorem 4 of Section 2 and also the conditions:*

1. *$a(t, x)$, $b_k(t, x)$ are continuous with respect to t.*

2. $$\lim_{t_1 - t_2 \to 0} \int_{R^{(m)}} |f(t_1, x, u) - f(t_2, x, u)|^2 \frac{du}{|u|^{m+1}} = 0.$$

Then the process $\xi(t)$ has an Itō differential equal to

$$i(a(t, \xi(t)), z) - \frac{1}{2} \sum_{k=1}^{m} (b_k(t, \xi(t)), z)$$

$$+ \int_{R^{(m)}} [e^{i(z, f(t, \xi(t), u))} - 1 - i(z, f(t)\xi(t), u)] \frac{du}{|u|^{m+1}} \tag{4.11}$$

at the point t.

Proof. We note that

$$\xi(t + \Delta) = \xi(t) + \int_t^{t+\Delta} a(s, \xi(s))\, ds + \sum_{k=1}^{m} \int_t^{t+\Delta} b_k(s, \xi(s))\, dw_k(s)$$

$$+ \int_t^{t+\Delta} \int_{R^{(m)}} f(s, \xi(s), u) q\, (ds \times du),$$

because

$$\xi(t + \Delta) - \xi(t) = \int_t^{t+\Delta} a(t, \xi(t))\, ds + \sum_{k=1}^{m} \int_t^{t+\Delta} b_k(t, \xi(t))\, dw_k(s)$$

$$+ \int_t^{t+\Delta} \int_{R^{(m)}} f(t, \xi(t), u) q\, (ds \times du) + \sum_{1}^{m+2} \Theta_i\, (\Delta),$$

where

$$\Theta_1 = \int_t^{t+\Delta} [a(s, \xi(s)) - a(t, \xi(t))]\, ds,$$

$$\Theta_i = \int_t^{t+\Delta} [b_{i-1}(s, \xi(s)) - b_{i-1}(t, \xi(t))]\, dw_{i-1}(s),$$

$$i = 2, \ldots, m + 1,$$

$$\Theta_{m+2} = \int_t^{t+\Delta} \int_{R^{(m)}} [f(s, \xi(s), u) - f(t, \xi(t), u)] q\, (ds \times du).$$

For future use, we shall need certain facts about infinitesimal random variables. Suppose that $\alpha(\Delta)$ is a particular quantity dependent on Δ. We shall write $\alpha(\Delta) = o(\Delta)$ if for every $\epsilon > 0$, the following conditions are fulfilled:

$$\mathbf{P}\{|\alpha(\Delta)| > \epsilon\} = o(\Delta), \qquad \mathbf{M}\psi_\epsilon(\alpha(\Delta))\alpha(\Delta) = o(\Delta),$$
$$\mathbf{M}\psi_\epsilon(\alpha(\Delta))(\alpha(\Delta))^2 = o(\Delta).$$

We denote by $\psi_\epsilon(x)$ the function that is equal to unity for $|x| \leq \epsilon$ and equal to zero for $|x| > \epsilon$. It is easy to verify the following assertions:

(a) if $\alpha(\Delta) = o(\Delta)$ and $\beta(\Delta) = o(\Delta)$, then $\alpha(\Delta) + \beta(\Delta) = o(\Delta)$;

(b) if $\mathbf{M}\alpha(\Delta) = o(\Delta)$ and $\mathbf{D}\alpha(\Delta) = o(\Delta)$, then $\alpha(\Delta) = o(\Delta)$;

(c) if $\alpha(\Delta) \to 0$ in probability, then $\alpha(\Delta) = o(\Delta)$;

(d) if for every bounded, twice continuously differentiable function with bounded derivatives the limit $\lim_{\Delta\to 0} (1/\Delta)\mathbf{M}g(\alpha(\Delta))$ exists, and if $\beta(\Delta) = o(\Delta)$, then

$$\lim_{\Delta\to 0} \frac{1}{\Delta}\mathbf{M}[g(\alpha(\Delta) + \beta(\Delta)) - g(\alpha(\Delta))] = 0.$$

In the same manner as in the case of Theorem 2, we can show that

$$\log \mathbf{M} \exp\left\{i\left(z, \int_t^{t+\Delta} a(t, \xi(t))\, ds\right)\right.$$
$$\left. + \sum_{k=1}^{m} \int_t^{t+\Delta} b_k(t, \xi(t))\, dw_k(s) + \int_t^{t+\Delta} \int_{R^{(m)}} f(t, \xi(t), u) q\,(ds \times du)\right\}$$
$$= \Delta\Big\{i(z, a(t, \xi(t))\} - \frac{1}{2}\sum_{k=1}^{m} (b_k(t, \xi(t)), z)^2$$
$$\left. + \int_{R^{(m)}} [e^{i(z, f(t,\xi(t),u))} - 1 - i(z, f(t, \xi(t), u))]\, \frac{du}{|u|^{m+1}}\right\}.$$

Thus to prove the theorem, from Property (d) it is sufficient to show that $\sum_1^{m+2} \Theta_i(\Delta) = o(\Delta)$. This will be shown if we can show that $\Theta_i(\Delta) = o(\Delta)$ for $i = 1, \ldots, m + 2$. From Remark 2 following Theorem 3 of Section 2, we know that there exists an R_1 for which

$$\mathbf{M} \sup_{t\leq s\leq t+\Delta} |\xi(s) - \xi(t)|^2 \leq R_1\Delta;$$

consequently $\sup_{t\leq s\leq t+\Delta} |\xi(s) - \xi(t)| \to 0$ in probability as $\Delta \to 0$. Therefore $(\Theta_1/\Delta) \to 0$ in probability as $\Delta \to 0$; therefore $\Theta_1(\Delta) = o(\Delta)$.

For $i = 2, \ldots, m+1$,

$$\mathbf{M}\Theta_i = 0, \qquad \mathbf{D}\Theta_i = \int_t^{t+\Delta} \mathbf{M}|b_{i-1}(s, \xi(s)) - b_{i-1}(t, \xi(t))|^2\, ds = o(\Delta),$$

because $|b_k(s, \xi(s)) - b_k(t, \xi(t))| \to 0$ in probability as a result of the continuity of $b_k(s, x)$ and, in addition, $|b(s, \xi(s)) - b(t, \xi(t))|^2 \leq 4K\,(\sup_s |\xi(s)|^2 + 1)$ and hence $\mathbf{M}\sup_s |\xi(s)|^2 < \infty$. This means that $\Theta_i(\Delta) = o(\Delta)$ for $i = 2, \ldots, m+1$. In an analogous manner, we can show that $\Theta_{m+2}(\Delta) = o(\Delta)$. This proves the theorem.

5. The dependence of the solutions of stochastic equations on initial conditions. Let us examine the dependence of the solution of Eq. (1.8) on the parameter appearing in the initial condition. To study this dependence, we need a limit theorem. Let us suppose that the set of σ-algebras $\mathbf{F}_t$, the process $w_1(t), \ldots, w_l(t)$, and the measure q are defined as in Section 2. Let us consider the sequence of equations:

$$\xi_n(t) = \varphi_n(t) + \int_{t_0}^t A_n(s, \xi_n(s))\, ds + \sum_{k=1}^{l} \int_{t_0}^t B_k^{(n)}(s, \xi_n(s))\, dw_k(s)$$
$$+ \int_{t_0}^t \int_{R^{(m)}} F_n(s, \xi_n(s), u) q\,(ds \times du), \qquad n = 0, 1, 2, \ldots, \tag{5.1}$$

where $\varphi_n(t)$, $A_n(t, x)$, $B_1^{(n)}(t, x), \ldots, B_l^{(n)}(t, x)$, $F_n(t, x, u)$ are particular random functions that satisfy the measurability conditions given at the beginning of Section 2.

THEOREM 1. *Assume that the following conditions are fulfilled:*

1. *There exists a K such that for all n and x,*

$$(T - t_0)|A_n(t, x)|^2 + \sum_{k=1}^{l} |B_k^{(n)}(t, x)|^2$$
$$+ \int_{R^{(m)}} |F_n(t, x, u)|^2 \frac{du}{|u|^{m+1}} \leq K(|x|^2 + 1).$$

2. *For every $C > 0$, there exists an L_C such that for every $|x| \leq C$, $|y| \leq C$ and $t \in [t_0, T]$,*

$$(T - t_0)|A_n(t, x) - A_n(t, y)|^2 + \sum_{k=1}^{l} |B_k^{(n)}(t, x) - B_k^{(n)}(t, y)|^2$$
$$+ \int_{u \in R^{(m)}} |F_n(t, x, u) - F_n(t, y, u)|^2 \frac{du}{|u|^{m+1}} \leq L_C^2 |x - y|^2.$$

3. *$\varphi_n(t)$ does not have discontinuities of the second kind with probability 1 and*

$$\sup_n \mathbf{M} \sup_t |\varphi_n(t)|^2 < \infty.$$

4. *For every t and x,*

$$|\varphi_n(t) - \varphi_0(t)| + |A_n(t, x) - A_0(t, x)| + \sum_{k=1}^{l} |B_k^{(n)}(t, x) - B_0^{(n)}(t, x)| + \int_{R^{(m)}} |F_n(t, x, u) - F_0(t, x, u)|^2 \frac{du}{|u|^{m+1}} \to 0$$

in probability. Then for every t, the functions $\xi_n(t) \to \xi_0(t)$ in probability.

Proof. Define $\psi_n^{(N)}(t) = 1$ if $|\varphi_n(s)| + |\varphi_0(s)| + |\xi_n(s)| + |\xi_0(s)| \leq N$ for $s \in [t_0, t]$ and $\varphi_n^{(N)}(t) = 0$ if for some $s \in [t_0, t]$, $\varphi_n(s)| + |\varphi_0(s)| + |\xi_n(s)| + |\xi_0(s)| > N$. Then

$$\begin{aligned}
&(\xi_n(t) - \xi_0(t))\psi_n^{(N)}(t) \\
&= [\varphi_n(t) - \varphi_0(t)]\psi_n^{(N)}(t) + \psi_n^{(N)}(t)\Big[\int_{t_0}^{t} [A_n(s, \xi_n(s)) - A_n(s, \xi_0(s))]\, ds \\
&\quad + \sum_{k=1}^{l} \int_{t_0}^{t} [B_k^{(n)}(s, \xi_n(s)) - B_k^{(n)}(s, \xi_0(s))]\, dw_k(s) \\
&\quad + \int_{t_0}^{t} \int_{R^{(m)}} [F_n(s, \xi_n(s), u) - F_n(s, u, \xi_0(s))]q\, (ds \times du) \\
&\quad + \int_{t_0}^{t} (A_n(s, \xi_0(s)) - A_0(s, \xi_0(s)))\, ds \\
&\quad + \sum_{k=1}^{l} \int_{t_0}^{t} (B_k^{(n)}(s, \xi_0(s)) - B_k^{(0)}(s, \xi_0(s)))\, dw_k(s) \\
&\quad + \int_{t_0}^{t} \int_{R^{(m)}} [F_n(s, \xi_0(s), u) - F_0(s, \xi_0(s), u)]q\, (ds \times du)\Big].
\end{aligned}$$

Taking the mathematical expectations of the squares of both sides of the inequality and applying Condition 2, we see that for some H,

$$\mathbf{M}|(\xi_n(t) - \xi_0(t))\psi_n^{(N)}(t)|^2 \leq H\int_{t_0}^{t} \mathbf{M}|(\xi_n(s) - \xi_0(s)|^2\psi_n^{(N)}(s)\, ds + \alpha_N^{(n)}(t), \tag{5.2}$$

where

$$\alpha_N^{(n)}(t) = H\Big[\mathbf{M}(\psi_n^{(N)}(t)|\varphi_n(t) - \varphi_0(t)|^2)$$

$$+ \int_{t_0}^{t} \mathbf{M}\Big\{\psi_n^{(N)}(s)\Big(|A_n(s, \xi_0(s)) - A_0(s, \xi_0(s))|^2$$

$$+ \sum_{k=1}^{l} |B_k^{(n)}(s, \xi_0(s)) - B_k^{(0)}(s, \xi(s))|^2$$

$$+ \int_{R^{(m)}} |F_n(s, \xi_0(s), u) - F_0(s, \xi_0(s), u)|^2 \frac{du}{|u|^{m+1}}\Big)\Big\} \times ds\Big].$$

By using Condition 4 and also the inequality

$$|\varphi_n(t) - \varphi_0(t)|\psi_n^{(N)}(t) \leq 2N,$$

$$((T - t_0)|A_n(s, \xi_0(s)) - A_0(s, \xi_0(s))|^2$$

$$+ \sum_{1}^{l} |B_k^{(n)}(s, \xi_0(s)) - B_k^{(0)}(s, \xi_0(s))|^2$$

$$+ \int_{R^{(m)}} |F_n(s, \xi_0(s), u) - F_0(s, \xi_0(s), u)|^2 \frac{du}{|u|^{m+1}}\Big) \varphi_k^{(N)}(t) \leq 4KN^2,$$

on the basis of Lebesgue's theorem on passage to the limit under the integral sign we can show that $\alpha_N^{(n)}(t) \to 0$ uniformly with respect to t as $n \to \infty$. This allows us to deduce easily from (5.2) the relation

$$\lim_{n\to\infty} \mathbf{M}|\xi_0(t) - \xi_n(t)|^2\psi_n^{(N)}(t) = 0;$$

consequently

$$\mathbf{P}\{|\xi_n(t) - \xi_0(t)| > \epsilon\} \leq \frac{\mathbf{M}\psi_n^{(N)}(t)|\xi_0(t) - \xi_n(t)|^2}{\epsilon^2} + \mathbf{P}\{\psi_n^{(N)}(t) = 0\}.$$

But

$$\mathbf{P}\{\psi_n^{(N)}(t) = 0\} \leq \mathbf{P}\Big\{\sup_t |\varphi_n(t)| > \frac{N}{4}\Big\} + \mathbf{P}\Big\{\sup_t |\varphi_0(t)| > \frac{N}{4}\Big\}$$

$$+ \mathbf{P}\Big\{\sup_t |\xi_n(t)| > \frac{N}{4}\Big\} + \mathbf{P}\Big\{\sup_t |\xi_0(t)| > \frac{N}{4}\Big\}.$$

Here, every term on the right-hand side approaches zero uniformly with respect to n as $N \to \infty$ because of Condition 3 and Remark 1 following

Theorem 3 of Section 2. Therefore

$$\lim_{N\to\infty} \overline{\lim_{n\to\infty}} \mathbf{P}\{|\xi_n(t) - \xi_0(t)| > \epsilon\} \le \frac{1}{\epsilon^2} \lim_{N\to\infty} \overline{\lim_{n\to\infty}} \mathbf{M}\psi_n^{(N)}(t)|\xi_0(t) - \xi_n(t)|^2$$
$$+ \lim_{N\to\infty} \overline{\lim_{n\to\infty}} \mathbf{P}\{\psi_n^{(N)}(t) = 0\} = 0.$$

This proves the theorem.

Now let $\xi_\alpha(t)$ be a solution of the equation

$$\xi_\alpha(t) = x_\alpha + \int_{t_0}^t a(s, \xi_\alpha(s))\, ds + \sum_{k=1}^{l} \int_{t_0}^t b_k(s, \xi_\alpha(s))\, dw_k(s)$$
$$+ \int_{t_0}^t \int_{R^{(m)}} f(s, \xi_\alpha(s), u) q\,(ds \times du), \tag{5.3}$$

where α is a numerical parameter that varies in some interval.

THEOREM 2. *Suppose that $a(t, x)$, $b_k(t, x)$, $f(t, x, u)$ satisfy the conditions of Theorem 3 of Section 2 and that, in addition, the following conditions are fulfilled:*

1. *First- and second-order derivatives with respect to the space variables (the coordinates of a point x) of the functions $a(t, x)$ and $b_k(t, x)$ exist and are bounded and continuous with respect to x.*

2. *If $(x^1, x^2, \ldots, x^m)$ are the coordinates of a point x, then for all i and j, the derivatives $f'_{x^i}(t, x, u)$, $f''_{x^ix^j}(t, x, u)$ also exist; further,*

$$\int_{R^{(m)}} |f'_{x^i}(t, x, u)|^2 \frac{du}{|u|^{m+1}}, \qquad \int_{R^{(m)}} |f''_{x^ix^j}(t, x, u)|^2 \frac{du}{|u|^{m+1}}$$

are bounded, and for every x_0, $t \in [t_0, T]$,

$$u\int_{R^{(m)}} \frac{|f'_{x^i}(t, x, u) - f'_{x^i}(t, y, u)|^4}{|x - y|^4} \frac{du}{|u|^{m+1}},$$

$$\lim_{x\to x_0} \int_{R^{(m)}} |f''_{x^ix^j}(t, x_0, u) - f''_{x^ix^j}(t, x, u)|^2 \frac{du}{|u|^{m+1}} = 0.$$

3. *The derivatives $(d/d\alpha)x_\alpha$ and $(d^2/d\alpha^2)x_\alpha$ exist and are continuous. Then $\xi_\alpha(t)$ has a stochastically continuous second-order derivative with respect to α.* (The derivative is defined by means of passing to the limit in the sense of convergence in probability.)

Proof. To simplify the notation, we shall consider processes in a one-dimensional space and shall assume $l = 1$. It follows from Theorem 1 that $\xi_\alpha(t)$ is continuous in probability as a function of α. Suppose that

$\eta_\alpha(t)$ is a solution of the equation

$$\eta_\alpha(t) = \frac{d}{d\alpha}x_\alpha + \int_{t_0}^{t} \frac{\partial}{\partial x} a(s, \xi_\alpha(s))\eta_\alpha(s)\,ds + \int_{t_0}^{t} \frac{\partial}{\partial x} b(s, \xi_\alpha(s))\eta_\alpha(s)\,dw(s)$$
$$+ \int_{t_0}^{t}\int_{u\in R^{(1)}} \frac{\partial}{\partial x} f(s, \xi_\alpha(s), u)\eta_\alpha(s)q\,(ds \times du). \quad (5.4)$$

Then since $[\xi_{\alpha+\Delta\alpha}(t) - \xi_\alpha(t)]/\Delta\alpha = \eta_{\alpha,\Delta\alpha}(t)$ is a solution of the equation

$$\eta_{\alpha,\Delta\alpha}(t) = \frac{x_{\alpha+\Delta\alpha} - x_\alpha}{\Delta\alpha} + \int_{t_0}^{t} \frac{a(s, \xi_{\alpha+\Delta\alpha}(s)) - a(s, \xi_\alpha(s))}{\xi_{\alpha+\Delta\alpha}(s) - \xi_\alpha(s)} \cdot \eta_{\alpha,\Delta\alpha}(s)$$
$$+ \int_{t_0}^{t} \frac{b(s, \xi_{\alpha+\Delta\alpha}(s)) - b(s, \xi_\alpha(s))}{\xi_{\alpha+\Delta\alpha}(s) - \xi_\alpha(s)}\eta_{\alpha,\Delta\alpha}(s)\,dw(s)$$
$$+ \int_{t_0}^{t}\int_{R^{(1)}} \frac{f(s, \xi_{\alpha+\Delta\alpha}(s), u) - f(s, \xi_\alpha(s), u)}{\xi_{\alpha+\Delta\alpha}(s) - \xi_\alpha(s)}\eta_{\alpha,\Delta\alpha}(s)q\,(ds \times du), \quad (5.5)$$

by using Theorem 1, we can see that $\eta_{\alpha,\Delta\alpha}(t) \to \eta_\alpha(t)$ in probability as $\Delta\alpha \to 0$; therefore $\eta_\alpha(t) = (\partial/\partial\alpha)\,\xi_\alpha(t)$.

In an analogous manner, we can show that $\partial^2\xi_\alpha(t)/\partial\alpha^2 = \zeta_\alpha(t)$, where $\zeta_\alpha(t)$ is a solution of the equation

$$\zeta_\alpha(t) = \frac{d^2}{d\alpha^2}x_\alpha + \int_{t_0}^{t} \frac{\partial^2}{\partial x^2} a(s, \xi_\alpha(s))\eta_\alpha^2(s)\,ds$$
$$+ \int_{t_0}^{t} \frac{\partial^2}{\partial x^2} b(s, \xi_\alpha(s))\eta_\alpha^2(s)\,dw(s)$$
$$+ \int_{t_0}^{t}\int_{R^{(1)}} \frac{\partial^2}{\partial x^2} f(u, s, \xi_\alpha(s))\eta_\alpha^2(s)q\,(ds \times du)$$
$$+ \int_{t_0}^{t} \frac{\partial}{\partial x} a(s, \xi_\alpha(s))\zeta_\alpha(s)\,ds$$
$$+ \int_{t_0}^{t} \frac{\partial}{\partial x} b(s, \xi_\alpha(s))\zeta_\alpha(s)\,dw(s)$$
$$+ \int_{t_0}^{t}\int_{R^{(1)}} \frac{\partial}{\partial x} f(s, \xi_\alpha(s), u)\zeta_\alpha(s)q\,(ds \times du). \quad (5.6)$$

The stochastic continuity of $\eta_\alpha(t)$ and $\zeta_\alpha(t)$ also follows from Theorem 1. This proves the theorem.

Remark 1. Since the variables

$$x'_\alpha, \qquad \frac{a(s, x_1) - a(s, x_2)}{x_1 - x_2}, \qquad \frac{b(s, x_1) - b(s, x_2)}{x_1 - x_2},$$

and

$$\int_{u \in R^{(1)}} \left| \frac{f(s, x_1, u) - f(s, x_2, u)}{x_2 - x_1} \right|^2 \frac{du}{|u|^2}$$

are bounded, by applying Remark 1 following Theorem 3 of Section 2 to Eq. (5.5), we can see that

$$\mathbf{M} \sup_t |\eta_{\alpha,\Delta\alpha}(t)|^2$$

is a finite constant independent of α and $\Delta\alpha$.

Let us now consider a process $\zeta_{\alpha,\Delta\alpha}(t)$ satisfying the condition

$$\begin{aligned} \zeta_{\alpha,\Delta\alpha}(t) &= \frac{1}{\Delta\alpha}\left(\frac{d}{d\alpha} x_{\alpha+\Delta\alpha} - \frac{d}{d\alpha} x_\alpha\right) + \varphi_{\alpha,\Delta\alpha}(t) \\ &\quad + \int_{t_0}^t a'_x(s, \xi_\alpha(s))\zeta_{\alpha,\Delta\alpha}(s)\,ds + \int_{t_0}^t b'_x(s, \xi_\alpha(s))\zeta_{\alpha,\Delta\alpha}(s)\,dw(s) \\ &\quad + \int_{t_0}^t \int_{R^{(1)}} f'_x(s, \xi_\alpha(s), u)\zeta_{\alpha,\Delta\alpha}(s) q\,(ds \times du), \end{aligned} \tag{5.7}$$

where

$$\begin{aligned} \varphi_{\alpha,\Delta\alpha}(t) &= \int_{t_0}^t \frac{a'_x(s, \xi_{\alpha+\Delta\alpha}(s)) - a'_x(s, \xi_\alpha(s))}{\xi_{\alpha+\Delta\alpha}(s) - \xi_\alpha(s)} \, \frac{\xi_{\alpha+\Delta\alpha}(s) - \xi_\alpha(s)}{\Delta\alpha} \, \eta_{\alpha+\Delta\alpha}(s)\,ds \\ &\quad + \int_{t_0}^t \frac{b'_x(s, \xi_{\alpha+\Delta\alpha}(s)) - b'_x(s, \xi_\alpha(s))}{\xi_{\alpha+\Delta\alpha}(s) - \xi_\alpha(s)} \, \frac{\xi_{\alpha+\Delta\alpha}(s) - \xi_\alpha(s)}{\Delta\alpha} \, \eta_{\alpha+\Delta\alpha}(s)\,dw(s) \\ &\quad + \int_{t_0}^t \int_{R^{(1)}} \frac{f'_x(s, \xi_{\alpha+\Delta\alpha}(s), u) - f'_x(s, \xi_\alpha(s), u)}{\xi_{\alpha+\Delta\alpha}(s) - \xi_\alpha(s)} \, \frac{\xi_{\alpha+\Delta\alpha}(s) - \xi_\alpha(s)}{\Delta\alpha} \\ &\quad \times \eta_{\alpha+\Delta\alpha}(s) q\,(ds \times du). \end{aligned}$$

By using the boundedness of the quantities

$$\frac{b'_x(s, x_1) - b'_x(s, x_2)}{x_1 - x_2}, \qquad \frac{a'_x(s, x_1) - a'_x(s, x_2)}{x_1 - x_2},$$

and

$$\int_{R^{(1)}} \left| \frac{f'_x(s, x_1, u) - f'_x(s, x_2, u)}{x_1 - x_2} \right|^2 \frac{du}{|u|^2},$$

we see that for some H_1,

$$\mathbf{M} \sup |\varphi_{\alpha,\Delta\alpha}(t)|^4 \leq H_1 \int_{t_0}^{T} \mathbf{M}\left[\left(\frac{\xi_{\alpha+\Delta\alpha}(s) - \xi_\alpha(s)}{\Delta\alpha}\right)^4 + (\eta_{\alpha+\Delta\alpha}(s))^4\right] ds, \tag{5.8}$$

provided the right-hand side is finite. Let us show that under our assumptions

$$\mathbf{M}\left(\frac{\xi_{\alpha+\Delta\alpha}(s) - \xi_\alpha(s)}{\Delta\alpha}\right)^4$$

is uniformly bounded with respect to α and $\Delta\alpha$. Hence, since η_α is the limit in the sense of convergence in probability of the quantity

$$\left(\frac{\xi_{\alpha+\Delta\alpha}(s) - \xi_\alpha(s)}{\Delta\alpha}\right),$$

the uniform boundedness of $\mathbf{M}(\eta_\alpha(s))^4$ follows.

LEMMA 1. *Suppose that $f(s, u)$ is measurable with respect to F_s and that*

$$\int_{t_0}^{T} \int_{R^{(1)}} \mathbf{M}|f(s, u)|^4 \frac{ds\, du}{|u|^2} < \infty.$$

Then there exists a constant $\overline{H}$, dependent only on $(T - t_0)$, for which

$$\mathbf{M}\left[\int_{t_0}^{T} \int_{R^{(1)}} f(s, u) q\,(ds \times du)\right]^4 \leq \overline{H} \int_{t_0}^{T} \int_{R^{(1)}} \mathbf{M}|f(s, u)|^4 \frac{ds\, du}{|u|^2}.$$

Proof. It is easy to see that for step functions,

$$\mathbf{M}\left[\int_{t}^{t+h} \int_{R^{(1)}} f(s, u) q\,(ds \times du)\right]^4 = \int_{t}^{t+h} \int_{R^{(1)}} \mathbf{M}|f(s, u)|^4 \frac{ds\, du}{|u|^2} + o(h),$$

$$\mathbf{M}\left[\int_{t}^{t+h} \int_{R^{(1)}} f(s, u) q\,(ds \times du)\right]^3 = \int_{t}^{t+h} \int_{R^{(1)}} \mathbf{M}(f(s, y))^3 \frac{ds \times du}{|u|^2} + o(h).$$

Therefore if $t_0 < t_1 < \cdots < t_n = T$, then

$$\mathbf{M}\left|\int_{t_0}^{T}\int_{R^{(1)}} f(u, s)q\,(ds \times du)\right|^4 = \sum_{k=1}^{n}\left(\mathbf{M}\left|\int_{t_0}^{t_k}\int_{R^{(1)}} f(s, u)q\,(ds \times du)\right|^4 - \mathbf{M}\left|\int_{t_0}^{t_{k-1}}\int_{R^{(1)}} f(s, u)q\,(ds \times du)\right|^4\right)$$

$$= 6\sum_{k=1}^{n}\mathbf{M}\left(\int_{t_0}^{t_{k-1}}\int_{R^{(1)}} f(s, u)q\,(ds \times du)^2 \int_{t_{k-1}}^{t_k}\int_{R^{(1)}} |f(t, v)|^2 \frac{dv\,dt}{|v|^2}\right)$$

$$+ 4\sum_{k=1}^{n}\mathbf{M}\int_{t_0}^{t_{k-1}}\int_{R^{(1)}} f(s, u)q\,(ds \times du)\left(\int_{t_{k-1}}^{t_k}\int_{R^{(1)}} (f(t, v))^3 \frac{dt\,dv}{v^2}\right)$$

$$+ o(t_{k+1} - t_k) + \int_{t_0}^{T} \mathbf{M}|(f(s, u))|^4 \frac{ds\,du}{|u|^2} + o\left(\max_k (t_{k+1} - t_k)\right).$$

Taking the limit as $\max_k (t_{k+1} - t_k) \to 0$, we obtain

$$\mathbf{M}\left|\int_{t_0}^{T}\int_{R^{(1)}} f(s, u)q\,(ds \times du)\right|^4$$

$$\leq 6\int_{t_0}^{T}\mathbf{M}\left(\int_{t_0}^{t}\int_{R^{(1)}} f(s, u)q\,(ds \times du)^2 \int_{R^{(1)}} |f(t, v)|^2 \frac{dv}{|v|^2}\right) dt$$

$$+ 4\int_{t_0}^{T}\mathbf{M}\left(\int_{t_0}^{t}\int_{R^{(1)}} f(s, u)q\,(ds \times du)\int_{R^{(1)}} |f(t, v)|^3 \frac{dv}{|v|^2}\right) dt$$

$$+ \int_{t_0}^{T}\int_{R^{(1)}} \mathbf{M}|f(s, u)|^4 \frac{ds\,du}{|u|^2}.$$

By using Hölder's inequality, we obtain

$$\mathbf{M}\left(\int_{t_0}^{T}\int_{R^{(1)}} f(s, u)q\,(ds \times du)\right)^4 \leq \int_{t_0}^{T}\int_{R^{(1)}} \mathbf{M}|f(s, u)|^4 \frac{ds\,du}{|u|^2}$$

$$+ 6\left[\left\{\int_{t_0}^{T}\mathbf{M}\left(\int_{t_0}^{t}\int_{R^{(1)}} f(s, u)q\,(ds \times du)\right)^4 dt\right\} \times \left(\int_{t_0}^{T}\int_{R^{(1)}} \mathbf{M}|f(s, u)|^4 \frac{ds\,du}{|u|^2}\right)\right]^{1/2}$$

$$+ 4\left[\int_{t_0}^{T}\mathbf{M}\left(\int_{t_0}^{t}\int_{R^{(1)}} f(s, u)q\,(ds \times du)\right)^4 dt\right]^{1/4} \times \left[\int_{t_0}^{T}\int_{R^{(1)}} \mathbf{M}|f(s, u)|^4 \frac{ds\,du}{|u|^2}\right]^{3/4}.$$

If we define

$$A = \sup_{t_0 \le t \le T} \mathbf{M}\left[\int_{t_0}^{t} \int_{R^{(1)}} f(s, u) q\,(ds \times du)\right]^4,$$

$$B = \int_{t_0}^{T} \int_{R^{(1)}} \mathbf{M}|f(s, u)|^4 \frac{ds \times du}{|u|^2},$$

then from the preceding inequality, we obtain

$$\frac{A}{B} \le 6(T - t_0)^{1/2}\left(\frac{A}{B}\right)^{1/2} + 4(T - t_0)^{1/4}\left(\frac{A}{B}\right)^{1/4} + 1 \le 8(T - t_0)^{1/2}\left(\frac{A}{B}\right)^{1/2} + 3$$

because

$$2(T - t_0)^{1/4}(A/B)^{1/4} \le (T - t_0)^{1/2}(A/B)^{1/2} + 1).$$

Consequently

$$\left(\frac{A}{B}\right)^{1/2} \le \sqrt{3 + 16(T - t_0)} - 4\sqrt{(T - t_0)}.$$

The proof of the lemma for arbitrary step functions follows. The general case can be obtained by passing to the limit with step functions.

By using the lemma just proved and Theorem 4 of Section 2, Chapter 2, we can easily see from (5.5) that there exists a constant C, independent of α and $\Delta\alpha$, such that

$$\mathbf{M}\left(\eta_{\alpha,\Delta\alpha}(t)\right)^4 \le C\left[\int_{t_0}^{t} \mathbf{M}\left(\eta_{\alpha,\Delta\alpha}(s)\right)^4 ds + (\sup_{\alpha} x'_\alpha)^4\right],$$

which means that

$$\mathbf{M}\left(\eta_{\alpha,\Delta\alpha}(t)\right)^4 \le (\sup_{\alpha} x'_\alpha)^4 e^{C(t-t_0)};$$

thus from (5.8), $\mathbf{M}\sup_t|\varphi_{\alpha,\Delta\alpha}(t)|^2$ is uniformly bounded with respect to α and $\Delta\alpha$. If we now apply Remark 1 following Theorem 3 of Section 2 to Eq. (5.7), we can see that Remark 2 holds.

Remark 2. If x_α, $a(t, x)$, $b(t, x)$, $f(t, x, u)$ satisfy the conditions of Theorem 2, and if $\xi_{\alpha,\Delta\alpha}(t)$ satisfies Eq. (5.7), then there exists a constant C_2 independent of α and $\Delta\alpha$ such that

$$\mathbf{M}|\zeta_{\alpha,\Delta\alpha}(t)|^2 \le C_2;$$

therefore the inequality

$$\mathbf{M}\left(\frac{\partial^2}{\partial\alpha^2}\,\xi_\alpha(t)\right)^2 \le C_2$$

will also be satisfied for

$$\frac{\partial^2}{\partial\alpha^2}\,\xi_\alpha(t) = \lim_{\Delta\alpha\to 0} \zeta_{\alpha,\Delta\alpha}(t)$$

(where the limit is in the sense of convergence in probability).

THEOREM 3. *If $g(x)$ is a twice continuously differentiable function, if it and its first two derivatives are bounded, and if $\xi_\alpha(t)$ is a solution of Eq. (5.1) with coefficients satisfying the conditions of Theorem 2, then the function*

$$\Phi(\alpha) = \mathbf{M}g\big(\xi_\alpha(t)\big)$$

will be a twice continuously differentiable function of α.

To prove this theorem, we use the following lemma.

LEMMA 2. *If $\xi_n \to \xi_0$ in probability and* $\sup_n \mathbf{M}\xi_n^2 < \infty$, *then*

$$\mathbf{M}\xi_n \to \mathbf{M}\xi_0.$$

Proof. We define $\psi_N(t) = 1$ for $|t| \le N$, $\psi(t) = N + 1 - |t|$ for $t \in [N, N+1]$, and $\psi_N(t) = 0$ for $|t| > N$. Then $\xi_n\psi_N(\xi_n)$ is bounded by the quantity $N + 1$ and it converges in probability to $\xi_0\psi_N(\xi_0)$. Therefore

$$\mathbf{M}\xi_n\psi_N(\xi_n) \to \mathbf{M}\xi_0\psi_N(\xi_0).$$

But

$$\begin{aligned}
|\mathbf{M}\xi_n - \mathbf{M}\xi_0| &\le |\mathbf{M}\xi_n\psi_N(\xi_n) - \mathbf{M}\xi_0\psi_N(\xi_0)| + \mathbf{M}|\xi_n|\big(1 - \psi_N(\xi_n)\big) \\
&\qquad + \mathbf{M}|\xi_0|\big(1 - \psi_N(\xi_0)\big) \\
&\le |\mathbf{M}\xi_n\psi_N(\xi_n) - \mathbf{M}\xi_0\psi_N(\xi_0)| + \sqrt{\mathbf{M}\xi_0^2\cdot\mathbf{M}\big(1 - \psi_N(\xi_0)\big)} \\
&\qquad + \sqrt{\mathbf{M}\xi_n^2\cdot\mathbf{M}\big(1 - \psi_N(\xi_n)\big)} \\
&\le |\mathbf{M}\xi_n\psi_N(\xi_n) - \mathbf{M}\xi_0\psi_N(\xi_0)| + \sqrt{\mathbf{M}\xi_n^2\cdot\mathbf{P}\{|\xi_0| > N\}} \\
&\qquad + \sqrt{\mathbf{M}\xi_n^2\cdot\mathbf{P}\{|\xi_n| > N\}} \\
&\le |\mathbf{M}\xi_n\psi_N(\xi_n) - \mathbf{M}\xi_0\psi_N(\xi_0)| + \frac{2}{N}\sup_n \mathbf{M}\xi_n^2.
\end{aligned}$$

The last inequality proves the lemma.

Applying the lemma just proved to the equation

$$\frac{\Phi(\alpha + \Delta\alpha) - \Phi(\alpha)}{\Delta\alpha} = \mathbf{M}\,\frac{g\big(\xi_{\alpha+\Delta\alpha}(t)\big) - g\big(\xi_\alpha(t)\big)}{\xi_{\alpha+\Delta\alpha}(t) - \xi_\alpha(t)}\;\frac{\xi_{\alpha+\Delta\alpha}(t) - \xi_\alpha(t)}{\Delta\alpha}$$

(the possibility of doing this being assured by Remark 1 following the preceding theorem), we obtain

$$\Phi'(\alpha) = \mathbf{M}g'_x(\xi_\alpha(t)) \frac{\partial}{\partial\alpha} \xi_\alpha(t).$$

Now, applying Lemma 2 to the inequality

$$\frac{\Phi'(\alpha + \Delta\alpha) - \Phi'(\alpha)}{\Delta\alpha} =$$

$$\mathbf{M} \frac{g_x(\xi_{\alpha+\Delta\alpha}(t)) - g_x(\xi_\alpha(t))}{\xi_{\alpha+\Delta\alpha}(t) - \xi_\alpha(t)} \frac{\xi_{\alpha+\Delta\alpha}(t) - \xi_\alpha(t)}{\Delta\alpha} \frac{\partial}{\partial\alpha} \xi_{\alpha+\alpha\Delta}(t)$$

$$+ \mathbf{M}g'_x(\xi_\alpha(t)) \frac{1}{\Delta\alpha} \left(\frac{\partial}{\partial\alpha} \xi_{\alpha+\Delta\alpha}(t) - \frac{\partial}{\partial\alpha} \xi_\alpha(t)\right)$$

(this being justified by Remark 2), we obtain

$$\Phi''(\alpha) = \mathbf{M}g'_{xx}(\xi_\alpha(t)) \left(\frac{\partial}{\partial\alpha} \xi_\alpha(t)\right)^2 + \mathbf{M}g'_x(\xi_\alpha(t)) \frac{\partial^2}{\partial\alpha^2} \xi_\alpha(t). \tag{5.9}$$

Since

$$\sup_\alpha [\mathbf{M}\{(g''_x(\xi_\alpha(t))\xi'_\alpha(t)^2)^2 + (g'_x(\xi_\alpha(t))\xi''_\alpha(t))^2\}] < \infty,$$

and since the expression under the mathematical expectation sign on the right-hand side of (5.9) is stochastically continuous with respect to α, we may conclude on the basis of Lemma 2 that $\Phi''(\alpha)$ is continuous with respect to α. This proves the theorem.

6. Differential equations defining the distributions of Markov processes. To define the distribution of a variable $\xi(t)$, it is sufficient to know $\mathbf{M}g(\xi(t))$ for the case in which the $g(x)$ are sufficiently smooth bounded functions. For $t_1 \in [t_0, t]$, let $\xi_{x,t_1}(t)$ denote a solution of the equation

$$\xi_{x,t_1}(t) = x + \int_{t_1}^t a(s, \xi_{x,t_1}(s))\, ds + \sum_{k=1}^l \int_{t_1}^t b_k(s, \xi_{x,t_1}(s))\, dw_k(s)$$

$$+ \int_{t_1}^t \int_{u\in R^{(m)}} f(s, \xi_{x,t_1}(s), u) q\, (ds \times du).$$

[We shall assume that the conditions of Theorem 3 of the preceding section are satisfied by the coefficients $a(t, x)$, $b_k(t, x)$, $f(t, x, u)$.] Then if $g(x)$ also satisfies the conditions of Theorem 3 of the preceding section, the function

$$\Phi(x, t_1) = \mathbf{M}g(\xi_{x,t_1}(t))$$

will be a twice continuously differentiable function of the space variables. Let us show that $\Phi(x, t_1)$ satisfies some integro-differential equation. As a preliminary, let us prove an auxiliary proposition.

LEMMA. *For every twice continuously differentiable function $g(x)$, such that it and its first two derivatives are bounded, the following relation holds:*

$$\lim_{t\to t_1} \frac{\mathbf{M}g(\xi_{x,t_1}(t)) - g(x)}{t - t_1} = (a(t_1, x), \nabla)g(t_1, x)$$

$$+ \sum_{k=1}^{l} (b_k(t_1, x), \nabla)^2 g(t_1, x)$$

$$+ \int_{R^{(m)}} (g(t_1, f(t_1, x, u)) - g(t_1, x) - (f, \nabla)g) \frac{du}{|u|^{m+1}}, \quad (6.1)$$

where ∇ is a vector with coordinates $(\partial/\partial x^1, \partial/\partial x^2, \ldots, \partial/\partial x^m)$ and $x^1, x^2, \ldots, x^m$ are the coordinates of the vector x.

Proof. Under our assumptions,

$$\xi_{x,t_1}(t) = x + \int_{t_1}^{t} a(t_1, x)\, ds + \sum_{k=1}^{l} \int_{t_1}^{t} b_k(t_1, x)\, dw_k(s)$$

$$+ \int_{t_1}^{t} \int_{u\in R^{(m)}} f(t_1, x, u) q\, (ds \times du) + \Theta(t - t_1),$$

where $\Theta(t - t_1)$ is a random variable such that $\Theta(t - t_1) = o(t - t_1)$. (This was established in the proof of Theorem 3 of Section 4.) On the basis of Property (d) of Section 4, we know that

$$\lim_{t\to t_1} \frac{1}{t - t_1} (g(\xi_{x,t_1}(t) - \Theta(t - t_1)) - g(x)) = \lim_{t\to t_1} (g(\xi_{x,t_1}(t)) - g(x)),$$

if this limit exists for every function $g(x)$ satisfying the condition of the lemma.

Let us show that

$$\lim_{t\to t_1} \frac{1}{t - t_1} (g(\xi_{x,t_1}(t) - \Theta(t - t_1)) - g(x))$$

coincides with the right-hand side of (6.1). First we find that

$$\lim_{t\to t_1} \frac{1}{t - t_1} \mathbf{M}\left[g\left(x + \int_{t_1}^{t} \int_{R^{(m)}} f(t_1, x, u) q\, (ds \times du)\right) - g(x)\right]$$

$$= \lim_{t\to t_1} \frac{1}{t - t_1} \mathbf{M}\left[g\left(x + \int_{t_1}^{t} \int_{R^{(m)}} f(t_1, x, u) q\, (ds \times du)\right) - g(x)\right.$$

$$\left. - \sum g'_{x^i}(x) \int_{t_1}^{t} \int f^i(t_1, x, u) q\, (ds \times du)\right],$$

since $\mathbf{M}\int_{t_1}^{t}\int_{R^{(m)}} f^i(t_1, x, u)q\ (ds \times du)$ and $f^i(t, x, u)$ are the coordinates of the vector $f(t, x, u)$. Let us first consider the case in which $f(t_1, x, u)$ is a step function with respect to u: $f(t_1, x, u)$ is different from zero only for $u \in \bigcup_{k=1}^{N} B_k$ and is constant on each of the sets B_k. Since

$$\mathbf{P}\left\{p\left(\bigcup_{k=1}^{N} B_k\right) > 1\right\} = o(t - t_1),$$

then for $u_k \in B_k$,

$$\begin{aligned}\mathbf{M}\Bigg[g\left(x + \int_{t_1}^{t}\int f(t_1, x, u)q\ (ds \times du)\right) - g(x)\\ - \sum_{i=1}^{m} \frac{\partial g(x)}{\partial x^i}\int_{t_1}^{t}\int f^i(t_1, x, u)q\ (ds \times du)\Bigg]\\ = \sum_{k=1}^{N}\Bigg\{g\left(x + \left[1 - (t - t_1)\int_{\bigcup_k B_k} \frac{du}{|u|^{m+1}}\right] f(t_1, x, u_k)\right) - g(x)\\ - \sum_{i=1}^{m} \frac{\partial g}{\partial x^i}(x)\left[1 - (t - t_1)\int_{\bigcup_k B_k}\frac{du}{|u|^{m+1}}\right] f^i(t_1, x, u_k)\Bigg\}\\ \times \int_{B_k} \frac{du}{|u|^{m+1}}(t - t_1)o(t - t_1).\end{aligned}$$

From this relation, it is easy to show that in the general case

$$\begin{aligned}\mathbf{M}\Bigg[g\left(x + \int_{t_0}^{t}\int f(t_1, x, u)q\ (ds \times du)\right) - g(x)\\ - \sum \frac{\partial g(x)}{\partial x^i}\int_{t_0}^{t}\int f^i(t_1, x, u)q\ (ds \times du)\Bigg]\\ = (t - t_1)\int_{R^{(m)}}\left[g(x + f(t_1, x, u)) - g(x) - \sum_{i=1}^{m} g'_{x^i}(x)f^i(t_1, x, u)\right]\\ \times \frac{du}{|u|^{m+1}} + o(t - t_1).\end{aligned}$$

To prove the lemma, we still need to show that

$$\begin{aligned}\lim_{t \to t_1} \frac{1}{t - t_1}\mathbf{M}\left[g\left(x + \int_{t_1}^{t} a(t_1, x)\, ds + \sum_{k=1}^{l}\int_{t_1}^{t} b_k(t_1, x)\, dw_k(s)\right) - g(x)\right]\\ = \sum_{i=1}^{m} a^i(t_1, x)\frac{\partial g(x)}{\partial x^i} + \sum_{1 \le i,j \le m}\sum_{k=1}^{l} b_k^i(t_1, x)b_k^j(t_1, x)g''_{x^ix^j}(x)\end{aligned}$$

uniformly with respect to x in every bounded region. This follows from the relation

$$\mathbf{M}\left[g\left(x+\int_{t_1}^{t} a(t_1, x)\,ds+\sum_{k=1}^{l}\int_{t_1}^{t} b_k(t_1, x)\,dw_k(s)\right)-g(x)\right]$$
$$=\frac{1}{[2\bar{\pi}(t-t_1)]^{l/2}}\int\left[g\left(x+a(t_1, x)(t-t_1)+\sum_{k=1}^{l} b_k(t_1, x)\alpha_k\right)-g(x)\right]$$
$$\times\exp\left[-\frac{\sum_{k=1}^{l}\alpha_k^2}{2(t-t_1)}\right]d\alpha_1\cdots d\alpha_l.$$

Combining these results, we obtain a proof of the lemma.

THEOREM. *Suppose that* $\Phi(t_1, x)=\mathbf{M}g(\xi_{t_1,x}(t))$, *where* $\xi_{t_1,x}(t)$ *and* g *satisfy the conditions stated at the beginning of this section. Then*

$$-\frac{\partial\Phi}{\partial t_1}=\sum_{i=1}^{m} a^i(t_1, x)\frac{\partial\Phi}{\partial x^i}+\frac{1}{2}\sum_{i,j=1}^{m}\sum_{k=1}^{l} b_k^i(t_1, x)b_k^j(t_1, x)\frac{\partial^2\Phi}{\partial x^i\,\partial x^j}$$
$$+\int_{R^{(m)}}\left[\Phi(t_1, x+f(t_1, x, u))-\Phi(t_1, x)-\sum_{i=1}^{m} f^i(t_1, x, u)\frac{\partial\Phi}{\partial x^i}\right]\frac{du}{|u|^{m+1}}. \tag{6.2}$$

Proof. For $t_1 < t_2 < t$,

$$\mathbf{M}g(\xi_{t_1,x}(t)/\xi_{t_1,x}(t_2))=\Phi(t_2, \xi_{t_1,x}(t_2)).$$

This means that

$$\Phi(t_1, x)=\mathbf{M}\Phi(t_2, \xi_{t_1,x}(t_2)).$$

Therefore

$$\frac{\Phi(t_1-\Delta, x)-\Phi_1(t_1, x)}{\Delta}=\mathbf{M}\frac{\Phi(t_1, \xi_{t_1-\Delta,x}(t_1))-\Phi(t_1, x)}{\Delta}. \tag{6.3}$$

We note, on the basis of Theorem 3 of Section 5, that the function $\Phi(t_1, x)$ is twice continuously differentiable. Thus we may apply the lemma of this section to (6.3), which proves the theorem.

CHAPTER 4

DIFFERENTIABILITY OF MEASURES CORRESPONDING TO MARKOV PROCESSES

1. Statement of the problem. Suppose that two measures μ_1 and μ_2 are given on a σ-algebra $\mathbf{B}$ of subsets of some set X with elements x. The measure μ_2 is said to be *absolutely continuous with respect to the measure* μ_1 if for every $A \in \mathbf{B}$ such that $\mu_1(A) = 0$, it is also true that $\mu_2(A) = 0$. If μ_1 and μ_2 are finite, the Radon-Nikodym theorem states that for every measure μ_2 that is absolutely continuous with respect to the measure μ_1, there exists a $\mathbf{B}$-measurable function $p(x)$ such that for every $A \in \mathbf{B}$,

$$\mu_2(A) = \int_A p(x)\, \mu_1(dx). \tag{1.1}$$

The function $p(x)$ is called the derivative or the density of the measure μ_2 with respect to μ_1, and is denoted by $(d\mu_2/d\mu_1)(x)$. The measure μ_2 is then said to be differentiable with respect to the measure μ_1.

Let us examine a random process $\xi(t)$ that is defined on $[t_0, T]$ and that takes its values in $R^{(m)}$. We denote by $\Phi_m[t_0, T]$ the set of all functions $x(t)$ that are defined on $[t_0, t]$ and that take values in $R^{(m)}$. If $F_m[t_0, T]$ is the minimal σ-algebra of subsets of $\Phi_m[t_0, T]$ that contains all cylindrical sets, then to the process $\xi(t)$ there corresponds on $F_m[t_0, T]$ some measure μ such that $\mu(\mathbf{A}) = \mathbf{P}\{\xi(t) \in \mathbf{A}\}$ for every cylindrical set $\mathbf{A} \in F_m[t_0, T]$. In this chapter, we are concerned with the following question: Under what conditions will the measure μ_2, corresponding to the process $\xi_2(t)$, be absolutely continuous with respect to the measure μ_1 corresponding to the process $\xi_1(t)$? Also, if μ_2 is absolutely continuous with respect to μ_1, we shall evaluate $d\mu_2/d\mu_1$. In that case, $d\mu_2/d\mu_1$ will be a functional with respect to $F_m[t_0, T]$ in the space $\Phi_m[t_0, T]$ and will be determined to within sets of measure zero under the measure μ_1. Therefore the variable $(d\mu_2/d\mu_1)(\xi_1(t))$ will be measurable with respect to the original field of probabilities. What is known about the variable $(d\mu_2/d\mu_1)(\xi_1(t))$ allows us to write Formula (1.1) in the following form: If $\psi_B(x(t))$ is the characteristic function of a set B in $F_m[t_0, T]$, then

$$\mu_2(B) = \mathbf{M}\psi_B(\xi_2(t)) = \mathbf{M}\psi_B(\xi_1(t))\, \frac{d\mu_2}{d\mu_1}\, (\xi_1(t)). \tag{1.2}$$

Therefore we shall always confine ourselves to the matter of evaluating the quantity $(d\mu_2/d\mu_1)(\xi_1(t))$. The processes $\xi_1(t)$ and $\xi_2(t)$ will henceforth be assumed to be Markov processes that are solutions of the stochastic equations of the form of Eq. (1.7) of Chapter 3.

The study of the absolute continuity of measures corresponding to random processes is of interest in connection with the following considerations.

In the first place, among the various properties of random processes, an especially important role is played by the properties which are satisfied by the process almost everywhere; if the measure μ_2 corresponding to the process $\xi_2(t)$ is absolutely continuous with respect to the measure μ_1 corresponding to the process $\xi_1(t)$, then $\xi_2(t)$ will satisfy almost everywhere all those properties that the process $\xi_1(t)$ satisfies almost everywhere.

In the second place, a knowledge of $(d\mu_2/d\mu_1)(\xi_1(t))$ allows us to evaluate by means of Formula (1.2) the probabilities of the various events related to the process $\xi_2(t)$, provided we can compute the distributions of the functionals of the process $\xi_1(t)$.

Finally, knowledge of $(d\mu_2/d\mu_1)(\xi_1(t))$ is necessary for solving various statistical problems by methods based on a plausibility relationship—problems dealing with choice among hypotheses and problems dealing with the estimation of an unknown distribution parameter.

2. Lemmas. Let us establish certain auxiliary propositions dealing with the absolute continuity of measures in general and measures corresponding to random processes in particular.

LEMMA 1. *Suppose that μ_1 and μ_2 are finite measures on* $\mathbf{B}$, *and that there exists a sequence B_n such that for every $A \in \mathbf{B}_n$, $\mu_2(A/B_n) \to 0$. If the measure $\mu_2^{(n)}(A) = \mu_2(A \cap B_n)$ is absolutely continuous with respect to the measure $\mu_1^{(n)}(A) = \mu_1(A \cap B_n)$ for every n, then μ_2 is absolutely continuous with respect to μ_1, and for $x \in B_n$,*

$$\frac{d\mu_2}{d\mu_1}(x) = \frac{d\mu_2^{(n)}}{d\mu_1^{(n)}}(x).$$

Proof. If $\mu_1(A) = 0$, then $\mu_1(A \cap B_n) = 0$, which implies that $\mu_2(A \cap B_n) = 0$. But $\mu_2(A) = \lim_{n\to\infty} \mu_2(A \cap B_n)$ and consequently $\mu_2(A) = 0$; that is, μ_2 is absolutely continuous with respect to μ_1. It is now easy to establish that for almost all x with respect to the measure μ_1 belonging to B_n, $(d\mu_2/d\mu_1)(x) = (d\mu_2^{(n)}/d\mu_1^{(n)})(x)$ because

$$\int_A \left(\frac{d\mu_2}{d\mu_1}(x) - \frac{d\mu_2^{(n)}}{d\mu_1^{(n)}}(x)\right)\mu_1(dx) = 0 \qquad \text{for all} \qquad A \subset B_n.$$

Let $\mathbf{B}_x$ be a σ-algebra of subsets of X; let $\mathbf{B}_y$ be a σ-algebra of subsets of Y; let $Z = X \times Y$ be the set of pairs (x, y), $x \in X$, $y \in Y$; let $\mathbf{B}_z$ be the minimal σ-algebra of subsets of Z that contains all the subsets $B_z = B_x \times B_y$, $B_x \in \mathbf{B}_x$, $B_y \in \mathbf{B}_y$; further, let μ be a measure on $\mathbf{B}_x$ and ν be a measure on $\mathbf{B}_y$. Let us denote by $\mu \times \nu$ a measure on $\mathbf{B}_z$ such that

$$\mu \times \nu(B_x \times B_y) = \mu(B_x)\nu(B_y).$$

We shall call this measure the *product of the measures* μ *and* ν. The following obvious lemma is valid for the products of measures.

LEMMA 2. *If* μ_1 *and* μ_2 *are measures on* $\mathbf{B}_x$ *and* ν_1 *and* ν_2 *are measures on* $\mathbf{B}_y$, *and if* μ_2 *and* ν_2 *are absolutely continuous with respect to* μ_1 *and* ν_1, *respectively, then the measure* $\mu_2 \times \nu_2$ *is absolutely continuous with respect to* $\mu_1 \times \nu_1$ *and*

$$\frac{d\mu_2 \times \nu_2}{d\mu_1 \times \nu_1}(x, y) = \frac{d\mu_2}{d\mu_1}(x)\frac{d\nu_2}{d\nu_1}(y).$$

Again let there be defined σ-algebras of the subsets $\mathbf{B}_x$ and $\mathbf{B}_y$ on the sets X and Y. A mapping T of the set X into the set Y is said to be $(\mathbf{B}_x \times B_y)$-measurable if the inverse image of each set in $\mathbf{B}_y$ belongs to $\mathbf{B}_x$.

LEMMA 3. *Suppose that* $\xi_1(t)$ *and* $\xi_2(t)$ *are processes defined on* $[t_0, T]$ *and taking values in* $R^{(m)}$; *suppose that* T *is an* $F_m[t_0, T] \times F_{m'}[t_0, T]$-*measurable mapping of* $\Phi_m[t_0, T]$ *into* $\Phi_{m'}[t_0, T]$. *Suppose that* $\eta_1(t) = T\xi_1(t)$, *that* $\eta_2(t) = T\xi_2(t)$, *that the* μ_i *are measures corresponding to the processes* $\xi_i(t)$ *on* $F_m[t_0, T]$, *and that the* ν_i *are measures corresponding to the processes* $\eta_i(t)$ *on* $F_{m'}[t_0, T]$. *Then if* μ_2 *is absolutely continuous with respect to* μ_1, *it follows that* ν_2 *is absolutely continuous with respect to* ν_1 *and*

$$\frac{d\nu_2}{d\nu_1}(\eta_1(t))\mathbf{M}\left(\frac{d\mu_2}{d\mu_1}(\xi_1(t))/\eta_1(t),\ t \in [t_0, T]\right).$$

Proof. For every set $\mathbf{A} \in F_{m'}[t_0, T]$, denote its characteristic function by $\psi_{\mathbf{A}}(x(t))$. Then

$$\begin{aligned}\mathbf{M}\psi_{\mathbf{A}}(\eta_1(t))\mathbf{M}&\left(\frac{d\mu_2}{d\mu_1}(\xi_1(t))/\eta_1(t),\ t \in [t_0, T]\right)\\ &= \mathbf{M}\Psi_{\mathbf{A}}(\eta_1(t))\frac{d\mu_2}{d\mu_1}(\xi_1(t)) = \mathbf{M}\Psi_{T^{-1}\mathbf{A}}(\xi_1(t))\frac{d\mu_2}{d\mu_1}(\xi_1(t))\\ &= \mathbf{M}\Psi_{T^{-1}\mathbf{A}}(\xi_2(t)) = \mathbf{M}\Psi_{\mathbf{A}}(\eta_2(t)).\end{aligned}$$

The proof of the lemma follows.

The basic method that we shall use in showing the absolute continuity of measures corresponding to random processes is outlined as follows: We shall define sequences of simpler processes $\xi_n^{(1)}(t)$ and $\xi_n^{(2)}(t)$ that converge to $\xi_1(t)$ and $\xi_2(t)$. We choose these processes in such a way that it is possible to compute $(d\mu_2^{(n)}/d\mu_1^{(n)})(\xi_n^{(1)}(t))$, where $\mu_i^{(n)}$ are measures corresponding to the processes $\xi_n^{(i)}(t)$. It is natural to expect that when $n \to \infty$, the value of $(d\mu_2^{(n)}/d\mu_1^{(n)})(\xi_n^{(1)}(t))$ will converge to $(d\mu_2/d\mu_1)(\xi_1(t))$. For this kind of approach, we need the following lemma.

LEMMA 4. *Suppose that a sequence of processes $\xi_n^{(1)}(t)$ converges in probability to $\xi^{(1)}(t)$ for every $t \in [t_0, T]$ and that the sequence $\xi_n^{(2)}(t)$ converges in probability to $\xi^{(2)}(t)$ for every $t \in [t_0, T]$. Suppose that the measure $\mu_2^{(n)}$ corresponding to the process $\xi_n^{(2)}(t)$ and the measure $\mu_1^{(n)}$ corresponding to the process $\xi_n^{(1)}(t)$ are absolutely continuous with respect to each other. Suppose also that the variables*

$$\log \frac{d\mu_1^{(n)}}{d\mu_2^{(n)}} (\xi_n^{(2)}(t)) \qquad \textit{and} \qquad \log \frac{d\mu_2^{(n)}}{d\mu_1^{(n)}} (\xi_n^{(1)}(t))$$

converge in probability to some particular variables ρ_{12} and ρ_{21}, respectively. Then the measures μ_1 and μ_2, corresponding to the processes $\xi^{(1)}(t)$ and $\xi^{(2)}(t)$ will be absolutely continuous with respect to each other, and

$$\rho_{12} = \log \frac{d\mu_1}{d\mu_2} (\xi^{(2)}(t)) \qquad \textit{and} \qquad \rho_{21} = \log \frac{d\mu_2}{d\mu_1} (\xi^{(1)}(t)).$$

Proof. Let us establish, for example, the absolute continuity of μ_2 with respect to μ_1. For every set $\mathbf{A} \in \mathbf{F}_m[t_0, T]$, we can write

$$\begin{aligned}\mu_1^{(n)}(\mathbf{A}) = \int_A \frac{d\mu_1^{(n)}}{d\mu_2^{(n)}} d\mu_2^{(n)} &\geq e^{-N} \mu_2^{(n)}(\mathbf{A}) \\ &+ \int_A \left(\frac{d\mu_1^{(n)}}{d\mu_2^{(n)}} - e^{-N}\right) d\mu_2^{(n)} \geq e^{-N} \mu_2^{(n)}(\mathbf{A}) \\ &+ \int\limits_{\left(\frac{d\mu_1^{(n)}}{d\mu_2^{(n)}} < e^{-N}\right)} \left(\frac{d\mu_1^{(n)}}{d\mu_2^{(n)}} - e^{-N}\right) d\mu_2^{(n)} \\ &\geq e^{-N} \mu_2^{(n)}(\mathbf{A}) - e^{-N} \mathbf{P}\left\{\left|\log \frac{d\mu_1^{(n)}}{d\mu_2^{(n)}} (\xi_2^{(n)}(t))\right| > N\right\}.\end{aligned}$$

Consequently

$$\mu_2^{(n)}(\mathbf{A}) \leq e^N \mu_1^{(n)}(\mathbf{A}) + \mathbf{P}\left\{\left|\log \frac{d\mu_1^{(n)}}{d\mu_2^{(n)}} (\xi_2^{(n)}(t))\right| > N\right\}.$$

This means that for any $\mathbf{A} \in F_m[t_0, T]$ for which $\mu_i^{(n)}(\mathbf{A}) \to \mu_i(\mathbf{A})$, we have

$$\mu_2(\mathbf{A}) \leq e^N \mu_1(\mathbf{A}) + \mathbf{P}\{|\rho_{12}| \geq N\}.$$

But for every set $\mathbf{A}$ for which $\mu_1(\mathbf{A}) = 0$, it is possible to find a monotonically increasing sequence of cylindrical sets C_k such that

$$\mu_i^{(n)}(C_k) \to \mu_i(C_k), \qquad \bigcup_{k=1} C_k \supset \mathbf{A} \qquad \text{and} \qquad \mu_1(C_m) = \epsilon,$$

where ϵ is a preassigned arbitrary small positive number. Then

$$\mu_2(\mathbf{A}) \leq e^N\epsilon + \mathbf{P}\{|\rho_{12}| \geq N\}.$$

From the fact that $\epsilon > 0$ and N are arbitrary, we conclude that $\mu_2(\mathbf{A}) = 0$. The absolute continuity of μ_2 with respect to μ_1 is established. By switching the indices 1 and 2 in this argument, we obtain the absolute continuity of μ_1 with respect to μ_2. Let us now show that

$$\rho_{21} = \log \frac{d\mu_2}{d\mu_1}\,(\xi_1(t)).$$

Let C be a cylindrical set for which $\psi_C(\xi_i^{(n)}(t)) \to \psi_C(\xi_i(t))$ in probability and let ψ_C be the characteristic function of the set C. Then

$$\mathbf{M}\psi_C(\xi_2^{(n)}(t)) = \mathbf{M}\psi_C(\xi_1^{(n)}(t))\,\frac{d\mu_2^{(n)}}{d\mu_1^{(n)}}\,(\xi_1^{(n)}(t)).$$

Since

$$\psi_C(\xi_2^{(n)}(t)) \to \psi_C(\xi_2(t)) \qquad \text{and} \qquad \psi_C(\xi_1^{(n)}(t)) \to \psi_C(\xi_1(t)),$$

it follows from Fatou's lemma that

$$\begin{aligned}\mathbf{M}\psi_C(\xi_2(t)) &= \lim_{n\to\infty} \mathbf{M}\psi_C(\xi_2^{(n)}(t)) \\ &= \lim_{n\to\infty} \mathbf{M}\psi_C(\xi_1^{(n)}(t))\left(\frac{d\mu_2^{(n)}}{d\mu_1^{(n)}}\,(\xi_1^{(n)}(t))\right) \\ &\geq \mathbf{M}\psi_C(\xi_1(t))e^{\rho_{21}}.\end{aligned}$$

Therefore

$$\mathbf{M}\psi_C(\xi_1(t))\left(\frac{d\mu_2}{d\mu_1}\,(\xi_1(t)) - e^{\rho_{21}}\right) \geq 0.$$

From this inequality, it is easy to show that for every set $\mathbf{A}$ in $F_m[t_0, T]$,

$$\mathbf{M}\psi_{\mathbf{A}}(\xi_1(t))\left(\frac{d\mu_2}{d\mu_1}\,(\xi_1(t)) - e^{\rho_{21}}\right) \geq 0.$$

Thus $e^{\rho_{21}} \leq (d\mu_2/d\mu_1)(\xi_1(t))$ almost everywhere. To show that $\rho_{21} = \log\,(d\mu_2/d\mu_1)(\xi_1(t))$, it is sufficient to establish that

$$\mathbf{M}e^{\rho_{21}} \geq \mathbf{M}\,\frac{d\mu_2}{d\mu_1}\,(\xi_1(t)) = 1.$$

We denote by $g_N(t)$ a function equal to t for $|t| \leq N$ and equal to N (sign t)

for $|t| > N$. Then

$$\lim_{n\to\infty} \mathbf{M}g_N\left(\frac{d\mu_2^{(n)}}{d\mu_1^{(n)}}\,(\xi_1^{(n)}(t))\right) = \mathbf{M}g_N(e^{\rho_{21}}).$$

But

$$1 - \mathbf{M}g_N\left(\frac{d\mu_2^{(n)}}{d\mu_1^{(n)}}\,(\xi_1^{(n)}(t))\right) \le \int\limits_{\frac{d\mu_2^{(n)}}{d\mu_1^{(n)}} > N} \frac{d\mu_2^{(n)}}{d\mu_1^{(n)}}\,d\mu_1^{(n)}$$

$$= \mathbf{P}\left\{\left|\frac{d\mu_2^{(n)}}{d\mu_1^{(n)}}\,(\xi_2^{(n)}(t))\right| > N\right\} \le \mathbf{P}\left\{\left|\log\frac{d\mu_1^{(n)}}{d\mu_2^{(n)}}\,(\xi_2^{(n)}(t))\right| > \log N\right\}\cdot$$

Taking the limit as $n \to \infty$, we obtain

$$1 - \mathbf{M}e^{\rho_{21}} \le 1 - \mathbf{M}g_N(e^{\rho_{21}}) \le \mathbf{P}\{|\rho_{12}| \ge \log N\}.$$

From this we get the inequality $1 - \mathbf{M}e^{\rho_{21}} \le 0$. Therefore

$$\rho_{21} = \log\frac{d\mu_2}{d\mu_1}\,(\xi_1(t)).$$

In an analogous manner, we can show that

$$\rho_{12} = \log\frac{d\mu_1}{d\mu_2}\,(\xi_2(t)).$$

This proves the lemma.

For future use, we note the following: Since, in this chapter, we are interested only in measures corresponding to random processes, we shall replace one process with another if the finite-dimensional distributions of these processes coincide.

3. Some sufficient conditions for the absolute continuity of measures corresponding to homogeneous processes with independent increments. Suppose that two homogeneous processes with independent increments $\xi_1(t)$ and $\xi_2(t)$ are given on an interval $[0, T]$. The characteristic functions of these processes take the form

$$\mathbf{M}e^{i(z,\xi_j(t))} = \exp\left\{t\left[i(z, a_j) - \tfrac{1}{2}(A_jz, z) + \int_{|x|\le 1} (e^{i(z,x)} - 1 - i(z, x))\Pi_j\,(dx) + \int_{|x|>1} (e^{i(z,x)} - 1)\Pi_j\,(dx)\right]\right\}, \qquad j = 1,2. \tag{3.1}$$

Let us assume that a process $\xi_1(t)$ is of the form

$$\xi_1(t) = \bar{a}t + \sum_{k=1}^{l} b_k^{(1)} w_k(t) + \int_0^t \int_{|u| \le 1} f(u) q\,(ds \times du) + \int_0^t \int_{|u|>1} f(u) p\,(ds \times du). \tag{3.2}$$

[Theorem 2 of Section 4, Chapter 3, assures the existence of vectors $\bar{a}_1, b_1^{(1)}, b_2^{(1)}, \ldots, b_l^{(1)}$, such that the vectors $b_j^{(1)}$ are mutually orthogonal, and of a function $f(u)$ such that the process (3.2) will have the characteristic function (3.1) for $t > 0$.]

THEOREM. *Suppose that the following conditions are fulfilled:*

1. *The linear transformations A_1 and A_2 coincide.*

2. *Π_2 and Π_1 are absolutely continuous with respect to each other, and $\rho(x) = (d\Pi_2/d\Pi_1)(x)$ has the properties:*

(a) $$\int_{|\rho(x)-1| \le 1/2} (1 - \rho(x))^2 \Pi_1\,(dx) < \infty;$$

(b) $$\int_{|\rho(x)-1| > 1/2} |1 - \rho(x)| \Pi_1\,(dx) < \infty;$$

(c) *there exist numbers $\alpha_1, \alpha_2, \ldots, \alpha_l$ such that*

$$a_2 - a_1 + \int_{|x| \le 1} (1 - \rho(x)) \Pi_1\,(dx) = \sum_{k=1}^{l} \alpha_k b_k^{(1)}.$$

Then the measures μ_1 and μ_2 corresponding to the processes $\xi_1(t)$ and $\xi_2(t)$ are absolutely continuous with respect to each other and

$$\log \frac{d\mu_2}{d\mu_1}(\xi_1(t)) = \bar{\alpha}T + \sum_{k=1}^{l} \alpha_k w_k(T) + \int_0^T \int_{|\rho(f(u))-1| \le 1/2} \log \rho(f(u)) q\,(ds \times du) + \int_0^T \int_{|\rho(f(u))-1| > 1/2} \log \rho(f(u)) p\,(ds \times du), \tag{3.3}$$

where

$$\bar{\alpha} = -\frac{1}{2} \sum_{k=1}^{l} \alpha_k^2 + \int_{|\rho(f(u))-1| > 1/2} (1 - \rho(f(u))) \frac{du}{|u|^{m+1}} + \int_{|\rho(f(u))-1| \le 1/2} (1 - \rho(f(u)) + \log \rho(f(u))) \frac{du}{|u|^{m+1}}. \tag{3.4}$$

Before proving the theorem, let us note that Formulas (3.3) and (3.4) are meaningful because

$$\int\limits_{|\rho(f(u))-1|\le 1/2} |\log \rho(f(u))|^2 \frac{du}{|u|^{m+1}} \le 4 \int\limits_{|\rho(f(u))-1|\le 1/2} |\rho(f(u)) - 1|^2 \frac{du}{|u|^{m+1}} = 4 \int\limits_{|\rho(x)-1|\le 1/2} (\rho(x) - 1)^2 \Pi_1 (dx) < \infty,$$

$$\int\limits_{|\rho(f(u))-1|>1/2} \frac{du}{|u|^{m+1}} \le 2 \int\limits_{|\rho(f(u))-1|>1/2} |\rho(f(u)) - 1| \frac{du}{|u|^{m+1}} = 2 \int\limits_{|\rho(x)-1|>1/2} |\rho(x) - 1| \Pi_1 (dx) < \infty$$

and

$$\left| \int\limits_{|\rho(f(u))-1|\le 1/2} (1 - \rho(f(u)) + \log \rho(f(u))) \frac{du}{|u|^{m+1}} \right| \le 2 \int\limits_{|\rho(f(u))-1|\le 1/2} (1 - \rho(f(u)))^2 \frac{du}{|u|^{m+1}} = 2 \int\limits_{|\rho(x)-1|\le 1/2} (1 - \rho(x))^2 \Pi_1 (dx) < \infty.$$

[In all three integrals, the change of variable $f(u) = x$ was made and the relation $\Pi_1(A) = \int_{f(u)\in A} du/|u|^{m+1}$ was used.]

The proof of the theorem will follow from the following lemmas, in which the hypotheses of the theorem will be assumed.

LEMMA 1. *Let $\overline{p}$ be a Poisson measure with independent values that is defined for all Borel sets A in $[t_0, T] \times R^{(m)}$ for which*

$$\int_{(t,u)\in A} \rho(f(u)) \frac{dt\, du}{|u|^{m+1}} < \infty$$

and

$$\mathbf{M}\overline{p}(A) = \int_{(t,u)\in A} \rho(f(u)) \frac{dt\, du}{|u|^{m+1}}.$$

We define $\overline{q}(A) = \overline{p}(A) - \mathbf{M}\overline{p}(A)$ and

$$\overline{a}_2 = a_2 + \int\limits_{\substack{|u|\le 1\\ |f(u)|>1}} \rho(f(u))f(u) \frac{du}{|u|^{m+1}} - \int\limits_{\substack{|f(u)|\le 1\\ |u|>1}} \rho(f(u))f(u) \frac{du}{|u|^{m+1}}.$$

Then the process

$$\tilde{\xi}_2(t) = \bar{a}_2 t + \sum_{k=1}^{l} b_k w_k(t) + \int_0^t \int_{|u| \le 1} f(u)\bar{q}\,(ds \times du)$$

$$+ \int_0^t \int_{|u|>1} f(u)\bar{p}\,(ds \times du) \qquad (3.5)$$

will have the same distributions as the process $\xi_2(t)$.

To prove this lemma, it is necessary to compute the characteristic function of the process $\tilde{\xi}_2(t)$, which is carried out just as with Theorem 2 of Section 4, Chapter 3. We obtain

$$\mathbf{M} \exp \{i(z, \tilde{\xi}_2(t))\} = \exp\Bigg[t\Bigg(i(\bar{a}_2, z) - \frac{1}{2}\sum_{k=1}^{l} (b_k, z)^2$$

$$+ \int_{|u| \le 1} (e^{i(z,f(u))} - 1 - i(z, f(u))\rho(f(u)))\,\frac{du}{|u|^{m+1}}$$

$$+ \int_{|u|>1} (e^{i(z,f(u))} - 1)\rho(f(u))\,\frac{du}{|u|^{m+1}}\Bigg)\Bigg].$$

Then we make the substitution $f(u) = x$ in the integrals, recalling that $\rho(f(u))(du/|u|^{m+1})$ now becomes $\rho(x)\Pi_1\,(dx) = \Pi_2(dx)$.

LEMMA 2. *Let* $\epsilon > 0$ *be arbitrary and set*

$$\xi_1^{(\epsilon)}(t) = \int_0^t \int_{|u|>\epsilon} f(u)p\,(ds \times du); \qquad \xi_2^{(\epsilon)} = \int_0^t \int_{|u|>\epsilon} f(u)\bar{p}\,(ds \times du). \qquad (3.6)$$

If $\mu_1^{(\epsilon)}$ *and* $\mu_2^{(\epsilon)}$ *are measures corresponding to the processes* $\xi_1^{(\epsilon)}(t)$ *and* $\xi_2^{(\epsilon)}(t)$, *then these measures are absolutely continuous with respect to each other and*

$$\log \frac{d\mu_2^{(\epsilon)}}{d\mu_1^{(\epsilon)}}\,(\xi_1^{(\epsilon)}(t)) = \int_0^T \int_{|u|>\epsilon} \log \rho(f(u))p\,(ds \times du)$$

$$+ \int_{|u|>\epsilon} T(1 - \rho(f(u)))\,\frac{du}{|u|^{m+1}}, \qquad (3.7)$$

$$\log \frac{d\mu_1^{(\epsilon)}}{d\mu_2^{(\epsilon)}}\,(\xi_2^{(\epsilon)}(t)) = -\int_0^T \int_{|u|>\epsilon} \log \rho(f(u))\bar{p}\,(ds \times du)$$

$$-T\int_{|u|>\epsilon} (1 - \rho(f(u)))\,\frac{du}{|u|^{m+1}}. \qquad (3.8)$$

Proof. We introduce the variables

$$\xi_n^{(k)} = \int_{(kT/n)}^{((k+1)/n)T} \int_{|u|<\epsilon} f(u)p\,(ds \times du) \times g_1\left(\int_{(kT/n)}^{((k+1)/n)T} \int_{|u|>\epsilon} p\,(ds \times du)\right),$$

$$\eta_n^k = \int_{(kT/n)}^{((k+1)/n)T} \int_{|u|>\epsilon} f(u)\overline{p}\,(ds \times du) \times g_1\left(\int_{(kT/n)}^{((k+1)/n)T} \int_{|u|>\epsilon} \overline{p}\,(ds \times du)\right),$$

where $g_1(t) = t$ for $t \leq 1$ and $g_1(t) = 0$ for $t > 1$. Since

$$\mathbf{P}\left\{\left|\eta_n^{(k)} - \int_{(kT/n)}^{((k+1)/n)T} \int_{|u|>\epsilon} f(u)\overline{p}\,(ds \times du)\right| > 0\right\} = O\left(\frac{1}{n^2}\right),$$

$$\mathbf{P}\left\{\left|\xi_n^{(k)} - \int_{(kT/n)}^{((k+1)/n)T} \int_{|u|>\epsilon} f(u)\overline{p}\,(ds \times du)\right| > 0\right\} = O\left(\frac{1}{n^2}\right),$$

we have

$$\mathbf{P}\left\{\sup_k\left|\sum_{j=0}^{k-1} \xi_n^{(j)} - \int_0^{(kT/n)} \int_{|u|>\epsilon} f(u)\overline{p}\,(ds \times du)\right| > 0\right\} = O\left(\frac{1}{n}\right),$$

$$\mathbf{P}\left\{\sup_k\left|\sum_{j=0}^{k-1} \eta_n^{(j)} - \int_0^{(kT/n)} \int_{|u|>\epsilon} f(u)\overline{p}\,(ds \times du)\right| > 0\right\} = O\left(\frac{1}{n}\right).$$

Consequently if $\eta_n(t) = \sum_{k<nt} \eta_n^{(k)}$ and $\xi_n(t) = \sum_{k<nt} \xi_n^{(k)}$, then $\eta_n(t) \to \xi_2^{(\epsilon)}(t)$ and $\xi_n(t) \to \xi_1^{(\epsilon)}(t)$ in probability. The variables $\xi_n^{(k)}$ are independent and identically distributed. We have

$$\mathbf{P}\,\{\xi_n^{(k)} \in A\} = \frac{T}{n}\int_{f(u)\in A} \frac{du}{|u|^{m+1}} \exp\left\{-\frac{T}{n}\int_{f(u)\in A} \frac{du}{|u|^{m+1}}\right\},$$

and if $0 \in A$,

$$\mathbf{P}\,\{\xi_n^{(k)} = 0\} = 1 - \frac{T}{n}\int_{|u|>\epsilon} \frac{du}{|u|^{m+1}} \exp\left\{-\frac{T}{n}\int_{|u|>\epsilon} \frac{du}{|u|^{m+1}}\right\}.$$

The variables $\eta_n^{(k)}$ are also independent and identically distributed. We have

$$\mathbf{P}\{\eta_n^{(k)} \in A\} = \frac{T}{n}\int_{f(u)\in A} \rho(f(u))\,\frac{du}{|u|^{m+1}} \exp\left\{-\frac{T}{n}\int_{f(u)\in A} \rho(f(u))\,\frac{du}{|u|^{m+1}}\right\},$$

and if $0 \in A$,

$$\mathbf{P}\{\eta_n^{(k)} = 0\} = 1 - \frac{T}{n}\int_{|u|>\epsilon} \rho(f(u)) \frac{du}{|u|^{m+1}} \times \exp\left\{-\frac{T}{n}\int_{|u|<\epsilon} \rho(f(u)) \frac{du}{|u|^{m+1}}\right\}.$$

We denote by $\varphi_n(x)$ the distribution density of the variable $\eta_n^{(k)}$ with respect to the distribution of the variable $\xi_n^{(k)}$. From the relationships above, it follows that

$$\varphi_n(x) = \rho(x) \exp\left\{-\frac{T}{n}\int_{|u|>\epsilon} (\rho(f(u)) - 1) \frac{du}{|u|^{m+1}}\right\}$$

for $x \neq 0$, and

$$\varphi_n(0) = \frac{1 - \frac{T}{n}\int_{|u|>\epsilon} \rho(f(u)) \frac{du}{|u|^{m+1}} \exp\left\{-\frac{T}{n}\int_{|u|>\epsilon} \rho(f(u)) \frac{du}{|u|^{m+1}}\right\}}{1 - \frac{T}{n}\int_{|u|>\epsilon} \frac{du}{|u|^{m+1}} \exp\left\{-\frac{T}{n}\int_{|u|>\epsilon} \frac{du}{|u|^{m+1}}\right\}}.$$

Since $\xi_n(t)$ is completely defined by $\xi_n^{(k)}$, and since the $\xi_n^{(k)}$ are uniquely determined by $\xi_n(t)$, and the $\xi_n^{(k)}$ are mutually independent, it follows from Lemma 3 of Section 2 that

$$\frac{d\nu_n}{d\mu_n}(\xi_n(t)) = \prod_{k=0}^{n-1} \varphi_n(\xi_n^{(k)}),$$

which implies that

$$\log \frac{d\nu_n}{d\mu_n}(\xi_n(t)) = \sum_{k=0}^{n-1} \log \varphi_n(\xi_n^{(k)}).$$

[ν_n and μ_n are the measures corresponding to the processes $\eta_n(t)$ and $\xi_n(t)$.] Thus if we define $\rho(0) = 1$ and introduce the function $\Psi_0(x) = 0$ for $x \neq 0$ and $\Psi_0(x) = 1$ for $x = 0$, we obtain

$$\log \frac{d\nu_n}{d\mu_n}(\xi_n(t)) = \sum_{k=0}^{n-1} \log \rho(\xi_n^{(k)}) - \frac{T}{n}\int_{|u|>\epsilon} (\rho(f(u)) - 1) \frac{du}{|u|^{m+1}}$$

$$\times \sum_{k=0}^{n-1} (1 - \Psi_0(\xi_n^{(k)})) + \sum_{k=1}^{n} \Psi_0(\xi_n^{(k)})$$

$$\times \log \frac{1 - \frac{T}{n}\int_{|u|>\epsilon} \rho(f(u)) \frac{du}{|u|^{m+1}} \exp\left\{-\frac{T}{n}\int_{|u|>\epsilon} \rho(f(u)) \frac{du}{|u|^{m+1}}\right\}}{1 - \frac{T}{n}\int_{|u|>\epsilon} \frac{du}{|u|^{m+1}} \exp\left\{-\frac{T}{n}\int_{|u|>\epsilon} \frac{du}{|u|^{m+1}}\right\}}.$$

Taking into consideration the quantities $\mathbf{P}\{\xi_n^{(k)} = 0\}$, we easily see that $(1/n)\sum_{k=0}^{n-1} \Psi_0(\xi_n^{(k)}) \to 1$ in probability and

$$n \log \frac{1 - \dfrac{T}{n}\displaystyle\int_{|u|>\epsilon} \rho(f(u)) \frac{du}{|u|^{m+1}} \exp\left\{-\dfrac{T}{n}\displaystyle\int_{|u|>\epsilon} \rho(f(u)) \frac{du}{|u|^{m+1}}\right\}}{1 - \dfrac{T}{n}\displaystyle\int_{|u|>\epsilon} \frac{du}{|u|^{m+1}} \exp\left\{-\dfrac{T}{n}\displaystyle\int_{|u|>\epsilon} \frac{du}{|u|^{m+1}}\right\}}$$
$$\to T\int_{|u|>\epsilon} (1 - \rho(f(u))) \frac{du}{|u|^{m+1}}.$$

We note further that

$$\log \rho(\xi_n^{(k)}) = \int_{(T/n)k}^{(T/n)(k+1)} \int_{|u|>\epsilon} \log \rho(f(u)) p\,(ds \times du)$$

if
$$\int_{(T/n)k}^{(T/n)(k+1)} \int_{|u|>\epsilon} \log p\,(ds \times du) \le 1.$$

Thus

$$\mathbf{P}\left\{\left|\sum_0^{n-1} \log \rho(\xi_n^{(k)}) - \int_0^T \int_{|u|>\epsilon} \log \rho(f(u)) p\,(ds \times du)\right| > 0\right\} = O\left(\frac{1}{n}\right).$$

Therefore

$$\sum_0^{n-1} \log \rho(\xi_n^{(k)}) \to \int_0^T \int_{|u|>\epsilon} \log \rho(f(u)) p\,(ds \times du)$$

in probability also. We have shown that

$$\log \frac{d\nu_n}{d\mu_n}(\xi_n(t)) \to \int_0^T \int_{|u|>\epsilon} \log \rho(f(u)) p\,(ds \times du)$$
$$+ T\int_{|u|>\epsilon} (1 - \rho(f(u))) \frac{du}{|u|^{m+1}} \tag{3.9}$$

in probability. By the same considerations, we can show that

$$\log \frac{d\mu_n}{d\nu_n}(\eta_n(t)) \to -\int_0^T \int_{|u|>\epsilon} \log \rho(f(u)) \overline{p}\,(ds \times du)$$
$$- T\int_{|u|>\epsilon} (1 - \rho(f(u))) \frac{du}{|u|^{m+1}} \tag{3.10}$$

in probability.

On the basis of Lemma 4 of Section 2, we conclude from (3.9) and (3.10) that the measures $\mu_1^{(\epsilon)}$ and $\mu_2^{(\epsilon)}$ are absolutely continuous with respect to each other and that Formulas (3.7) and (3.8) are valid.

LEMMA 3. *Suppose that $w(t)$ is a Brownian process on the interval $[0, T]$. Suppose that $\eta_1(t) = w(t) + \gamma_1(t)$ and $\eta_2(t) = w(t) + \gamma_2(t)$, where γ_1 and γ_2 are real constants. If μ_1^* and μ_2^* are measures corresponding to the processes $\eta_1(t)$ and $\eta_2(t)$, they are absolutely continuous with respect to each other and*

$$\log \frac{d\mu_2^*}{d\mu_1^*}(\eta_1(t)) = (\gamma_2 - \gamma_1)w(T) - \tfrac{1}{2}(\gamma_2 - \gamma_1)^2 T.$$

Proof. Let $\eta_1^{(n)}(t) = \eta_1(kT/n)$ if $t \in ((kT/n), ((k+1)/n)T)$, and let $\eta_2^{(n)}(t) = \eta_2(kT/n)$ if $t \in ((kT/n)((k+1)/n)T)$. We denote by $\mu_1^{(n)}$ and $\mu_2^{(n)}$ the measures corresponding to the processes $\eta_1^{(n)}(t)$ and $\eta_2^{(n)}(t)$. Since $\eta_1^{(n)}(t)$ and $\eta_2^{(n)}(t)$ are identical functionals of

$$\Delta_{nk}^{(1)} = \eta_1^{(n)}\left(\frac{k+1}{n}T\right) - \eta_1^{(n)}\left(\frac{k}{n}T\right),$$

$$\Delta_{nk}^{(2)} = \eta_2^{(n)}\left(\frac{k+1}{n}T\right) - \eta_2^{(n)}\left(\frac{k}{n}T\right),$$

and since $\Delta_{nk}^{(1)}$ and $\Delta_{nk}^{(2)}$ are independent random variables having normal distributions with means $\gamma_1 T/n$, $\gamma_2 T/n$ and both with dispersion T/n, we have

$$\frac{d\mu_2^{(n)}}{d\mu_1^{(n)}}(\eta_1^{(n)}(t)) = \prod_{k=0}^{n-1} \exp\left\{-\frac{\left(\Delta_{nk}^{(1)} - \frac{\gamma_2 T}{n}\right)^2}{2\frac{T}{n}} + \frac{\left(\Delta_{nk}^{(1)} - \frac{\gamma_1 T}{n}\right)^2}{2\frac{T}{n}}\right\}$$

$$= \prod_{k=0}^{n} \exp\left\{(\gamma_2 - \gamma_1)\,\Delta_{nk}^{(1)} + \frac{\gamma_1^2 T}{2n} - \frac{\gamma_2^2 T}{2n}\right\}$$

$$= \exp\left\{(\gamma_2 - \gamma_1)w(T) - \frac{(\gamma_2 - \gamma_1)^2}{2}T\right\}.$$

Therefore

$$\log \frac{d\mu_2^{(n)}}{d\mu_1^{(n)}}(\eta_1^{(n)}(t)) = (\gamma_2 - \gamma_1)w(T) - \frac{(\gamma_2 - \gamma_1)^2}{2}T.$$

In an analogous fashion, it can be shown that

$$\log \frac{d\mu_1^{(n)}}{d\mu_2^{(n)}}(\eta_2^{(n)}(t)) = (\gamma_1 - \gamma_2)w(T) - \frac{(\gamma_1 - \gamma_2)^2}{2}T.$$

To prove the lemma, it only remains to use Lemma 4 of Section 2.

Now let us proceed to prove the theorem. Suppose that $\xi_1(t)$ is defined by Formula (3.2), and $\bar{\xi}_2(t)$ by Formula (3.5). Set

$$\bar{\xi}_1^{(\epsilon)}(t) = \bar{a}_1 t + \sum_{k=1}^{l} b_k^{(1)} w_k(t) + \int_0^t \int_{|u|>\epsilon} f(u) p\,(ds \times du) - t \int_{1 \geq |u| > \epsilon} f(u) \frac{du}{|u|^{m+1}},$$

$$\bar{\xi}_2^{(\epsilon)}(t) = \bar{a}_2 t + \sum_{k=1}^{l} b_k^{(1)} w_k(t) + \int_0^t \int_{|u|>\epsilon} f(u) \bar{p}\,(ds \times du) - t \int_{1 \geq |u| > \epsilon} f(u) \rho(f(u)) \frac{du}{|u|^{m+1}} + t \int_{|u| \leq \epsilon} (1 - \rho(f(u))) f(u) \frac{du}{|u|^{m+1}}.$$

By using condition (c), we can write the expression for $\bar{\xi}_2^{(\epsilon)}(t)$ in the following form:

$$\bar{\xi}_2^{(\epsilon)}(t) = \bar{a}_1 t + \sum_{k=1}^{l} b_k^{(1)} (w_k(t) + \alpha_k t) + \int_0^t \int_{|u|>\epsilon} f(u) \bar{p}\,(ds \times du) - t \int_{\epsilon < |u| \leq 1} f(u) \frac{du}{|u|^{m+1}}.$$

Suppose that $\zeta_1(t)$ is a process in an $(l + m)$-dimensional space with components

$$\left[w_1(t), \ldots, w_l(t), \int_0^t \int_{|u|>\epsilon} f(u) p(ds \times du) \right]$$

and that $\zeta_2(t)$ is a process in the same space with components

$$\left[w_1(t) + \alpha_1 t, \ldots, w_l(t) + \alpha_l t, \int_0^t \int_{|u|>\epsilon} f(u) \bar{p}\,(ds \times du) \right].$$

(In both processes, the last component is m-dimensional and the remaining components are one-dimensional.) Since the components of the processes are independent and the measures corresponding to the same component of these processes are absolutely continuous with respect to each other, if we denote by μ_{ζ_i} the measure corresponding to the process $\zeta_i(t)$ and by $\nu_1, \ldots, \nu_l, \nu_1', \ldots, \nu_l', \mu_1^{(\epsilon)}, \mu_2^{(\epsilon)}$ the measures corresponding to the processes $w_1(t), \ldots, w_l(t), w_1(t) + \alpha_1 t, \ldots, w_l(t) + \alpha_l t$ (concerning $\mu_i^{(\epsilon)}$, see the condition of Lemma 2), on the basis of Lemma 2 of Section 2 and Lemmas

2 and 3 of this section we have

$$\frac{d\mu_{\zeta_2}}{d\mu_{\zeta_1}}(\zeta_1(t)) = \exp\left\{\sum_{k=1}^{l} \alpha_k w_k(T) - \frac{T}{2}\sum_{k=1}^{l} \alpha_k^2 + \int_0^T \int_{|u|>\epsilon} \log \rho(f(u))p\,(ds \times du) + T\int_{|u|>\epsilon} (1 - \rho(f(u)))\frac{du}{|u|^{m+1}}\right\}.$$

Since $\bar{\xi}_2^{(\epsilon)}(t)$ and $\bar{\xi}_1^{(\epsilon)}(t)$ are obtained by the same transformation of $\zeta_2(t)$ and $\zeta_1(t)$, if $\bar{\mu}_1^{(\epsilon)}$ and $\bar{\mu}_2^{(\epsilon)}$ are the measures corresponding to the processes $\bar{\xi}_1^{(\epsilon)}(t)$ and $\bar{\xi}_2^{(\epsilon)}(t)$, then on the basis of Lemma 3 of Section 2,

$$\frac{d\bar{\mu}_2^{(\epsilon)}}{d\mu_1^{(\epsilon)}}(\bar{\xi}_1^{(\epsilon)}(t)) = \mathbf{M}\left(\exp\left\{\sum_{k=1}^{l}\alpha_k w_k(T) - \frac{T}{2}\sum_{k=1}^{l}\alpha_k^2 + T\int_{|u|>\epsilon}(1 - \rho(f(u)))\frac{du}{|u|^{m+1}} + \int_0^T\int_{|u|>\epsilon}\log\rho(f(u))p\,(ds \times du)\right\} \Big| f(\bar{\xi}_1^{(\epsilon)}(t)),\, t \in (0, T)\right). \tag{3.11}$$

But the expression under the conditional mathematical expectation sign is completely determined by the values of $\bar{\xi}_1^{(\epsilon)}(t)$, because by adding the jumps of $\bar{\xi}_1^{(\epsilon)}(t)$ it is possible to obtain

$$\int_0^t \int_{|u|>\epsilon} f(u)p\,(ds \times du),$$

and from Lemma 2,

$$\int_0^T \int_{|u|>\epsilon} \log \rho(f(u))p\,(ds \times du)$$

is a functional of the process

$$\int_0^t \int_{|u|>\epsilon} f(u)p\,(ds \times du);$$

by computing the jumps and the fixed function from $\bar{\xi}_1^{(\epsilon)}(t)$, we obtain $\sum_{k=1}^{l} b_k^{(1)} w_k(t)$. From this expression, in view of the orthogonality of $b_k^{(1)}$, it is possible to determine $w_k(t)$. Consequently the conditional mathematical expectation of the quantity in (3.11) coincides with the quantity itself. Thus

$$\log \frac{d\bar{\mu}_2^{(\epsilon)}}{d\bar{\mu}_1^{(\epsilon)}} (\bar{\xi}_1^{(\epsilon)}(t)) = \sum_{k=1}^{l} \alpha_k w_k(T) - \frac{T}{2} \sum_{k=1}^{l} \alpha_k^2$$

$$+ \int_0^T \int_{\substack{|u|>\epsilon \\ |\rho(f(u))-1|>(1/2)}} \log \rho(f(u)) p\,(ds \times du)$$

$$+ \int_0^T \int_{\substack{|u|>\epsilon \\ |\rho(f(u))-1|\leq(1/2)}} \log \rho(f(u)) q\,(ds \times du)$$

$$+ T \int_{\substack{|u|>\epsilon \\ |\rho(f(u))-1|>(1/2)}} (1 - \rho(f(u))) \frac{du}{|u|^{m+1}}$$

$$+ T \int_{\substack{|u|>\epsilon \\ |\rho(f(u))-1|\leq(1/2)}} [1 - \rho(f(u)) + \log \rho(f(u))] \frac{du}{|u|^{m+1}} \cdot \quad (3.12)$$

An analogous calculation shows that

$$\log \frac{d\bar{\mu}_1^{(\epsilon)}}{d\bar{\mu}_2^{(\epsilon)}} (\bar{\xi}_2^{(\epsilon)}(t)) = - \sum_{k=1}^{l} \alpha_k w_k(T) - \frac{T}{2} \sum_{k=1}^{l} \alpha_k^2$$

$$- \int_0^T \int_{\substack{|u|>\epsilon \\ |\rho(f(u))-1|>(1/2)}} \log \rho(f(u)) \bar{p}\,(ds \times du)$$

$$- \int_0^T \int_{\substack{|u|>\epsilon \\ |\rho(f(u))-1|\leq(1/2)}} \log \rho(f(u)) \bar{q}\,(ds \times du)$$

$$+ T \int_{\substack{|u|>\epsilon \\ |\rho(f(u))-1|>(1/2)}} (\rho(f(u)) - 1) \frac{du}{|u|^{m+1}}$$

$$+ T \int_{\substack{|u|>\epsilon \\ |\rho(f(u))-1|\leq(1/2)}} [\rho(f(u)) - 1 - \log \rho(f(u))] \frac{du}{|u|^{m+1}} \cdot \quad (3.13)$$

Since $\bar{\xi}_1^{(\epsilon)}(t)$ and $\bar{\xi}_2^{(\epsilon)}(t)$ converge in probability to $\xi_1(t)$ and $\xi_2(t)$ as $\epsilon \to 0$, and since the expressions (3.12) and (3.13) also have a limit as $\epsilon \to 0$ in the sense of convergence in probability [also, (3.12) converges in probability to (3.3)], by applying Lemma 4 of Section 2 we complete the proof of the theorem.

4. The absolute continuity of measures corresponding to Markov processes. Suppose that $\xi_1(t)$ and $\xi_2(t)$ are solutions of the stochastic equations

$$\xi_i(t) = \xi_i(t_0) + \int_{t_0}^{t} a_i(s, \xi_i(s))\, ds + \sum_{k=1}^{l} \int_{t_0}^{t} b_k(s, \xi_i(s))\, dw_k(s)$$
$$+ \int_{t_0}^{t} \int_{|u| \leq 1} f_i(s, \xi_i(s), u) q\,(ds \times du)$$
$$+ \int_{t_0}^{t} \int_{|u| > 1} f_i(s, \xi_i(s), u) p\,(ds \times du), \quad i = 1, 2, \tag{4.1}$$

the coefficients of which satisfy the conditions of the existence and uniqueness theorems of Section 3, Chapter 3. Let us examine two families of homogeneous processes with independent increments

$$\eta_{t,x}^{(i)}(s) = \int_{t}^{s} a_i(t, x)\, d\tau + \sum_{k=1}^{l} b_k(t, x)[w_k(s) - w_k(t)]$$
$$+ \int_{t}^{s} \int_{|u| \leq 1} f_i(t, x, u) q\,(d\tau \times du)$$
$$+ \int_{t}^{s} \int_{|u| > 1} f_i(t, x, u) p\,(d\tau \times du), \qquad i = 1, 2, \tag{4.2}$$

that are defined for every $t \in [t_0, T]$ and $x \in R^{(m)}$ on $[t, T]$.

Let us subdivide the interval $[t_0, T]$: $t_0 < t_1 < \cdots < t_N = T$; let us denote by $\xi_i^{(N)}(t)$ the processes defined by the relations

$$\xi_i^{(N)}(t_0) = \xi_i(t_0),$$
$$\xi_i^{(N)}(t) = \xi_i^{(N)}(t_k) + \eta_{t_k, \xi_i^{(N)}(t_k)}^{(i)}(t)$$

for $t_k \leq t \leq t_{k+1}$.

Suppose that the following conditions are fulfilled:

1. There exists a function $\rho(t, x, u) > 0$ such that for every $A \in R^{(m)}$,

$$\int_{f_2(t,x,u) \in A} \frac{du}{|u|^{m+1}} = \int_{f_1(t,x,u) \in A} \rho(t, x, f_1(t, x, u))\, \frac{du}{|u|^{m+1}},$$

and that for every t and x,

$$\text{(a)}\ \delta(t, x) = \int_{|1-\rho(t,x,f_1(t,x,u))|\leq 1/2} [1 - \rho(t, x, f_1(t, x, u))]^2 \frac{du}{|u|^{m+1}} < \infty;$$

$$\text{(b)}\ \gamma(t, x) = \int_{|1-\rho(t,x,f_1(t,x,u))|>1/2} |1 - \rho(t, x, f_1(t, x, u))| \frac{du}{|u|^{m+1}} < \infty.$$

2. There exist $\alpha_1(t, x), \ldots, \alpha_l(t, x)$ such that

$$a_2(t, x) - a_1(t, x) \neq \int_{|u|\leq 1} (f_1(t, x, u) - f_2(t, x, u)) \frac{du}{|u|^{m+1}} = \sum_{k=1}^{l} \alpha_k(t, x) b_k(t, x).$$

3. The distributions of $\xi_1(t_0)$ and $\xi_2(t_0)$ are absolutely continuous with respect to one another, and the density of the distribution of $\xi_2(t_0)$ with respect to the distribution of $\xi_1(t_0)$ is $p_0(x)$.

We shall henceforth denote by $\mu_{x,t_k}^{(i)}$ the measure corresponding to the process $\eta_{t_k,x}^{(i)}(t)$ on $F_m[t_k, t_{k+1}]$. Then under the assumptions stated, the measures $\mu_{x,t_k}^{(1)}$ and $\mu_{x,t_k}^{(2)}$, will, on the basis of the theorem of Section 3, be absolutely continuous with respect to one another. On the basis of the same theorem,

$$\begin{aligned}
\log \frac{d\mu_{x,t_k}^{(2)}}{d\mu_{x,t_k}^{(1)}} (\eta_{t_k,x}^{(1)}(t)) &= \Bigg[-\frac{1}{2} \sum_{j=1}^{l} \alpha_j^2(t_k, x) + \int_{R^{(m)}} [1 - \rho(t_k, x, f_1(t_k, x, u)) \\
&\quad + \psi(t_k, x, u) \log \rho(t_k, x, f_1(t_k, x, u))] \frac{du}{|u|^{m+1}} \Bigg] \times (t_{k+1} - t_k) \\
&\quad + \int_{t_k}^{t_{k+1}} \int_{R^{(m)}} \log \rho(t_k, x, f_1(t_k, x, u)) \psi(t_k, x, u)) q\, (ds \times du) \\
&\quad + \sum_{j=1}^{l} \alpha_j(t_k, x)[w_j(t_{k+1}) - w_j(t_k)] \\
&\quad + \int_{t_k}^{t_{k+1}} \int_{R^{(m)}} \log \rho(t_k, x, f_1(t_k, x, u))(1 - \psi(t_k, x, u)) p\, (ds \times du);
\end{aligned}$$

$$\psi(t, x, u) = \begin{cases} 1 & \text{if} \quad |\rho(t, x, f_1(t, x, u)) - 1| \leq \frac{1}{2}, \\ 1 - 4(|\rho(t, x, f_1(t, x, u)) - 1| - \frac{1}{2}) & \\ & \text{if} \quad \frac{1}{2} < |\rho(t, x, f_1(t, x, u)) - 1| \leq \frac{3}{4}, \\ 0 & \text{if} \quad |\rho(t, x, f_1(t, x, u)) - 1| \geq \frac{3}{4}. \end{cases} \tag{4.3}$$

The quantity

$$\log \frac{d\mu_{x,t_k}^{(1)}}{d\mu_{x,t_k}^{(2)}} (\eta_{t_k,x}^{(2)}(t))$$

can be written with the help of the same formula except that instead of $\rho(t_k, x, f_1(t_k, x, u))$ we need to substitute $1/[\rho(t_k, x, f_2(t_k, x, u))]$, and instead of $\alpha_j(t_k, x)$ we need to substitute $-\alpha_j(t_k, x)$. We denote by $\alpha(t_k, x)$ the coefficient of $(t_{k+1} - t_k)$ on the right-hand side of Eq. (4.3) and by $\bar{\alpha}(t_k, x)$ the corresponding quantity in the expression for

$$\log \frac{d\mu_{x,t_k}^{(1)}}{d\mu_{x,t_k}^{(2)}} (\eta_{t_k,x}^{(2)}(t)).$$

Let us establish several auxiliary propositions.

LEMMA 1. *Suppose that $\mu_1^{(N)}$ and $\mu_2^{(N)}$ are measures corresponding to the processes $\xi_1^{(N)}(t)$ and $\xi_2^{(N)}(t)$. Then*

$$\log \frac{d\mu_2^{(N)}}{d\mu_1^{(N)}} (\xi_1^{(N)}(t)) = \log p_0(\xi_1(t_0)) + \sum_{k=0}^{N-1} \alpha(t_k, \xi_1^{(N)}(t_k))(t_{k+1} - t_k)$$

$$+ \sum_{k=0}^{N-1} \sum_{j=1}^{l} \alpha_j(t_k, \xi_1^{(N)}(t_k))[W_j(t_{k+1}) - W_j(t_k)]$$

$$+ \sum_{k=0}^{N-1} \int_{t_k}^{t_{k+1}} \int_{R^{(m)}} \log \rho(t_k, \xi_1^{(N)}(t_k), f_1(t_k, \xi_1^{(N)}(t_k), u))\psi(t_k, \xi_1^{(N)}(t_k), u)$$

$$\times q(ds \times du) + \sum_{k=0}^{N-1} \int_{t_k}^{t_{k+1}} \int_{R^{(m)}} \log \rho(t_k, \xi_1^{(N)}(t_k), f_1(t_k, \xi_1^{(N)}(t_k), u))$$

$$\times (1 - \psi(t_k, \xi_1^{(N)}(t_k), u))p(ds \times du). \tag{4.4}$$

The formula for $\log (d\mu_1^{(N)}/d\mu_2^{(N)})(\xi_1^{(N)}(t))$ *is obtained from Formula* (4.4) *if we substitute* p_0 *for* $1/p_0$, $\rho[(t_k, x, f_1(t_k, x, u))$ *for* $1/[\rho(t_k, x, f_2(t_k, x, u))]$, *and* $\xi_1^{(N)}(t)$ *for* $\xi_2^{(N)}(t)$.

Proof. Suppose that $\mathbf{A}_k$ are arbitrary cylindrical sets in $F_m[t_k, t_{k+1}]$. Suppose that for every Borel set $A \subset R^{(m)}$,

$$\mu_{x,t_k}^{(i)}(\mathbf{A}_k; A) = \mathbf{P}\{\eta_{t_k,x}^{(i)}(t) \in \mathbf{A}_k; \eta_{t_k,x}^{(i)}(t_{k+1}) \in A\}.$$

If $\mathbf{A} = \cap_{k=0}^{N-1} \mathbf{A}_k$, then the fact that $\xi_i^{(N)}(t)$ are Markov processes implies that

$$\mu_i^{(N)}(\mathbf{A}) = \int_{R^{(m)}} \mathbf{P}\{\xi_i(t_0) \in dx_0\} \int_{R^{(m)}} \mu_{x_0,t_0}^{(i)}(\mathbf{A}_0, dx_1) \cdots$$

$$\times \int_{R^{(m)}} \mu_{x_{N-2},t_{N-2}}(\mathbf{A}_{N-2}, dx_N)\ \mu_{x_{N+1},t_{N_1}}(\mathbf{A}_{N_1}). \tag{4.5}$$

From (4.5), by using the absolute continuity of the measures $\mathbf{P}\{\xi_1(t_0) \in A\}$ and $\mathbf{P}\{\xi_2(t_0) \in A\}$ with respect to one another and of the measures $\mu^{(1)}_{x_k,t_k}$ and $\mu^{(2)}_{x_k,t_k}$, it is easy to show that

$$\frac{d\mu_2^{(N)}}{d\mu_1^{(N)}}(x(t)) = p_0(x(t_0)) \prod_{k=0}^{N-1} \frac{d\mu^{(2)}_{x_k,t_k}}{d\mu^{(1)}_{x_k,t_k}}(x(t)), \qquad \text{where} \qquad x_k = x(t_k).$$

Consequently

$$\log \frac{d\mu_2^{(N)}}{d\mu_1^{(N)}}(x(t)) = \log p_0(x(t_0)) + \sum_{k=0}^{N-1} \log \frac{d\mu^{(2)}_{x(t_k),t_k}}{d\mu^{(1)}_{x(t_k),t_k}}(x(t)). \tag{4.6}$$

In order to obtain (4.4), it is sufficient to substitute the process $\xi_1^{(N)}(t)$ for $x(t)$ in (4.6) and to use Formula (4.3).

LEMMA 2. *Suppose that* $\bar{\xi}_i^{(N)}(t) = \xi_i^{(N)}(t_k)$ *for* $t \in [t_k, t_{k+1}]$. *Then* $\bar{\xi}_i^{(N)}(t) - \xi_i^{(N)}(t) \to 0$ *in probability as* $\max_k (t_{k+1} - t_k) \to 0$.

Proof. From the definition of $\xi_i^{(N)}(t)$, it follows that the $\xi_i^{(N)}(t_k)$ satisfy the relations (3.2) of Chapter 3 if in these relations we substitute $t_k^{(n)}$ for t_k, $a(t, x)$ for $a_i(t, x)$, $f(t, x, u)$ for $f_i(t, x, u)$, and $\xi_k^{(n)}$ for $\xi_i^{(N)}(t_k)$. Therefore, on the basis of the corollary to the existence theorem of Section 3, Chapter 3, we may assert that the finite-dimensional distributions of $\bar{\xi}_i^{(N)}(t)$ will converge to the finite-dimensional distributions of the process $\xi_i(t)$ as $\max_k (t_{k+1} - t_k) \to 0$. Therefore $\bar{\xi}_i^{(N)}(t)$ is uniformly bounded in probability with respect to N and t. Since for $t \in [t_k, t_{k+1}]$,

$$\begin{aligned} \xi_i^{(N)}(t) - \bar{\xi}_i^{(N)}(t) &= a_i(t_k, \bar{\xi}_i^{(N)}(t))(t - t_k) \\ &\quad + \sum_{j=1}^{l} b_j(t_k, \bar{\xi}_i^{(N)}(t))[w_j(t) - w_j(t_k)] \\ &\quad + \int_{t_k}^{t} \int_{|u| \le 1} f_i(t_k, \bar{\xi}_i^{(N)}(t), u) q\,(ds \times du) \\ &\quad + \int_{t_k}^{t} \int_{|u| > 1} f_i(t_k, \bar{\xi}_i^{(N)}(t), u) p\,(ds \times du), \end{aligned} \tag{4.7}$$

we have that $\xi_i^{(N)}(t) - \bar{\xi}_i^{(N)}(t) \to 0$ in probability as $t - t_k \to 0$. The convergence of the first two terms to zero is obvious. Also,

$$\begin{aligned} &\mathbf{P}\left\{\left|\int_{t_k}^{t} \int_{|u|>1} f_i(t_k, \bar{\xi}_i^{(N)}(t), u) p\,(ds \times du)\right| < 0\right\} \\ &\qquad \le \mathbf{P}\{p([t_k, t] \times \{|u| > 1\}) > 0\} \to 0 \end{aligned}$$

as $t - t_k \to 0$. Finally, if $g_N(x) = 1$ for $|x| \leq N$ and $g_N(x) = 0$ for $|x| > N$, then

$$\mathbf{M}\left(\int_{t_k}^{t}\int_{|u|\leq 1} f_i(t_k, \bar{\xi}_i^{(N)}(t), u), q_N(\bar{\xi}_i^{(N)}(t))q\,(ds \times du)\right)^2$$

$$\leq (t - t_k) \sup_{|x|\leq N}\int_{|u|\leq 1} |f_i(t_k, x, u)|^2 \frac{du}{|u|^{m+1}} \to 0$$

as $t - t_k \to 0$, and the probability

$$\mathbf{P}\left\{\left|\int_{t_k}^{t}\int_{|u|\leq 1} f_i(t_k, \bar{\xi}_i^{(N)}(t), u)\left(1 - g_N(\bar{\xi}_i^{(N)}(t))\right)q\,(ds \times du)\right| > 0\right\}$$
$$\leq \mathbf{P}\{|\bar{\xi}_i^N(t)| > N\}$$

can be made arbitrarily small by a sufficiently large choice of N. This proves the lemma.

Let us now establish a theorem giving sufficient conditions for the absolute continuity of the measures μ_2 and μ_1 corresponding to the processes $\xi_2(t)$ and $\xi_1(t)$ introduced at the beginning of this section.

THEOREM. *Suppose that the following conditions are fulfilled:*

1. $\rho(t, x, y) > 0$ *and is continuous over the set of the variables.*

2. $\alpha_1(t, x), \ldots, \alpha_l(t, x), \alpha(t, x), \bar{\alpha}(t, x)$ *are continuous over the set of variables.*

3. *For every* $C > 0$, $\sup\limits_{\substack{t_0\leq t\leq T\\ |x|\leq C}} \delta(t, x) < \infty$, *and the coefficients of Eqs.* (4.1) *satisfy the conditions of the existence and uniqueness theorems of Section 3, Chapter 3.*

Then the measures μ_1 *and* μ_2 *will be absolutely continuous with respect to one another and*

$$\log\frac{d\mu_2}{d\mu_1}(\xi_1(t)) = \log p_0(\xi_1(t_0)) + \int_{t_0}^{T} \alpha(t, \xi_1(t))\,dt$$

$$+ \sum_{k=1}^{l}\int_{t_0}^{T} \alpha_j(t, \xi_1(t))\,dw_j(t)$$

$$+ \int_{t_0}^{T}\int_{R^{(m)}} \log\rho(t, \xi_1(t), f_1(t, \xi_1(t), u))\psi_1(t, \xi_1(t), u)q\,(dt \times du)$$

$$+ \int_{t_0}^{T}\int_{R^{(m)}} \log\rho(t, \xi_1(t), f_1(t, \xi_1(t), u))(1 - \psi_1(t, \xi_1(t), u))p\,(ds \times du). \tag{4.8}$$

An analogous formula with p_0 substituted for $1/p_0$, α_j substituted for $-\alpha_j$, α substituted for $\bar{\alpha}$, and ρ, f_1 for $1/\rho$, f_2 holds for $(d\mu_1/d\mu_2)(\xi_2(t))$.

Proof. As was shown in the existence theorem of Section 3, Chapter 3, if $\max_k (t_{k+1} - t_k) \to 0$, it is possible to find a subsequence N' and processes

$$\tilde{\xi}_i^{(N')}(t),\ \xi_i^{(N')}(t),\ \tilde{w}_1^{(N')}(t), \ldots, \tilde{w}_l^{(N')}(t),\ \tilde{\zeta}^{(N')}(t)$$

which have for every N' the same joint finite-dimensional distributions as the processes

$$\bar{\xi}_i^{(N')}(t),\ \xi_i^{(N')}(t),\ w_1(t), \ldots, w_l(t),$$

$$\zeta(t) = \int_{t_0}^{t} \int_{|u| \leq 1} uq\,(ds \times du) + \int_{t_0}^{t} \int_{|u|>1} up\,(ds \times du),$$

and which converge to certain particular processes

$$\tilde{\xi}_i(t),\ \xi_i(t),\ \tilde{w}_1(t), \ldots, \tilde{w}_l(t),\ \tilde{\zeta}(t)$$

as $N' \to \infty$. Also, the $\tilde{\xi}_i(t)$ will have the same finite-dimensional distributions as the $\xi_i(t)$. It follows from Lemma 2 that $\tilde{\xi}_i(t) = \xi_i(t)$.

Since $\xi_i^{(N')}(t)$, $\tilde{w}_j^{(N')}(t)$, $\tilde{\zeta}^{(N')}(t)$ and $\xi_i^{(N')}(t)$, $(w_j(t), \zeta(t))$ have identical joint finite-dimensional distributions, the value of

$$\log \frac{d\mu_2^{(N')}}{d\mu_1^{(N')}} (\tilde{\zeta}_1^{(N')}(t))$$

will, from Lemma 1, be determined by Formula (4.4) if in this formula we replace $\xi_1^{(N')}(t)$ by $\xi_1^{(N')}(t)$, $w_j(t)$ by $\tilde{w}_j^{(N')}(t)$, and p and q by $\tilde{p}^{(N')}$ and $\tilde{q}^{(N')}$; $\tilde{p}^{(N')}$ and $\tilde{q}^{(N')}$ are defined in terms of $\tilde{\zeta}^{(N')}(t)$ in exactly the same way that p_0 and q_0 are defined in terms of $\zeta_0(t)$ in Lemma 4 of Section 3, Chapter 3. As $N' \to \infty$,

$$\sum_{k=0}^{N'-1} \alpha(t_k, \xi_1^{(N')}(t_k))(t_{k+1} - t_k) \to \int_{t_0}^{T} \alpha(s, \xi_1(s))\,ds$$

in probability in view of the continuity of $\alpha(t, x)$. Using the theorem of Section 3, Chapter 2, and the continuity of $\alpha_j(t, x)$, it is easy to show that

$$\sum_{k=0}^{N'-1} \alpha_j(t_k, \xi_1^{(N')}(t_k))[\tilde{w}_j^{(N')}(t_{k+1}) - \tilde{w}_j^{(N')}(t_k)] \to \int_{t_0}^{T} \alpha_j(s, \xi_1(s)\,d\tilde{w}_j(s)$$

in probability.

Finally, by using Lemma 5 of Section 3, Chapter 3, it is possible in the same way as in the proof of the existence theorem of Section 3, Chapter 3, to establish the convergence of the right-hand side of (3.7) to the right-hand side of the equation that $\xi(t)$ satisfies; then we can show that

$$\sum_{k=0}^{N'-1} \int_{t_0}^{t_{k+1}} \int_{R^{(m)}} \log \rho\big(t_k, \tilde{\xi}_1^{(N')}(t_k), f_1\big(t_k, \tilde{\xi}_1^{(N')}(t_k), u\big)\big)\psi_1\big(t_k, \tilde{\xi}_1^{(N')}(t_k), u\big)$$

$$\times\, \tilde{q}^{(N')}(ds \times du) \to \int_{t_0}^{T} \int_{R^{(m)}} \log \rho\big(t, \tilde{\xi}_1(t), f_1\big(t, \tilde{\xi}_1(t), u\big)\big)\psi_1\big(t, \tilde{\xi}_1(t), u\big) \times \tilde{q}\,(ds \times du),$$

$$\sum_{k=0}^{N'-1} \int_{t_k}^{t_{k+1}} \int_{R^{(m)}} \log \rho\big(t_k, \tilde{\xi}_1^{(N')}(t_k), f_1\big(t_k, \xi^{(N')}(t_k), u\big)\big)$$

$$\times\, \big(1 - \psi_1\big(t_k, \tilde{\xi}_1^{(N')}(t_k), u\big)\big)\, \tilde{p}^{(N')}(ds \times du)$$

$$\to \int_{t_0}^{T} \int_{R^{(m)}} \log \rho\big(t, \tilde{\xi}_1(t), f_1\big(t, \tilde{\xi}_1(t), u\big)\big)\big(1 - \psi_1\big(t, \tilde{\xi}_1(t), u\big)\big)\tilde{p}\,(dt \times du)$$

in probability, where $\tilde{p}$ and $\tilde{q}$ are defined in terms of $\zeta(t)$ in the same way as in Lemma 4 of Section 3, Chapter 3; p_0 and q_0 are defined in terms of $\zeta_0(t)$. Thus, $\log\,(d\mu_2^{(N')}/d\mu_1^{(N')})(\tilde{\xi}_1^{(N')}(t))$ converges in probability to the right-hand side of (4.8), if we there replace $\xi_1(t)$ by $\tilde{\xi}_1(t)$, $w_j(t)$ by $\tilde{w}_j(t)$, and p and q by $\tilde{p}$ and $\tilde{q}$.

In an analogous fashion, we could show that $\log\,(d\mu_1^{(N')}/d\mu_2^{(N')})\,(\tilde{\xi}_2^{(N')}(t))$ also converges in probability to a random variable that is obtainable from (4.8) by replacing $\xi_1(t)$ by $\tilde{\xi}_2(t)$, $w_j(t)$ by $\tilde{w}_j(t)$, p and q by $\tilde{p}$ and $\tilde{q}$, ρ by $1/\rho$, p_0 by $1/p_0$, α_j by $-\alpha_j$, and α by $\bar{\alpha}$.

If we apply Lemma 4 of Section 2 and take into account the fact that $\xi_i(t)$, $w_j(t)$, p, q, and $\tilde{\xi}_i(t)$, $\tilde{w}_j(t)$, $\tilde{p}$, $\tilde{q}$ have identical distributions, we obtain the proof of the theorem.

CHAPTER 5

ONE-DIMENSIONAL DIFFUSION PROCESSES

1. Preliminary remarks. In Section 1, Chapter 3, we examined stochastic equations for diffusion processes. If the diffusion and transition coefficients of a diffusion process are $\sigma^2(t, x)$ and $a(t, x)$, then the diffusion process will be in the form of a solution to the equation

$$d\xi(t) = a\big(t, \xi(t)\big)\, dt + \sigma\big(t, \xi(t)\big)\, dw(t); \tag{1.1}$$

if we examine a process on the interval $[t_0, T]$ and if an initial condition $\xi(t_0)$ is given at the point t_0, then Eq. (1.1) can be written in integral form:

$$\xi(t) = \xi(t_0) + \int_{t_0}^{t} a\big(s, \xi(s)\big)\, ds + \int_{t_0}^{t} \sigma\big(s, \xi(s)\big)\, dw(s). \tag{1.2}$$

From Theorem 3 of Section 2, Chapter 2, it follows that

$$\int_{t_0}^{t} \sigma\big(s, \xi(s)\big)\, dw(s)$$

is with probability 1 a continuous process provided $\sigma\big(s, \xi(s)\big)$ is with probability 1 a bounded process [for example, in the case in which $\sigma(s, x)$ is a bounded function or $\sigma(s, x)$ is bounded in every bounded region of variation of x and $\xi(t)$ is with probability 1 a bounded function of t]. In the future, we shall examine only those functions for which

$$\sup_{|x| \leq C,\, s \in [t_0, T]} \big(|a(s, x)| + |\sigma(s, x)|\big) < \infty,$$

for every $C > 0$, and the solutions to Eq. (1.2) will be assumed to be bounded. Then it follows from Formula (1.2) that such solutions are with probability 1 continuous. The following existence and uniqueness theorems concerning the solutions to Eq. (1.2) follow from the theorems proven in Chapter 3.

THEOREM 1. *Suppose that the following conditions are satisfied:*

1. $a(s, x)$ *and* $\sigma(s, x)$ *are defined and measurable with respect to their variables where* $s \in [t_0, T]$, $x \in (-\infty, \infty)$.

2. *There exists a* K *such that*

$$\big(a(s, x)\big)^2 + \big(\sigma(s, x)\big)^2 \leq K(1 + x^2).$$

3. *For every $C > 0$, there exists an L_C such that for $|x| \leq C$, $|y| \leq C$,*

$$|a(s, x) - a(s, y)| + |\sigma(s, x) - \sigma(s, y)| \leq L_C|x - y|.$$

Then no matter what the random variable $\xi(t_0)$, independent of $w(t)$, may be, Eq. (1.2) has a unique continuous solution with probability 1.

This theorem follows from Theorem 4 of Section 2, Chapter 3.

THEOREM 2. *Suppose that $a(s, x)$ and $\sigma(s, x)$ are continuous with respect to their variables for $s \in [t_0, T]$, $x \in (-\infty, \infty)$ and that Condition 2 of Theorem 1 is satisfied. Then no matter what the random variable $\xi(t_0)$, independent of $w(t)$, may be, Eq. (1.2) has a continuous solution with probability 1.*

This theorem is a consequence of the existence theorem of Section 3, Chapter 3.

Let $\xi_1(t)$ and $\xi_2(t)$ be two processes satisfying the equation

$$\xi_i(t) = \xi_i(t_0) + \int_{t_0}^{t} a_i(s, \xi_i(s))\, ds + \int_{t_0}^{t} \sigma(s, \xi_i(s))\, dw(s), \qquad i = 1,2 \tag{1.3}$$

and let μ_i be the measure corresponding to the process $\xi_i(t)$ on $\mathbf{F}_1[t_0, T]$. Then from the theorem of Section 4, Chapter 4, we have Theorem 3:

THEOREM 3. *If $a_i(s, x)$ and $\sigma(s, x)$ satisfy the conditions of Theorem 2, if the distributions of $\xi_1(t_0)$ and $\xi_2(t_0)$ are absolutely continuous with respect to one another, and if the density of the distribution of $\xi_2(t_0)$ relative to the distribution of $\xi_1(t_0)$ is equal to $\varphi(x)$, then the measures μ_1 and μ_2 are absolutely continuous with respect to one another and*

$$\log \frac{d\mu_2}{d\mu_1}(\xi_1(t)) = \log \varphi(\xi, (t_0)) + \int_{t_0}^{T} \frac{a_2(t, \xi_1(t)) - a_1(t, \xi_1(t))}{\sigma(t, \xi_1(t))}\, dw(t)$$
$$- \frac{1}{2}\int_{t_0}^{T} \left(\frac{a_2(t, \xi_1(t)) - a_1(t, \xi_1(t))}{\sigma(t, \xi_1(t))}\right)^2 dt. \tag{1.4}$$

Let us examine Eq. (1.2), the coefficients in which are bounded in every bounded region of variation of x. Suppose that $\xi(t)$ is a solution of this equation and that $f(t, x)$ is a function with continuous derivatives $f_t'(t, x)$, $f_x'(t, x)$, and $f_{xx}''(t, x)$, and is monotonic with respect to x for every t. We define $\eta(t) = f(t, \xi(t))$. Then from Theorem 5 of Section 2, Chapter 2, $\eta(t)$ must satisfy the relation

$$d\eta(t) = [f_t'(t, \xi(t)) + f_x'(t, \xi(t))a(t, \xi(t)) + \tfrac{1}{2}f_{xx}''(t, \xi(t))\sigma^2(t, \xi(t))]\, dt$$
$$+ f_x'(t, \xi(t))\sigma(t, \xi(t))\, dw(t).$$

If $\xi(t) = g(t, \eta(t))$ [the existence of such a function $g(t, x)$ follows from the monotonicity of $f(t, x)$], then $\eta(t)$ is a solution to the equations

$$d\eta(t) = \bar{a}(t, \eta(t))\,dt + \bar{\sigma}(t, \eta(t))\,dw(t), \tag{1.5}$$

$$\bar{a}(t, x) = f_t'(t, g(t, x)) + f_x'(t, g(t, x))a(t, g(t, x)) + \tfrac{1}{2}f_{xx}''(t, g(t, x))\sigma^2(t, g(t, x)), \tag{1.6}$$

$$\bar{\sigma}(t, x) = f_x'(t, g(t, x))\sigma(t, g(t, x)). \tag{1.7}$$

Since a one-to-one correspondence is established between the solutions of Eq. (1.1) and (1.5), the questions of existence and uniqueness of a solution to Eqs. (1.1) and (1.4) are settled simultaneously.

THEOREM 4. *Suppose that $a(t, x)$ and $\sigma(t, x)$ satisfy Condition 2 of Theorem 1 and that $\xi(t)$ is a solution of (1.2) that is bounded with probability 1. Then the process*

$$\eta(t) = -\ln\left(\sqrt{1 + \xi^2(t)} - \xi(t)\right)$$

will be a solution of the equation

$$d\eta(s) = \bar{a}(s, \eta(s))\,ds + \bar{\sigma}(s, \eta(s))\,dw(s)$$

with bounded coefficients, where

$$\bar{a}(s, x) = a(s, \sinh x)\frac{1}{\sqrt{1 + (\sinh x)^2}} - \frac{1}{4}\frac{\sinh x}{(1 + (\sinh x)^2)^{3/2}}\sigma^2(s, \sinh x),$$

$$\bar{\sigma}(s, x) = \sigma(s, \sinh x)(1 + (\sinh x)^2)^{-1/2},$$

and

$$\sinh x = \tfrac{1}{2}(e^x - \bar{e}^x).$$

The proof follows immediately from Formulas (1.5), (1.6), and (1.7) if we take $f(t, x) = -\ln(\sqrt{1 + x^2} - x)$ and $g(t, x) = \sinh x$.

Remark. Henceforth, when we study questions of existence and uniqueness of a solution to Eq. (1.2), we shall replace Condition 2 of Theorem 1 with the condition of boundedness of the coefficients of the equation.

2. The absolute continuity of measures corresponding to diffusion processes. Theorem 3 of the preceding section deals only with the case in which the coefficients of the equation satisfy the conditions of the existence and uniqueness theorems. We now examine a generalization of this theorem for the case in which the conditions of the uniqueness theorem are not fulfilled.

THEOREM. *Suppose that the coefficients $a_i(s, x)$ and $\sigma(s, x)$ of Eq. (1.3) are continuous with respect to their variables, that for some $K > 0$ the inequality*

$$|a_1(s, x)| + |a_2(s, x)| + |\sigma(s, x)| \leq K\sqrt{1 + x^2}$$

is satisfied, and that $\sigma(s, x) > 0$. Suppose also that the distributions of $\xi_i(t_0)$ are absolutely continuous with respect to each other and that the distribution density of $\xi_j(t_0)$ with respect to the distribution of $\xi_j(t_0)$ is equal to $\varphi(x)$. Then there exist solutions $\xi_1(t)$ and $\xi_2(t)$ to Eq. (1.3) such that the measures μ_1 and μ_2 corresponding to the processes $\xi_1(t)$ and $\xi_2(t)$ are absolutely continuous with respect to one another and Formula (1.4) is satisfied.

Proof. Let us examine a sequence of partitions of the interval $[t_0, T]$: $t_0 = t_0^{(n)} < t_1^{(n)} < \cdots < t_n^{(n)} = T$. We define $\Delta t_k^{(n)} = t_{k+1}^{(n)} - t_k^{(n)}$, $\Delta w_k^{(n)} = w(t_{k+1}^{(n)}) - w(t_k^{(n)})$. Suppose that $\lim_{N\to\infty} \max_k \Delta t_k^{(n)} = 0$, that $\xi_i^{(n)}(t) = \xi_i^{(n)}(t_k^{(n)})$ for $t \in [t_k^{(n)}, t_{k+1}^{(n)}]$, that $\xi_i^{(n)}(t_0) = \xi_i(t_0)$, and that

$$\xi_i^{(n)}(t_{k+1}^{(n)}) = a_i\big(t_k^{(n)}, \xi_i^{(n)}(t_k^{(n)})\big)\,\Delta t_k^{(n)} + \sigma\big(t_k^{(n)}, \xi_i^{(n)}(t_k^{(n)})\big)\,\Delta w_k^{(n)}. \tag{2.1}$$

We note that the difference equation (2.1) is a special case of Eq. (3.2) of Chapter 3, which was examined at the time of the proof of the existence theorem of Section 3, Chapter 3. It follows from Lemma 3 of Section 3, Chapter 3, that the processes $\xi_i^{(n)}(t)$ satisfy the conditions of Remark 2 of Section 6, Chapter 1; therefore, on the basis of Corollary 2 of Section 6, Chapter 1, we can choose a sequence n' and construct processes

$$\tilde{w}_n'(t),\ \xi_1^{(n')}(t),\ \xi_2^{(n')}(t)$$

that have for every n' the same common distributions as do the processes $w(t)$, $\xi_1^{(N')}(t)$, and $\xi_2^{(N')}(t)$, and that converge in probability to certain processes $\tilde{w}(t)$, $\xi_1(t)$, and $\xi_2(t)$ as $n' \to \infty$. As was shown in the proof of the existence theorem of Section 3, Chapter 3, the processes $\xi_i(t)$ satisfy the equation

$$\xi_i(t) = \xi_i(t_0) + \int_{t_0}^{t} a_i\big(s, \xi_i(s)\big)\,ds + \int_{t_0}^{t} \sigma\big(s, \xi_i(s)\big)\,d\tilde{w}(s). \tag{2.2}$$

Let $\tilde{\mu}_i(n')$ be a measure corresponding to the process $\xi_i^{(n')}(t)$. Using relation (2.1) and Formula (4.4) of Chapter 4 (here, $f_1 = f_2 = 0$ and therefore we consider ρ equal to 1), we can write

$$\begin{aligned}
\log \frac{d\tilde{\mu}_i^{(n')}}{d\tilde{\mu}_j^{(n')}}\,(\xi_j^{(n')}(t)) &= \log \varphi_{i,j}(\xi_j^{(n')}(t_0)) \\
&\quad + \sum_{k=0}^{n'-1} \frac{a_i\big(t_k^{(n')}, \xi_j^{(n')}(t_k^{(n')})\big) - a_j\big(t_k^{(n')}, \xi_j^{(n')}(t_k^{(n')})\big)}{\sigma\big(t_k^{(n')}, \xi_j^{(n')}(t_k^{(n')})\big)}\,\Delta\tilde{w}_k^{(n')} \\
&\quad - \frac{1}{2}\sum_{k=0}^{n'-1} \left(\frac{a_i\big(t_k^{(n')}, \xi_j^{(n')}(t_k^{(n')})\big) - a_j\big(t_k^{(n')}, \xi_j^{(n')}(t_k^{(n')})\big)}{\sigma\big(t_k^{(n')}, \xi_j^{(n')}(t_k^{(n')})\big)}\right)^2 \Delta t_k^{(n')}.
\end{aligned} \tag{2.3}$$

On the basis of the theorem of Section 3, Chapter 2, we obtain

$$\log \frac{d\tilde{\mu}_i^{(n')}}{d\tilde{\mu}_j^{(n')}} (\xi_j^{(n')}(t)) \to \log \psi(\xi_j(t_0)) + \int_{t_0}^{T} \frac{a_i(s, \xi_j(s)) - a_j(s, \xi_j(s))}{\sigma(s, \xi_j(s))}\, d\tilde{w}(s) - \frac{1}{2}\int_{t_0}^{T} \left(\frac{a_i(s, \xi_j(s)) - a_j(s, \xi_j(s))}{\sigma(s, \xi_j(s))}\right)^2 ds; \quad (2.4)$$

thus, by using Lemma 4 of Section 2, Chapter 3, we conclude that the measures $\tilde{\mu}_1$ and $\tilde{\mu}_2$ corresponding to the processes $\xi_1(t)$ and $\xi_2(t)$ are absolutely continuous with respect to one another and $\log (d\tilde{\mu}_i/d\tilde{\mu}_j)(\tilde{\xi}_j(t))$ is defined by the right-hand side of the relation (2.4).

Since $\tilde{\xi}_i(t_0)$ and $\tilde{w}(s)$ have exactly the same common distributions as do $\xi_i(t_0)$ and $w(s)$, there exists a process $\xi_i(s)$ such that the common distributions of $\xi_i(s)$, $\xi_i(t_0)$, and $w(s)$ are the same as the common distributions of $\tilde{\xi}_i(s)$, $\tilde{\xi}_i(t_0)$, and $\tilde{w}(s)$. [Since $\tilde{\xi}_i(s)$ is some measurable function of $\tilde{\xi}_i(t_0)$ and $\tilde{w}(t)$, it follows that $\xi_i(s)$ must be equal to this function if we substitute $\xi_i(t_0)$ for $\tilde{\xi}_i(t_0)$ and $w(t)$ for $\tilde{w}(t)$. Then $\xi_i(s)$ will satisfy Eq. (1.3), $\tilde{\mu}_i$ will coincide with μ_i, and from (2.4), $\log (d\mu_2/d\mu_1)(\xi_1(t))$ will be given by Formula (1.4). This proves the theorem.

3. A comparison theorem for diffusion processes. We shall show that under certain assumptions a diffusion process is a monotonic function of the transition coefficient. This fact will be used for establishing more general conditions for the uniqueness of the solution to Eq. (1.2).

THEOREM. *Suppose that $a_1(t, x)$, $a_2(t, x)$, and $\sigma(t, x)$ satisfy the following conditions:*

1. *$a_1(t, x)$, $a_2(t, x)$, and $\sigma(t, x)$ are continuous in their variables for $t \in [t_0, T]$, $x \in (-\infty, \infty)$.*

2. *$\sigma(t, x) > 0$, and for every $C > 0$, there exists $\alpha > \frac{1}{2}$ and $L > 0$ such that for $|x| \leq C$, $|y| \leq C$,*

$$|\sigma(t, x) - \sigma(t, y)| \leq L|x - y|^{\alpha}.$$

Suppose further that $\xi_1(t)$ and $\xi_2(t)$ are with probability 1 continuous solutions of the equation

$$\xi_i(t) = \xi_i(t_0) + \int_{t_0}^{t} a_i(s, \xi_i(s))\, ds + \int_{t_0}^{t} \sigma(s, \xi_i(s))\, dw(s), \qquad i = 1, 2.$$

Under these conditions, if $a_1(t, x) < a_2(t, x)$ for every t and x and if $\mathbf{P}\{\xi_1(t_0) \leq \xi_2(t_0)\} = 1$, then $\xi_1(t) \leq \xi_2(t)$ for every t with probability 1.

The proof of the theorem depends on the following lemma.

LEMMA. *Suppose that the conditions of the theorem are met, and that τ is a random variable such that the event $\{\tau > s\}$ does not depend on $w(t) - w(s)$ for $t > s$ and $\xi_1(\tau) = \xi_2(\tau)$ with probability 1. Then there exists a τ_1 such that $\tau_1 > \tau$ with probability 1 and for $s \in (\tau, \tau_1)$ the inequality $\xi_1(s) < \xi_2(s)$ is fulfilled with probability 1.*

Proof. Let $\chi_a^b(s)$ denote the characteristic function of the interval (a, b). Define $\psi(s) = 0$ for $s < \tau$ and for those $s > \tau$ for which

$$\inf_{\tau \le u \le s} \big(a_2(u, \xi_2(u)) - a_1(u, \xi_1(u))\big) \le \tfrac{1}{2}\big(a_2(\tau, \xi_2(\tau)) - a_1(\tau, \xi_1(\tau))\big),$$

and define $\psi(s) = 1$ if $s \ge \tau$ and

$$\inf_{\tau \le u \le s} \big(a_2(u, \xi_2(u)) - a_1(u, \xi_1(u))\big) > \tfrac{1}{2}\big(a_2(\tau, \xi_2(\tau)) - a_1(\tau, \xi_1(\tau))\big).$$

Also define

$$\psi_h^{(C)}(s) = \chi_{-C}^{C}\big(\sup_{t_0 \le u \le s} (|\xi_1(u)| + |\xi_2(u)|)\big)\chi_\tau^{\tau+h}(s)\psi(s).$$

Let us show that with probability 1

$$\lim_{h \to 0} \frac{1}{h} \int_{t_0}^{T} \psi_h^{(C)}(s)[\sigma(s, \xi_2(s)) - \sigma(s, \xi_1(s))]\, dw(s) = 0 \tag{3.1}$$

is fulfilled. In fact, by using Hölder's inequality, we obtain

$$\begin{aligned}
\mathbf{M}\Big(\int_{t_0}^{T} & \psi_h^{(C)}(s)[\sigma(s, \xi_2(s)) - \sigma(s, \xi_1(s))]\, dw(s)\Big)^2 \\
&= \int_{t_0}^{T} \mathbf{M}\psi_h^{(C)}(s)[\sigma(s, \xi_1(s)) - \sigma(s, \xi_2(s))]^2\, ds \\
&\le L^2 \int_{t_0}^{T} \mathbf{M}\psi_h^{(C)}(s)|\xi_1(s) - \xi_2(s)|^{2\alpha}\, ds \\
&\le L^2\Big[\int_{t_0}^{T} \mathbf{M}\psi_h^{(C)}(s)\, ds\Big]^{1-\alpha} \cdot \Big[\mathbf{M}\int_{t_0}^{T} \psi_h^{(C)}(s)|\xi_2(s) - \xi_1(s)|^2\, ds\Big]^{\alpha} \\
&\le L^2 h^{1-\alpha}\Big[\int_{t_0}^{T} \mathbf{M}\psi_h^{(C)}(s)|\xi_2(s) - \xi_1(s)|^2\, ds\Big]^{\alpha}.
\end{aligned}$$

Since

$$\begin{aligned}
\psi_h^{(C)}(s)(\xi_2(s) - \xi_1(s)) &= \psi_h^{(C)}(s)\int_{t_0}^{S} \psi_h^{(C)}(u)[a_2(u, \xi_2(u)) - a_1(u, \xi_1(u))]\, du \\
&\quad + \psi_h^{(C)}(s)\int_{t_0}^{S} \psi_h^{(C)}(u)[\sigma(u, \xi_2(u)) - \sigma(u, \xi_1(u))]\, dw(u),
\end{aligned}$$

and for some H, the inequality $\chi_{-C}^{C}(x)(|a_1(s, x)| + |a_2(s, x)|) \leq H$ holds, it follows that

$$\mathbf{M}\left(\int_{t_0}^{T} \psi_h^{(C)}(s)[\sigma(s, \xi_2(s)) - \sigma(s, \xi_1(s))]\, dw(s)\right)^2$$
$$\leq L^2 h^{1-\alpha}\Bigg[\int_{t_0}^{T} \mathbf{M}\psi_h^{(C)}(s)\Big[8H^2h^2$$
$$+ 2\left\{\int_{t_0}^{S} \psi_h^{(C)}(u)(\sigma(u, \xi_2(u)) - \sigma(u, \xi_1(u)))\, dw(u)\right\}^2\Big] ds\Bigg]^{\alpha}$$
$$\leq L^2 h^{1-\alpha}\Big[8H^2h^3$$
$$+ 2\int_{t_0}^{T} \mathbf{M}\psi_h^{(C)}(s)\left\{\int_{t_0}^{S} \psi_h^{(C)}(u)(\sigma(u, \xi_2(u)) - \sigma(u, \xi_1(u)))\, dw(u)\right\}^2 ds\Big]^{\alpha}$$
$$\leq C_1 h^{1+2\alpha} + C_2 h^{1-\alpha}\left(\int_{t_0}^{T} \mathbf{M}\psi_h^{(s)}(s)\left\{\int_{t_0}^{s} \psi_h^{(C)}(u)(\sigma(u, \xi_2(u))\right.\right.$$
$$\left.\left. - \sigma(u, \xi_1(u)))\, dw(u)\right\}^2 ds\right)^{\alpha},$$

where C_1 and C_2 are constants. But

$$\int_{t_0}^{T} \psi_h^{(C)}(s)\left[\int_{t_0}^{S} \psi_h^{(C)}(u)(\sigma(u, \xi_2(u)) - \sigma(u, \xi_1(u)))\, dw(u)\right]^2 ds$$
$$\leq h \max_s \left[\int_{t_0}^{S} \psi_h^{(C)}(u)(\sigma(u, \xi_2(u)) - \sigma(u, \xi_1(u)))\, dw(u)\right]^2;$$

therefore, on the basis of Property 5 of martingales of Section 5, Chapter 1,

$$\mathbf{M}\int_{t_0}^{T} \psi_h^{(C)}(s)\left[\int_{t_0}^{S} \psi_h^{(C)}(u)(\sigma(u, \xi_2(u)) - \sigma(u, \xi_1(u)))\, dw(u)\right]^2 ds$$
$$\leq 4h\int_{t_0}^{T} \mathbf{M}\psi_h^{(C)}(u)(\sigma(u, \xi_2(u)) - \sigma(u, \xi_1(u)))^2\, du.$$

Thus if

$$\nu(h) = \int_{t_0}^{T} \mathbf{M}\psi_h^{(C)}(u)(\sigma(u, \xi_2(u)) - \sigma(u, \xi_1(u)))^2\, du,$$

then for some C_1 and C_3, the inequality

$$\nu(h) \leq C_1 h^{1+2\alpha} + C_3 h[\nu(h)]^{\alpha}$$

is fulfilled. Multiplying both sides of the inequality by $h^{-1-2\alpha}$, we obtain

$$(\nu(h)h^{-1-2\alpha}) \leq C_1 + C_3 h^{\alpha(2\alpha-1)}(\nu(h)h^{-1-2\alpha})^{\alpha}.$$

From this inequality, it follows that for $h \leq 1$, there exists a constant D, independent of h, such that

$$\nu(h)h^{-1-2\alpha} \leq D.$$

(We can assume that $\alpha < 1$, and if it were true that $\nu(h)h^{-1-2\alpha} \to \infty$, we would have the two contradictory relationships

$$(\nu(h)h^{-1-2\alpha})^\alpha = o(\nu(h)h^{-1-2\alpha}),$$
$$\nu(h)h^{-1-2\alpha} = O(\nu(h)h^{-1-2\alpha})^\alpha.)$$

This means that for some D,

$$\mathbf{M}\Big[\int_{t_0}^T \psi_h^{(C)}(s)\big(\sigma(s, \xi_2(s)) - \sigma(s, \xi_1(s))\big)\, dw(s)\Big]^2 \le Dh^{1+2\alpha}.$$

We now note that

$$\int_{t_0}^T \psi_h^{(C)}(s)[\sigma(s, \xi_2(s)) - \sigma(s, \xi_1(s))]\, dw(s) = \int_{t_0}^T f(s)\chi_\tau^{\tau+h}(s)\, dw(s),$$

where $f(s)$ is some function that is measurable with respect to the minimal σ-algebra with respect to which $\xi_i(t_0)$ and $w(t)$ are measurable for $t \le s$. Since

$$\lambda(h) = \int_{t_0}^T f(s)\chi_\tau^{\tau+h}(s)\, dw(s) = \zeta(\tau + h) - \zeta(\tau),$$

where $\zeta(t) = \int_{t_0}^t f(s)\, dw(s)$ and $\zeta(s)$ are martingales, on the basis of Property 7 of martingales of Section 5, Chapter 1, $\zeta(\tau + h)$, and hence $\lambda(h) = \zeta(\tau + h) - \zeta(\tau)$, will be a martingale. Therefore, applying Property 5 of martingales of Section 5, Chapter 1, to $\lambda(h)$, we obtain

$$\mathbf{M} \sup_{0<h\le h_0}\Big[\int_{t_0}^T \psi_h^{(C)}(s)[\sigma(s, \xi_2(s)) - \sigma(s, \xi_1(s))]\, dw(s)\Big]^2 \le 4Dh_0^{1+2\alpha}.$$

Consequently

$$\mathbf{P}\left\{\sup_{1/2^{k+1}\le h\le 1/2^k}\left|\frac{1}{h}\int_{t_0}^T \psi_h^{(C)}(s)[\sigma(s, \xi_2(s)) - \sigma(s, \xi(s))]\, dw(s)\right| > \frac{1}{k}\right\}$$
$$\le \mathbf{P}\left\{\sup_{0\le h\le 1/2^k}\left|\int_{t_0}^T \psi_h^{(C)}(s)[\sigma(s, \xi_2(s)) - \sigma(s, \xi_1(s))]\, dw(s)\right| > \frac{1}{k2^{k+1}}\right\}$$
$$\le 4Dk^2 2^{2k+2}\left(\frac{1}{2^k}\right)^{1+2\alpha} \le \frac{16Dk^2}{2^{(2\alpha-1)k}}.$$

Since

$$\sum_{k=1}^\infty \frac{k^2}{2^{(2\alpha-1)k}} < \infty$$

with probability 1, there exists a k_0 such that for $k > k_0$, the relation

$$\sup_{0<h<1/2^k} \frac{1}{h}\left|\int_{t_0}^{T} \psi_h^{(C)}(s)[\sigma(s, \xi_2(s)) - \sigma(s, \xi_1(s))]\, dw(s)\right| < \frac{1}{k}$$

will be satisfied. (We make use of the Borel-Cantelli lemma.) Equation (3.1) follows from the last inequality.

To prove the lemma, we use the relations

$$\begin{aligned} &\psi_h^{(C)}(\tau + h)[\xi_2(\tau + h) - \xi_1(\tau + h)] \\ &\qquad = \psi_h^{(C)}(\tau + h)\int_{t_0}^{T} \psi_h^{(C)}(u)(a_2(u, \xi_2(u)) - a_1(u, \xi_1(u)))\, du \\ &\qquad\quad + \int_{t_0}^{T} \psi_h^{(C)}(u)(\sigma(u, \xi_2(u)) - \sigma(u, \xi_1(u)))\, dw(u) \\ &\qquad > h\psi_h^{(C)}(\tau + h)[\tfrac{1}{2}(a_2(\tau, \xi_2(\tau)) - a_1(\tau, \xi_1(\tau))) \\ &\qquad\quad + \frac{1}{h}\int_{t_0}^{T} \psi_h^{(C)}(u)(\sigma(u, \xi_2(u)) - \sigma(u, \xi_1(u)))\, dw(u)]\cdot \end{aligned}$$

From (3.1), it follows that there exists an h' such that for $0 < h < h'$,

$$\frac{1}{h}\left|\int_{t_0}^{T} \psi_h^{(C)}(u)[\sigma(u, \xi_2(u)) - \sigma(u, \xi_1(u))]\, dw(u)\right| < \tfrac{1}{4}(a_2(\tau, \xi_2(\tau)) - a_1(\tau, \xi_1(\tau)))$$

and hence

$$\begin{aligned} &\psi_h^{(C)}(\tau + h)[\xi_2(\tau + h) - \xi_1(\tau + h)] \\ &\qquad \geq \frac{h}{4}\psi_h^{(C)}(\tau + h)(a_2(\tau, \xi_2(\tau)) - a_1(\tau, \xi_1(\tau))) \end{aligned}$$

for $h < h'$. The proof of the lemma follows from the last relation because for almost all elementary events ω there exist C and h' such that $h < h'$ and $\psi_h^{(C)}(\tau + h) = 1$.

Proof of the theorem. Let τ_1 be the first zero of the difference $\xi_2(t) - \xi_1(t)$. Then the event $\{\tau_1 > s\}$ depends only on the behavior of $w(t)$ in the interval $[t_0, s]$ and does not depend on $w(t) - w(s)$ for $t > s$. On the basis of the lemma, we may assert that the next zero of the difference $\xi_2(t) - \xi_1(t)$ following τ_1 will be at a positive distance from τ_1; if we denote this zero by τ_2, then $\xi_2(t) > \xi_1(t)$ for $t \in (\tau_1, \tau_2)$. Clearly, the event $\{\tau_2 > s\}$ likewise does not depend on $w(t) - w(s)$ for $t > s$. By analogous reasoning, we conclude that after every zero of the difference

$\xi_2(s) - \xi_1(s)$ there exists a zero immediately following. Thus we can write these zeros as a transfinite increasing sequence τ_α, where α is an ordinal number. Then for every α, the event $\{\tau_\alpha > s\}$ does not depend on $w(t) - w(s)$ for $t > s$, and for $s \in (\tau_\alpha, \tau_{\alpha+1})$, $\xi_2(s) > \xi_1(s)$. If there exists a maximal zero $\bar{\tau}$, it must also be a member of the transfinite sequence. Consequently again, we conclude by using the lemma, that for some h (random) $\xi_2(s) > \xi_1(s)$ if $t \in (\bar{\tau}, \bar{\tau} + h)$. Since $\xi_2(s) - \xi_1(s)$ does not change sign in the interval $(\bar{\tau}, T)$, we have $\xi_2(s) > \xi_1(s)$ for $s \in (\bar{\tau}, T)$. Therefore for all $t \in [t_0, T]$, the inequality $\xi_1(s) \leq \xi_2(s)$ is satisfied. This proves the theorem.

4. A theorem on the uniqueness of the solution of a stochastic equation for diffusion processes.

THEOREM. *Suppose that $a(t, x)$ and $\sigma(t, x)$ satisfy the following conditions:*

1. *$a(t, x)$ and $\sigma(t, x)$ are defined and continuous for $t \in [t_0, T]$, $x \in (-\infty, \infty)$.*

2. *For some K, the inequality*

$$|a(t, x)|^2 + |\sigma(t, x)|^2 \leq K(1 + x^2)$$

is satisfied.

3. *$\sigma(t, x) > 0$, and for every $C > 0$, there exist $L > 0$ and $\alpha > \frac{1}{2}$ such that for $|x| \leq C$, $|y| \leq C$,*

$$|\sigma(t, x) - \sigma(t, y)| \leq L|x - y|^\alpha.$$

Then for every $\xi(t_0)$ not depending on $w(t)$, Eq. (1.2) has a unique continuous solution with probability 1.

Proof. As was shown in the remark to Theorem 4 of Section 1, instead of Condition 2 we may assume that the condition

$$|a(s, x)| + |\sigma(s, x)| \leq K$$

is fulfilled. Let $\xi_1(t)$ and $\xi_2(t)$ be two solutions to Eq. (2.1). For $\epsilon \in (0, 1)$, set

$$a_\epsilon^*(s, x) = a(s, x) - \epsilon\sigma(s, x),$$
$$a_\epsilon^{**}(s, x) = a(s, x) + \epsilon\sigma(s, x);$$

let $\xi_\epsilon^*(t)$ and $\xi_\epsilon^{**}(t)$ be solutions of the equations

$$\xi_s^*(t) = \xi(t_0) + \int_{t_0}^t a_\epsilon^*(s, \xi_\epsilon^*(s))\, ds + \int_{t_0}^t \sigma(s, \xi_\epsilon^*(s))\, dw(s), \tag{4.1}$$

$$\xi_\epsilon^{**}(t) = \xi(t_0) + \int_{t_0}^t a_\epsilon^{**}(s, \xi_\epsilon^{**}(s))\, ds + \int_{t_0}^t \sigma(s, \xi_\epsilon^{**}(s))\, dw(s) \tag{4.2}$$

such that

$$\log \frac{d\mu_\epsilon^{**}}{d\mu_\epsilon^{*}}(\xi_\epsilon^*(t)) = \int_{t_0}^{T} \left(\frac{a_\epsilon^{**}(s, \xi_\epsilon^*(s)) - a_\epsilon^*(s, \xi_\epsilon^*(s))}{\sigma(s, \xi_\epsilon^*(s))} \right) dw(s)$$

$$- \frac{1}{2} \int_{t_0}^{T} \left(\frac{a_\epsilon^{**}(s, \xi_\epsilon^*(s)) - a_\epsilon^*(s, \xi_\epsilon^*(s))}{\sigma(s, \xi_\epsilon^*(s))} \right)^2 ds = 2\epsilon(w(T) - w(t_0))$$

$$- 2\epsilon^2(T - t_0), \tag{4.3}$$

where μ_ϵ^* and μ_ϵ^{**} denote the measures corresponding to the processes $\xi_\epsilon^*(t)$ and $\xi_\epsilon^{**}(t)$. The existence of solutions of (4.1) and (4.2) that satisfy (4.3) follows from the theorem of Section 2. From the theorem of Section 3, it follows that with probability 1 the inequalities

$$\xi_\epsilon^*(t) \leq \xi_1(t) \leq \xi_\epsilon^{**}(t),$$
$$\xi_\epsilon^*(t) \leq \xi_2(t) \leq \xi_\epsilon^{**}(t)$$

hold. Therefore $|\xi_1(t) - \xi_2(t)| \leq \xi_\epsilon^{**}(t) - \xi_\epsilon^*(t)$. We note that in view of the boundedness of $a(t, x)$ and $\sigma(t, x)$, for all $\epsilon \in (0, 1)$ the relations

$$\mathbf{M}(\xi_\epsilon^*(t) - \xi_\epsilon^*(t_0))^2 \leq K^2(T - t_0)^2 + K^2(T - t_0),$$
$$\mathbf{M}(\xi_\epsilon^{**}(t) - \xi_\epsilon^{**}(t_0))^2 \leq K^2(T - t_0)^2 + K^2(T - t_0)$$

are valid. From Formula (4.3), it follows that

$$\mathbf{M}(\xi_\epsilon^{**}(t) - \xi_\epsilon^{**}(t_0)) = \mathbf{M}(\xi_\epsilon^*(t) - \xi_\epsilon^*(t_0))e^{2\epsilon[w(T) - w(t_0) - \epsilon T + \epsilon t_0]};$$

consequently

$$\mathbf{M}|\xi_1(t) - \xi_2(t)| \leq \mathbf{M}(\xi_\epsilon^{**}(t) - \xi_\epsilon^*(t))$$
$$\leq \mathbf{M}(\xi_\epsilon^*(t) - \xi(t_0))(e^{2\epsilon[w(T) - w(t_0) - \epsilon T + \epsilon t_0]} - 1)$$
$$\leq \sqrt{\mathbf{M}(\xi_\epsilon^*(t) - \xi(t_0))^2} \sqrt{\mathbf{M}(e^{2\epsilon[w(T) - w(t_0) - \epsilon T + \epsilon t_0]} - 1)^2} \to 0$$

as $\epsilon \to 0$. Thus for all $t \in [t_0, T]$, we have $\mathbf{P}\{\xi_1(t) = \xi_2(t)\} = 1$. When we take into account the continuity of $\xi_1(t)$ and $\xi_2(t)$, we conclude that $\xi_1(t) = \xi_2(t)$ for all t with probability 1. This proves the theorem.

CHAPTER 6

THE LIMITING TRANSITION FROM A MARKOV CHAIN TO A MARKOV PROCESS WITH CONTINUOUS TIME

1. Statement of the problem. Let us examine a sequence of Markov chains: $\xi_0^{(n)}, \xi_1^{(n)}, \ldots, \xi_n^{(n)}$. Let $t_0 = t_0^{(n)} < t_1^{(n)} < \cdots < t_{n+1}^{(n)} = T$ be a sequence of partitions of the interval $[t_0, T]$. Let us associate with each of these Markov chains a random process $\xi^{(n)}(t) = \xi_k^{(n)}$ for $t \in [t_k^{(n)}, t_{k+1}^{(n)}]$. The process $\xi^{(n)}(t)$ is a step function having stochastic discontinuities at the points $t_k^{(n)}$. If $\xi_{k+1}^{(n)} - \xi_k^{(n)}$ converges uniformly to zero in probability with respect to k as $n \to \infty$ and $\max_k (t_{k+1}^{(n)} - t_k^{(n)}) \to 0$, then the process $\xi^{(n)}(t)$ can be considered approximately stochastically continuous. We shall be concerned with the question as to when this process can be considered an approximate solution to the stochastic equation examined in Chapter 3, that is, as to what the conditions are for which $\xi^{(n)}(t)$ will in some defined sense converge as $n \to \infty$ to a process $\xi(t)$ that is a solution to the stochastic equation.

We shall study below, first of all, the conditions under which the finite-dimensional distributions of the processes $\xi^{(n)}(t)$ will converge to the finite-dimensional distributions of the process $\xi(t)$ of the form shown; second, we shall study the limiting behavior of those characteristics of the processes $\xi^{(n)}(t)$ that are not expressed by knowledge of $\xi^{(n)}(t)$ at a finite number of points [for example, $\int_t^T f(t, \xi^{(n)}(t))\, dt$, $\sup_t \xi^{(n)}(t)$, etc.]. Therefore we are interested in finding the conditions under which the distributions of certain functionals of $\xi_n(t)$ will converge to the distribution of the corresponding functionals of $\xi(t)$. It is obvious that such conditions must depend on the class of functionals $F(x(t))$ for which the limiting distributions $F(\xi^{(n)}(t))$ must coincide with the distribution of $F(\xi(t))$. We shall find a rather wide class of functionals possessing a property under which, in the case of convergence of Markov processes to solutions of the stochastic equations of the type examined, the fulfillment of the conditions for convergence of finite-dimensional distributions implies the convergence of the distributions of the functionals of this class.

Since the processes $\xi^{(n)}(t)$ and $\xi(t)$ are solutions of stochastic equations and hence with probability 1 do not have discontinuities of the second kind, since the $\xi^{(n)}(t)$ are continuous from the right, and since the $\xi(t)$ can be considered to be such, we shall examine functionals that are defined on the space of functions that have no discontinuities of the second kind and that are continuous from the right. The class of functionals will be defined in terms of the continuity of the functionals with respect to some convergence defined for functions without discontinuities of the second kind that are continuous from the right. This convergence will be studied in Section 2. In Section 3, we shall prove a general limiting theorem for processes that with probability 1 have no discontinuities of the second kind. Afterwards, in Section 4, we shall study the conditions for convergence of finite-dimensional distributions of the processes $\xi^{(n)}(t)$ to a process $\xi(t)$ that is a solution to Eq. (1.8) of Chapter 3. Finally, in Section 5, we shall prove a limiting theorem for distributions of functionals of Markov processes of the type shown.

2. A type of convergence of functions without discontinuities of the second kind. Let us examine the space $D_n[t_0, T]$ of functions $x(t)$ that are defined for $t \in [t_0, T]$, that take values in $R^{(m)}$, and that possess the following properties:

1. At every point $t \in [t_0, T]$, $x(t-0)$ exists and $x(T-0) = x(T)$.
2. At every point $t \in [t_0, T]$, $x(t+0)$ exists and $x(t+0) = x(t)$ if $t_0 \leq t < T$. For every function $x(t) \in D_m[t_0, T]$, we set

$$x^{\epsilon}(t) = x(t) - \sum_{s \leq t} [x(s) - x(s-0)],$$

$$|x(s) - x(s-0)| > \epsilon.$$

Definition 1. A sequence $x_n(t)$ of functions in $D_m[t_0, T]$ is said to be **J**-convergent to the function $x_0(t)$ in $D_m[t_0, T]$ if

(a) for every $\epsilon > 0$ for which $x_0(t)$ does not have jumps in absolute value equal to ϵ, the quantity $x_n(t) - x_n^{\epsilon}(t) \to x_0(t) - x_0^{\epsilon}(t)$ for almost all $t \in [t_0, T]$;

(b) $\lim_{\epsilon \to 0} \overline{\lim_{n \to \infty}} \sup_{t_0 \leq t \leq T} |x_n^{\epsilon}(t) - x_0^{\epsilon}(t)| = 0.$

To indicate that $x_n(t)$ is **J**-convergent to $x_0(t)$, we shall write

$$x_n(t) \underset{\rightarrow}{\mathbf{J}} x_0(t) \qquad \text{or} \qquad x_0(t) = \mathrm{J}\lim_{n \to \infty} x_n(t).$$

It turns out that every **J**-convergent sequence can, by small deformations of the interval $[t_0, T]$, be transformed into a uniformly convergent sequence. The above definition is equivalent to the following:

Definition 2. A sequence $x_n(t)$ is **J**-convergent to $x_0(t)$ if there exists a sequence of continuous monotonically increasing functions $\lambda_n(t)$ such that $\lambda_n(t_0) = t_0$, $\lambda_n(T) = T$, and

$$\lim_{n \to \infty} \sup_{t} \left(|x_n(\lambda_n(t)) - x_0(t)| + |\lambda_n(t) - t|\right) = 0. \tag{2.1}$$

Suppose that $x(t) \in D_m[t_0, T]$. For every $C > 0$, we define

$$\Delta_c(x(t)) = \sup_{\substack{t_0 \le t' < t < t'' \le T \\ |t'-t''| < C}} \min [|x(t') - x(t)|; |x(t) - x(t'')|] + \sup_{0 < h \le C} (|x(t_0 + h) - x(t_0)| + |x(T) - x(T - h)|).$$

It is clear that every function $x(t)$ in $D_m[t_0, T]$ satisfies the condition

$$\lim_{C \to 0} \Delta_C(x(t)) = 0.$$

We now define the quantity $\rho_C(x(t), y(t))$ for all $x(t)$ and $y(t)$ in $D_m[t_0, T]$ and for $C > 0$ as follows:

$$\rho_C(x(t), y(t)) = \Delta_C(x(t)) + \Delta_C(y(t)) + \sup_{t_0 \le t \le T - C/2,} \inf_{s \in [t, t+C/2]} |x(s) - y(s)|.$$

The quantity $\rho_C(x_n(t), x_0(t))$ allows us to test the convergence of $x_n(t)$ to $x_0(t)$.

THEOREM. *In order that* $x_0(t) = \mathrm{J}\lim_{n\to\infty} x_n(t)$, *it is necessary that*

$$\lim_{C \to 0} \overline{\lim_{n \to \infty}} \rho_C(x_n(t), x_0(t)) = 0, \tag{2.2}$$

and it is sufficient that there exist a sequence c_n, *converging to zero, for which* $\rho_{c_n}(x_n(t), x_0(t)) \to 0$.

Proof of the necessity. Suppose that $x_n(t) \underset{\to}{\mathrm{J}} x_0(t)$. From the second definition of J-convergence, it is easy to see that $x_n(t) \to x_0(t)$ at every point of continuity of $x_0(t)$. Therefore for every $C > 0$, as $n \to \infty$,

$$\sup_{t_0 \le t \le T - C} \inf_{s \in [t, t+C]} |x_n(s) - x_0(s)| = 0.$$

Then $\Delta_C(x_0(t)) \to 0$ as $C \to 0$. Therefore we need only show that

$$\lim_{c \to 0} \overline{\lim_{n \to \infty}} \Delta_c(x_n(t)) = 0.$$

Suppose that the $\lambda_n(t)$ satisfy the conditions of Definition 2. Then if $\sup_t |\lambda_n(t) - t| < c/2$, the inequality

$$\Delta_c(x_n(t)) \le \Delta_{c/2}(x_0(t)) + 3 \sup_t |x_n(\lambda_n(t)) - x_0(t)|$$

holds. Consequently

$$\overline{\lim_{n \to \infty}} \Delta_c(x_n(t)) \le \Delta_{c/2}(x_0(t))$$

and thus we have

$$\lim_{c\to 0} \overline{\lim_{n\to\infty}} \Delta_c(x_n(t)) \leq \lim_{c\to 0} \Delta_{c/2}(x_0(t)) = 0.$$

This proves the necessity of (2.2).

To prove the sufficiency of the conditions of the theorem, we use the following two lemmas:

LEMMA 1. *Suppose that* $x(t), y(t) \in D_m[t_0, T]$ *and* $|x(t_1) - x(t_1 - 0)| > \epsilon$. *If, for some* $h > 0$, *it is true that* $\rho_h(x(t), y(t)) < \mu$ *and* $\mu < \epsilon/7$, *then there is a point* t' *in the interval* $(t_1 - h/2,\ t_1 + h/2)$ *such that* $|y(t' - 0) - x(t_1 - 0)| < 3\mu$, $|y(t') - x(t_1)| < 3\mu$, *and* $|y(t) - y(t' - 0)| \leq \mu$ *for* $t \in (t' - h, t')$ *and* $|y(t) - y(t')| < \mu$ *for* $t \in [t', t' + h]$.

Proof. There exist points $s_1 \in (t_1 - h/2, t_1)$ and $s_2 \in (t_1, t_1 + h/2)$ such that $|y(s_1) - x(s_1)| < \mu$ and $|y(s_2) - x(s_2)| < \mu$. Since $\Delta_h(x(t)) < \mu$, we have

$$\min\,[|x(s_1) - x(t_1 - 0)|;\, |x(t_1) - x(t_1 - 0)|] < \mu,$$
$$\min\,[|x(s_2) - x(t_1)|;\, |x(t_1) - x(t_1 - 0)|] < \mu.$$

Since $|x(t_1) - x(t_1 - 0)| > \epsilon > \mu$, we have $|x(s_1) - x(t_1 - 0)| < \mu$ and $|x(s_2) - x(t_1)| < \mu$. Therefore $|y(s_1) - x(t_1 - 0)| < 2\mu$ and $|y(s_2) - x(t_1) < 2\mu$; thus $|y(s_1) - y(s_2)| > \epsilon - 4\mu > 3\mu$. Since

$$\min_{s_1 \leq s' < t < s'' \leq s_2} [|y(s') - y(t)|;\, |y(t) - y(s'')|] < \mu,$$

there exists in the interval $[s_1, s_2]$ a point t' such that $|y(s_1) - y(t)| < \mu$ for $s_1 \leq t < t'$ and $|y(s_2) - y(t)| < \mu$ for $t' \leq t \leq s_2$. Consequently $|y(s_1) - y(t' - 0)| \leq \mu$ and $|y(s_2) - y(t')| < \mu$; also, $|y(t' - 0) - x(t_1 - 0)| < 3\mu$ and $|y(t') - x(t_1)| < 3\mu$.

Since $|y(t' - 0) - y(t')| > \epsilon - 6\mu < \mu$ and

$$\min_{t'-h<t<t'} [|y(t) - y(t' - 0)|;\, |y(t' - 0) - y(t')|] < \mu,$$

it follows that $|y(t) - y(t' - 0)| \leq \mu$. In the same manner, we could show that $|y(t) - y(t')| < \mu$ for $t \in [t', t' + h]$. This proves the lemma.

LEMMA 2. *If* $x(t)$ *and* $y(t)$ *do not have jumps exceeding in absolute value a number* ϵ, *and if for some* $h < (T_0 - t_0)/2$ *it is true that*

$$\rho_h(x(t), y(t)) < \mu,$$

then

$$\sup_t |x(t) - y(t)| < 2\epsilon + 5\mu.$$

Proof. First let us show that $|t_1 - t_2| < h$ implies $|x(t_1) - x(t_2)| < \epsilon + 2\mu$. Let $t_1 < t_2$ and $t' \in (t_1, t_2)$ be such that $|x(t_1) - x(s)| \leq \mu$ for

$s < t'$ and $|x(t_1) - x(t')| > \mu$. Then

$$|x(t_1) - x(t')| \leq |x(t_1) - x(t' - 0)| + |x(t' - 0) - x(t')| \leq \epsilon + \mu.$$

But

$$\min \big(|x(t_1) - x(t')|; |x(t') - x(t_2)|\big) < \mu,$$

which means that $|x(t') - x(t_2)| < \mu$. Therefore $|x(t_1) - x(t_2)| < \epsilon + 2\mu$. In an analogous manner, we could show that $|y(t_1) - y(t_2)| < \epsilon + 2\mu$ for $|t_1 - t_2| < h$. If $t \in [t_0, T]$, then there is a t' such that $|t - t'| < h/2$ and $|y(t') - x(t')| < \mu$. Therefore

$$|y(t) - x(t)| \leq |y(t') - x(t')| + |y(t') - y(t)| + |x(t') - x(t)| < 2\epsilon + 5\mu.$$

This proves the lemma.

Proof of the sufficiency. Suppose that $\rho_{C_n}(x_n(t), x_0(t)) \to 0$ and $C_n \to 0$. Let us choose $\epsilon > 0$ such that $x_0(t)$ does not have jumps equal in absolute value to ϵ. Then there exists $\mu > 0$ such that $\mu < \epsilon/7$ and $x_0(t)$ does not have jumps whose absolute values fall in the interval

$$[\epsilon - 6\mu, \epsilon + 6\mu].$$

We denote by $t_1 < t_2 < \cdots < t_k$ all points at which the jumps of $x_0(t)$ exceed ϵ in absolute value and we denote by δ the smallest of the numbers $t_1 - t_0, t_2 - t_1, \ldots, t_k - t_{k-1}, T - t_k$. We choose n sufficiently large that $C_n < \delta/2$ and $\rho_{C_n}(x_n(t), x_0(t)) < \mu$. Then in the intervals $(t_1 - (C_n/2), t_1 + (C_n/2)$, there exist (by Lemma 1) points $t_i^{(n)}$ such that $|x_n(t_i^{(n)} - 0) - x_0(t_i - 0)| < 3\mu$ and $|x_n(t_i^{(n)} - x_0(t_i)| < 3\mu$; consequently

$$|x_n(t_i^{(n)} - 0) - x_n(t_i^{(n)})| > |x_0(t_i) - x_0(t_i - 0)| - 6\mu > \epsilon.$$

It follows from Lemma 1 that $x_n(t)$ does not have jumps exceeding 2μ in absolute value in the intervals $(t_i^{(n)} - C_n, t_i^{(n)})$, $(t_i^{(n)}, t_i^{(n)} + C_n)$. Thus in each of the intervals $(t_i - C_n/2, t_i + C_n/2)$ there is only one jump in $x_n(t)$ the absolute value of which exceeds ϵ. $x_n(t)$ cannot have a jump exceeding ϵ in absolute value at a point t' different from the points $t_i^{(n)}$. In such a case, there would be (by Lemma 1) a point t'' such that $|t' - t''| < C_n/2$ and $|x_0(t'') - x_0(t'' - 0)| > \epsilon - 6\mu$. But we know from the choice of μ that the inequality $|x_0(t'') - x_0(t'' - 0)| > \epsilon - 6\mu$ implies the inequality $|x_0(t'') - x_0(t'' - 0)| > \epsilon$; thus for some j, we would have $t'' = t_j$. Then the interval $(t_j - (C_n/2), t_j + (C_n/2)$ would have two points t' and $t_j^{(n)}$ at which the jumps of $x_n(t)$ exceed ϵ in absolute value, which is impossible. Thus

$$x_n(t) - x_n^\epsilon(t) = \sum_{t_i^{(n)} \leq t} (x_n(t_i^{(n)}) - x_n(t_i^{(n)} - 0))$$

and

$$x_0(t) - x_0^\epsilon(t) = \sum_{ti \leq t} (x_0(t_i) - x_0(t_i - 0)).$$

From Lemma 2 and the condition that $\rho_{C_n}(x_n(t), x_0(t)) \to 0$, we know that $x_n(t) \to x_0(t)$ at every point of continuity of $x_0(t)$. Therefore, $|x_n(t_i^{(n)}) - x_0(t_i)| \to |x_n(t_i^{(n)} - 0) - x_0(t_i - 0)| \to 0$ as $n \to \infty$; and since $C_n \to 0$, it follows that $t_i^{(n)} \to t_i$. Consequently $x_n(t) - x_n^\epsilon(t) \to x_0(t) - x_0^\epsilon(t)$ as $t \neq t_i$. Therefore $x_n^\epsilon(t) \to x_0^\epsilon(t)$ for all t that are points of continuity of $x_0^\epsilon(t)$ [since this is fulfilled for $x_n(t)$]. Also, we may conclude that condition (a) of Definition 1 is fulfilled for the sequence $x_n(t)$.

Let us show, in the case in which $\Delta_C(x(t)) < \epsilon$, that $\Delta_C(x^\epsilon(t)) < 2\Delta_C(x(t))$. When $t_1 < t_2 < t_3$, $t_3 - t_1 < C$,

$$\min (|x^\epsilon(t_1) - x^\epsilon(t_2)|, |x^\epsilon(t_2) - x^\epsilon(t_3)|) \leq \Delta_C(x(t)),$$

provided that there is no jump in $x(t)$ exceeding ϵ in absolute value in the interval (t_1, t_3); if a jump exceeding ϵ exists at a point $t' \in (t_1, t_2)$, then

$$\begin{aligned} \min (|x^\epsilon(t_1) - x^\epsilon(t_2)|, |x^\epsilon(t_2)| - x^\epsilon(t_3)|) &\leq |x^\epsilon(t_2) - x^\epsilon(t_3)| \\ &\leq |x(t_2) - x(t')| + |x(t_3) - x(t')| \\ &= \min (|x(t' - 0) - x(t')|, |x(t') - x(t_2)|) \\ &\quad + \min (|x(t' - 0) - x(t')|, |x(t') - x(t_3)|) \leq 2\Delta_C(x(t)); \end{aligned}$$

an analogous inequality exists in the case in which $x(t)$ has a jump exceeding ϵ in absolute value in the interval (t_2, t_3). Therefore

$$\lim_{n \to \infty} \Delta_{c_n}(x_n^\epsilon(t)) = 0.$$

Let us now examine $\inf_{\bar{t} < t < \bar{t} + C_n/2} |x_n^\epsilon(t) - x_n^\epsilon(t)|$. If there is no point t_i, $t_i^{(n)}$ in $(\bar{t}, \bar{t} + (C_n/2))$, then

$$\begin{aligned} \inf_{\bar{t} \leq t < \bar{t} + c_n/2} |x_n^\epsilon(t) - x_0^\epsilon(t)| &< \inf_{\bar{t} \leq t < \bar{t} + c_n/2} |x_n(t) - x_0(t)| \\ &\quad + \sum_{i=1}^{k} (|x_n(t_i^{(n)}) - x_0(t_i)| + |x_n(t_i^{(n)} - 0) - x_0(t_i - 0)|) \\ &\leq (6k + 1)\rho_{C_n}(x_n(t), x_0(t)). \end{aligned}$$

If there exists a point t_i with $t_i^{(n)} < t_i$ in the interval $(\bar{t}, \bar{t} + (c_n/2))$, then since $|t_i^{(n)} - t_i| < (c_n/2)$ for $t' \in (t_i, \bar{t} + (c_n/2))$, we have $|t' - t_i| < (c_n/2))$, $t_i^{(n)} - t' \mid < c_n$, which means that

$$\begin{aligned} |x_n^\epsilon(t') - x_0^\epsilon(t')| &\leq |x_n(t') - x_0(t')| + 6k\rho_{c_n}(x_n(t), x_0(t)) \\ &\leq |x_n(t') - x_n(t_i^{(n)})| + |x_0(t') - x_0(t_i)| \\ &\quad + |x_n(t_i^{(n)}) - x_0(t_i)| + 6k\rho_{c_n}(x_n(t), x_0(t)) \\ &< (6k + 5)\rho_{c_n}(x_n(t), x_0(t)). \end{aligned}$$

By using the symmetry of $t_i^{(n)}$ and t_i, we may state that at all times

$$\inf_{\bar{t} \leq t \leq \bar{t} + c_n/2} |x_n^\epsilon(t) - x_0(t)| \leq (6k + 5)\rho_{c_n}(x_n(t), x_0(t)).$$

Therefore

$$\rho_{c_n}(x_n^\epsilon(t), x_0^\epsilon(t)) \leq (6k + 7)\rho_{c_n}(x_n(t), x_0(t))$$

for $c_n < \delta/2$, $\rho_{c_n}(x_n(t), x_0(t)) < \mu$. By applying Lemma 2, we obtain the inequality

$$\sup_t |x_n^\epsilon(t) - x_0^\epsilon(t)| \leq 2\epsilon + 5(6k + 7)\rho_{c_n}(x_n(t), x_0(t)),$$

so that condition (b) of Definition 1 is satisfied. Consequently

$$x_0(t) = \mathbf{J} \lim_{n \to \infty} x_n(t).$$

This proves the theorem.

3. A limit theorem for J-continuous functionals. Suppose that on an interval $[t_0, T]$ there are defined processes $\xi_n(t)$, $n = 0, 1, 2, \ldots$, having values in $R^{(m)}$, such that for every n, the process $\xi_n(t)$ as a function of t belongs to $D_m[t_0, T]$ with probability 1. We denote by $\mathbf{F}(D_m[t_0, T])$ the minimal σ-algebra of subsets of $D_m[t_0, T]$ that contains all subsets of the form $D_m[t_0, T] \cap \mathbf{A}$, where $\mathbf{A} \in \mathbf{F}_m[t_0, T]$ (see Section 1, Chapter 1). Suppose that μ_n is a measure defined on $\mathbf{F}(D_m[t_0, T]$ such that for every cylindrical set $\mathbf{A}$,

$$\mu_n(\mathbf{A}) = \mathbf{P}\{\xi_n(t) \in \mathbf{A} \cap D_m[t_0, T]\}.$$

We denote by $F(\mathbf{J}, \mu_0)$ the set of functionals $F(x(t))$ defined on $D_m[t_0, T]$ that are measurable with respect to $\mathbf{F}(D_m[t_0, T])$ and $\mathbf{J}$ and that are continuous at all points of some set $D' \in D_m[t_0, T]$ such that $\mu_0(D') = 1$ and for which $F(x(t))$ is $\mathbf{J}$-continuous at the point $x_0(t) \in D_m[t_0, T]$ if the relationship $x_n(t) \xrightarrow{\mathbf{J}} x_0(t)$ implies the relation $F(x_n(t)) \to F(x_0(t))$.

THEOREM. *Suppose that the following conditions are fulfilled:*

(a) *the finite-dimensional distributions of the processes* $\xi_n^{(t)}$ *converge to the finite-dimensional distributions of the process* $\xi_0(t)$;

(b) *for every* $\epsilon > 0$,

$$\lim_{c\to 0} \overline{\lim_{n\to\infty}} \mathbf{P}\{\Delta_c(\xi_n(t)) > \epsilon\} = 0;$$

(c) *for every* $\epsilon > 0$,

$$\lim_{h\to 0} \overline{\lim_{n\to\infty}} \sup_{|t_1-t_2|\le h} \mathbf{P}\{|\xi_n(t_1) - \xi_n(t_2)| > \epsilon\} = 0.$$

Then for an arbitrary functional $F(x(t))$ *in* $F(\mathbf{J}, \mu_0)$, *the distribution of* $F(\xi_n(t))$ *converges to the distribution of* $F(\xi_0(t))$.

Proof. As we know from the theorem of Section 6, Chapter 1, it is possible to construct a sequence of processes $x_n(t, \omega)$ the finite-dimensional distributions of which will coincide with the finite-dimensional distributions of $\xi_n(t)$. Then the $x_n(t, \omega)$ will for every t converge in probability to the process $x_0(t, \omega)$, the finite-dimensional distributions of which will coincide with the finite-dimensional distributions of the process $\xi_0(t)$. It follows from the coincidence of the finite-dimensional distributions of $x_n(t, \omega)$ and $\xi_n(t)$ that for every t and every sequence t_k satisfying the conditions $t_k \to t$ and $t_k > t$ with probability 1, there exists a limit $x_n(t_k, \omega)$ equal to $x_n(t, \omega)$. Let us choose a countable set N that is everywhere dense on $[t_0, T]$ and let us define the processes $\tilde{x}_n(t, \omega)$ by

$$\tilde{x}_n(t, \omega) = \lim_{\substack{s\to t \\ s>t,\, s\in N}} x_n(s, \omega).$$

The processes $\tilde{x}_n(t, \omega)$ will with probability 1 be continuous from the right and stochastically equivalent to the processes $x_n(t, \omega)$. From the separability of the processes $\tilde{x}_n(t, \omega)$ and from the fact that the $\xi_n(t)$ do not with probability 1 have discontinuities of the second kind, it is easy to see that the processes $x_n(t, \omega)$, and hence the $\tilde{x}_n(t, \omega)$, do not with probability 1 have discontinuities of the second kind. Consequently the $\tilde{x}_n(t, \omega)$ as functions of t with probability 1 belong to $D_m[t_0, T]$. Since for every cylindrical set $\mathbf{A}$, $\mu_n(\mathbf{A} \cap D_m[t_0, T]) = \mathbf{P}\{\tilde{x}_n(t, \omega) \in \mathbf{A} \cap D_m[t_0, T]\}$, the measure μ_n will correspond to the process $x_n(t, \omega)$ on $\mathbf{F}(D_m[t_0, T])$. Therefore, for any functional that is measurable with respect to

$$\mathbf{F}(D_m[t_0, T]),$$

the distributions of $F(\xi_n(t))$ and $F(\tilde{x}_n(t, \omega))$ coincide. Thus, to prove the theorem it is sufficient to show that $F(\tilde{x}_n(t, \omega))$ converges in probability to

$F(\tilde{x}_0(t, \omega))$, and to prove this it is sufficient to show that from every sequence n' it is possible to choose a subsequence n'_k such that

$$F(\tilde{x}_{n'_k}(t, \omega)) \to F(\tilde{x}_0(t, \omega))$$

with probability 1. Since $F(\tilde{x}_0(t, \omega))$ is continuous at the point $\tilde{x}_0(t, \omega)$ for almost all ω, the theorem will be proved if we establish the fact that from every sequence n' it is possible to choose a subsequence n'_k such that with probability 1, $\tilde{x}_{n_k}(t, \omega) \xrightarrow{J} (\tilde{x}_0(t, \omega))$.

From the fact that $\tilde{x}_n(t, \omega) \to \tilde{x}_0(t, \omega)$ in probability for all t, it follows that for every $c > 0$,

$$\sup_{t_0 \leq t \leq T - c/2} \inf_{t \leq s < t + c/2} |\tilde{x}_n(s, \omega) - \tilde{x}_0(s, \omega)| \to 0$$

in probability. Therefore for any $\epsilon > 0$ and $c > 0$,

$$\lim_{n\to\infty} \mathbf{P}\{ \sup_{t_0 \leq t \leq T - c/2} \inf_{t \leq s < t + c/2} |\tilde{x}_n(s, \omega) - \tilde{x}_0(s, \omega)| > \epsilon\} = 0.$$

It follows from condition (b) of the theorem that for every $\epsilon > 0$,

$$\lim_{c\to 0} \overline{\lim_{n\to\infty}} \mathbf{P}\{\Delta_c(\tilde{x}_n(s, \omega)) > \epsilon\} = 0.$$

Also, since $\tilde{x}_0(s, \omega) \in D_m[t_0, T]$, with probability 1 we have

$$\lim_{c\to 0} \mathbf{P}\{\Delta_c(\tilde{x}_0(t, \omega)) > \epsilon\} = 0.$$

Therefore for every $\epsilon > 0$,

$$\lim_{c\to 0} \overline{\lim_{n\to\infty}} \mathbf{P}\{\rho_c(\tilde{x}_n(t, \omega), \tilde{x}_0(t, \omega)) > \epsilon\} = 0.$$

Suppose that $\epsilon_m \to 0$. We denote by c_m a sequence converging to zero for which

$$\overline{\lim_{u\to\infty}} \mathbf{P}\{\rho_{c_m}(x_n(t, \omega), \tilde{x}_0(t, \omega)) > \epsilon_m\} < \frac{1}{m^2}\cdot$$

Let N_m be a number such that for $n > N_m$,

$$\mathbf{P}\{\rho_{c_m}(x_n(t, \omega), x_0(t, \omega)) > \epsilon_m\} < \frac{1}{m^2}\cdot$$

Let n_m be an arbitrary sequence for which $n_m > N_m$. Then, on the basis of the Borel-Cantelli Lemma, with probability 1 there exists a number m_0 such that for $m > m_0$, it is true that $\rho_{c_m}(\tilde{x}_{n_m}(t, \omega), \tilde{x}_0(t, \omega)) \leq \epsilon_m$, that is, $\rho_{c_m}(\tilde{x}_{n_m}(t, \omega), \tilde{x}_0(t, \omega)) \to 0$ with probability 1. It follows from the theorem of Section 2 that $\tilde{x}_{n_m}(t, \omega) \xrightarrow{J} \tilde{x}_0(t, \omega)$ with probability 1. This proves the theorem.

4. Conditions for convergence of finite-dimensional distributions. In this section, we shall study the conditions for convergence of finite-dimensional distributions of the processes $\xi^n(t)$, introduced in Section 1, to finite-dimensional distributions of the process $\xi(t)$ that is a solution to Eq. (1.8) of Chapter 3.

Let $P_{n,k}(x, A)$ be a transition probability in a Markov chain $\xi_0^{(n)}, \xi_1^{(n)}, \ldots, \xi_n^{(n)}$ on the kth step; that is, for every Borel set A in $R^{(m)}$ the relation

$$\mathbf{P}\{\xi_{k+1}^{(n)} \in A/\xi_k^{(n)}\} = \mathbf{P}_{n,k}(\xi_k^{(n)}, A)$$

is fulfilled with probability 1.

Let us examine the functions $f_k^{(n)}(x, u)$ for which the relations

$$\mathbf{P}_{n,k}(x, A) = (t_{k+1}^{(n)} - t_k^{(n)}) \int_{x + f_k^{(n)}(x,u) \in A} \frac{du}{|u|^{m+1}} \tag{4.1}$$

are fulfilled for every Borel set A in $R^{(m)}$ that does not contain the point zero, and $f_k^{(n)}(x, u) = 0$ for $|u| \leq r_k^{(n)}$, where $r_k^{(n)}$ is such that

$$(t_{k+1}^{(n)} - t_k^{(n)}) \int_{|u| \geq r_k^{(n)}} \frac{du}{|u|^{m+1}} = 1.$$

(The existence of such a function is assured by the lemma of Section 4, Chapter 3.)

Suppose also that $\mu_k^{(n)}$ are random measures in $R^{(m)}$ that are mutually independent for different values of k and that have the following property: For every Borel set A in $R^{(m)}$, $\mu_k^{(n)}(A)$ takes at least the values zero and one, and

$$\mathbf{P}\{\mu_k^{(n)}(A) = 1\} = (t_{k+1}^{(n)} - t_k^{(n)}) \int_{\substack{u \in A \\ |u| \geq r_k^{(n)}}} \frac{du}{|u|^{m+1}}.$$

Let us examine the quantity $\bar{\xi}_0^{(n)}$, which does not depend on $\mu_k^{(n)}$, and let us define $\bar{\xi}_k^{(n)}$ by the relation

$$\bar{\xi}_{k+1}^{(n)} = \bar{\xi}_k^{(n)} + \int f_k^{(n)}(\bar{\xi}_k^{(n)}, u)\mu_k^{(n)}(du). \tag{4.2}$$

Clearly, $\bar{\xi}_k^{(n)}$ will be a Markov chain with the same transition probabilities as $\xi_k^{(n)}$. Therefore the common distribution of the variables $\xi_k^{(n)}$ will coincide with the common distribution of the variables $\bar{\xi}_k^{(n)}$ provided that the distribution of $\xi_0^{(n)}$ coincides with the distribution of $\bar{\xi}_0^{(n)}$. In the future, therefore, instead of the variables $\xi_k^{(n)}$ we shall consider the variables $\bar{\xi}_k^{(n)}$, assuming that the distribution of $\bar{\xi}_0^{(n)}$ coincides with the distribution of $\xi_0^{(n)}$.

We choose δ_n such that $r_k^{(n)} < \delta_n$, and examine the quadratic form

$$B_k^{(n)}(z) = (t_{k+1}^{(n)} - t_k^{(n)}) \int_{|u|\leq\delta_n} (f_k^{(n)}(x, u), z)^2 \frac{du}{|u|^{m+1}}$$
$$- \left[(t_{k+1}^{(n)} - t_k^{(n)}) \int_{|u|\leq\delta_n} (f_k^{(n)}(x, u), z) \frac{du}{|u|^{m+1}} \right]^2 \tag{4.3}$$

[where (f, z) denotes the scalar product of f and z]. Suppose that $e_1^{(n)}(k, x)$, $e_2^{(n)}(k, x), \ldots, e_m^{(n)}(k, x)$ is a canonical basis of this quadratic form and that the $\lambda_1^{(n)}(k, x) \geq \cdots \geq \lambda_m^{(n)}(k, x)$ are canonical coefficients of this quadratic form. We define

$$b_i^{(n)}(k, x) = \sqrt{\lambda_i^{(n)}(k, x)}\, e_i^{(n)}(k, x),$$

$$\omega_n^{(l)}(k, x) = \frac{1}{\lambda_i^{(n)}(k, x)} \left(b_i^{(n)}(k, x), \quad \int_{|u|\leq\delta_n} f_k^{(n)}(x, u)\bar{\mu}_k^n (du)\right), \tag{4.4}$$

where $\bar{\mu}_k^{(n)}(A) = \mu_k^{(n)}(A) - \mathbf{M}\mu_k^{(n)}(A)$. It follows from (4.4) that

$$\int_{|u|\leq\delta_n} f_k^{(n)}(x, u)\bar{\mu}_k^{(n)} (du) = \sum_{i=1}^{m} b_i^{(n)}(k, x)\omega_n^{(i)}(k, x). \tag{4.5}$$

Since

$$\mathbf{M}\left(z, \int_{|u|\leq\delta_n} f_k^{(n)}(x, u)\bar{\mu}_k^{(n)} (du)\right)^2$$
$$= \mathbf{M}\left(\int_{|u|\leq\delta_n} (z, f_k^{(n)}(x, u))\bar{\mu}_k^{(n)} (du)\right)^2$$
$$= \mathbf{M}\left[\int_{|u|\leq\delta_n} (f_k^{(n)}(x, u), z)\mu_k^{(n)} (du)\right]^2$$
$$- \left[(t_{k+1}^{(n)} - t_k^{(m)}) \int_{|u|\leq\delta_n} (f_k^{(n)}(x, u), z) \frac{du}{|u|^{m+1}}\right]^2$$
$$= \mathbf{M}\left[\int_{|u|\leq\delta_n} \int_{|v|\leq\delta_n} (f_k^{(n)}(x, u), z)(f_k^{(n)}(x, v), z)\mu_k^{(n)} (du)\mu_k^{(n)} (dv)\right]$$
$$- \left[(t_{k+1}^{(n)} - t_k^{(n)}) \int_{|u|\leq\delta_n} (f_k^{(n)}(x, u), z) \frac{du}{|u|^{m+1}}\right]^2 = B_k^{(n)}(z),$$

$\mu_k^{(n)} (du) \cdot \mu_k^{(n)} (dv) = 0$ for nonintersecting du and dv,

$$[\mu_k^{(n)} (du)]^2 = \mu_k^{(n)} (du),$$

it follows from (4.5) that

$$B_k^{(n)}(z) = \sum_{i=1}^{m} \lambda_i^{(n)}(k, x)\left(e_i^{(n)}(k, x), z\right)^2$$

$$= \sum_{i,j=1}^{m} \left(b_i^{(n)}(k, x), z\right)\left(b_j^{(n)}(k, x), z\right)\mathbf{M}\omega_n^{(i)}(k, x)\omega_n^{(j)}(k, x)$$

$$= \sum_{i,j=1}^{m} \sqrt{\lambda_i^{(n)}(k, x)\lambda_j^{(n)}(k, x)}\left(e_i^{(n)}(k, x), z\right)\left(e_j^{(n)}(k, x), z\right)$$

$$\times \mathbf{M}\omega_n^{(i)}(k, x) \cdot \omega_n^{(j)}(k, x).$$

Since this is valid for arbitrary z, we have

$$\mathbf{M}\omega_n^{(i)}(k, x)\omega_n^{(j)}(k, x) = \begin{cases} 0 & \text{for} \quad i \neq j, \\ 1 & \text{for} \quad i = j. \end{cases} \tag{4.6}$$

We note that $\omega_n^{(i)}(k, x)$ is a random variable depending only on $\bar{\mu}_k^{(n)}$. We further define

$$b_i^{(n)}(t_k^{(n)}, x) = \frac{1}{\sqrt{i_{k+1}^{(n)} - t_k^{(n)}}}\, b_i^{(u)}(k, x),$$

$$a^{(n)}(t_k^{(n)}, x) = \int f_k^{(n)}(x, u)\,\frac{du}{|u|^{m+1}},$$

$$f^{(n)}(t_k^{(n)}, x, u) = \begin{cases} 0 & \text{for} \quad |u| \leq \delta_n, \\ f_k^{(n)}(x, u) & \text{for} \quad |u| > \delta_n. \end{cases}$$

Then Eq. (4.2) can be written in the form

$$\bar{\xi}_{k+1}^{(n)} = \xi_k^{(n)} + a^{(n)}(t_k^{(n)}, \bar{\xi}_k^{(n)})(t_{k+1}^{(n)} - t_k^{(n)}) + \sum_{i=1}^{m} b_i^{(n)}(t_k^{(n)}, \bar{\xi}_k^{(n)})$$

$$\times\, \omega_n^i(t_k^{(n)}) + \int f^{(n)}(t_k^{(n)}, \bar{\xi}_k^{(n)}, u)\bar{\mu}_k^{(n)}(du), \tag{4.7}$$

where

$$\omega_n^i(t_k^{(n)}) = \omega_n^i(t_k^{(n)}, \bar{\xi}_k^{(n)})\sqrt{t_{k+1}^{(n)} - t_k^{(n)}}.$$

For the variables $\omega_n^i(t_k^{(n)})$, Lemma 1 is valid.

LEMMA 1. *We denote by* $\mathbf{F}_k^{(n)}$ *the minimal* σ*-algebra of events relative to which* $\bar{\xi}_0^{(n)}$ *and* $\mu_r^{(n)}(A)$ *are measurable for every Borel set* A *and* $r < k$. *Then*

(a) $\mathbf{M}\left(\omega_n^i(t_k^{(n)})/\mathbf{F}_k^{(n)}\right) = 0,$

(b) $\mathbf{M}\left(\omega_n^i(t_k^{(m)})\omega_n^j(t_k^{(n)})/\mathbf{F}_k^{(n)}\right) = \begin{cases} t_{k+1}^{(n)} - t_k^{(n)} & \text{for} \quad i = j, \\ 0 & \text{for} \quad i \neq j. \end{cases}$

The proof of this lemma follows from Formula (4.6) if we take into account the fact that the measure $\mu_k^{(n)}$ does not depend on the σ-algebra $\mathbf{F}_k^n$. Let us show that in the case in which $a^{(n)}(t, x)$, $b_i^{(n)}(t, x)$, and $f^{(n)}(t, x, u)$ converge in the defined sense to $a(t, x)$, $b_i(t, x)$, and $f(t, x, u)$,

$$\max_k (t_{k+1}^{(n)} - t_k^{(n)}) \to 0,$$

the finite-dimensional distributions of the process $\bar{\xi}_k^{(n)}(t)$ defined by the relation $\bar{\xi}^{(n)}(t) = \bar{\xi}_k^{(n)}$ for $t \in [t_k^{(n)}, t_{k+1}^{(n)}]$ will converge to the finite-dimensional distributions of the solution to Eq. (1.8) of Chapter 3, if the distribution of $\xi_0^{(n)}$ converges to the distribution of $\xi(t_0)$.

As a preliminary, let us establish certain auxiliary propositions.

LEMMA 2. *Suppose that the following conditions are satisfied:*

1. *For some $K > 0$,*

$$(T - t_0)|a(t, x)|^2 + \sum_{i=1}^{m} |b_j(t, x)|^2 + \int |f(t, x, u)|^2 \frac{du}{|u|^{m+1}} \leq K^2(1 + |x|^2).$$

2. *There exists an $L > 0$ such that*

$$|a(t, x) - a(t, y)|^2 + \sum_{i=1}^{m} |b_i(t, x) - b_i(t, y)|^2 + \int |f(t, x, u) - f(t, y, u)|^2 \frac{du}{|u|^{m+1}} \leq L|x - y|^2.$$

3.

$$\lim_{n\to\infty} \left[\sum_{k=0}^{n-1} \left\{\mathbf{M}|a(t_k^{(n)}, \bar{\xi}_k^{(n)}) - a^{(n)}(t_k^{(n)}, \bar{\xi}_k^{(n)})|^2 + \sum_{i=1}^{m} \mathbf{M}|b_i(t_k^{(n)}, \bar{\xi}_k^{(n)}) - b_i^{(n)}(t_k^{(n)}, \bar{\xi}_k^{(n)})|^2 + \int \mathbf{M}|f(t_k^{(n)}, \bar{\xi}_k^{(n)}, u) - f^{(n)}(t_k^{(n)}, \bar{\xi}_k^{(n)}, u)|^2 \frac{du}{|u|^{m+1}}\right\} (t_{k+1}^{(u)} - t_k^{(n)})\right] = 0.$$

We define $\bar{\eta}^{(n)}(t) = \eta_k^{(n)}$ for $t \in [t_k^{(n)}, t_{k+1}^{(n)}]$,

$$\bar{\eta}_{k+1}^{(n)} = \bar{\eta}_k^{(n)} + a(t_k^{(n)}, \bar{\eta}_k^{(n)})(t_{k+1}^{(n)} - t_k^{(n)}) + \sum_{i=1}^{m} b(t_k^{(n)}, \bar{\eta}_k^{(n)})\omega_n^i(t_k^{(n)}) + \int f(t_k^{(n)}, \bar{\eta}_k^{(n)}, u)\bar{\mu}_k^{(n)}(du), \tag{4.8}$$

and $\bar{\eta}_0^{(n)} = \bar{\xi}_0^{(n)}$. *Under these conditions, if* $\sup_n \mathbf{M}|\bar{\xi}_0^{(n)}|^2 < \infty$, *then for all* t, *we have* $\mathbf{M}|\bar{\eta}^{(n)}(t) - \bar{\xi}^{(n)}(t)|^2 \to 0$ *for* $n \to \infty$.

Proof. It follows from (4.7) and (4.8) that

$$\begin{aligned}\mathbf{M}|\bar{\xi}_{k+1}^{(n)} - \bar{\eta}_{k+1}^{(n)}|^2 &\le \mathbf{M}|\bar{\xi}_k^{(n)} - \eta_k^{(n)}|^2 \\ &+ 2\mathbf{M}(\bar{\xi}_k^{(n)} - \bar{\eta}_k^{(n)}, a^{(n)}(t_k^{(n)}, \bar{\xi}_k^{(n)}) - a(t_k^{(n)}, \bar{\eta}_k^{(n)}))(t_{k+1}^{(n)} - t_k^{(n)}) \\ &+ (m+2)\Bigg[\sum_{i=1}^m \mathbf{M}|b_i(t_k^{(n)}, \bar{\xi}_k^{(n)}) - b_i(t_k^{(n)}, \bar{\eta}_k^{(n)})|^2(t_{k+1}^{(n)} - t_k^{(n)}) \\ &+ (t_{k+1}^{(n)} - t_k^{(n)})\int \mathbf{M}|f^{(n)}(t_k^{(n)}, \bar{\xi}_k^{(n)}, u) - f(t_k^{(n)}, \bar{\eta}_k^{(n)}, u)|^2 \frac{du}{|u|^{m+1}} \\ &+ \mathbf{M}|a^{(n)}(t_k^{(n)}, \bar{\xi}_k^{(n)}) - a(t_k^{(n)}, \bar{\eta}_k^{(n)})|^2(t_{k+1}^{(n)} - t_k^{(n)})^2\Bigg]. \end{aligned} \tag{4.9}$$

Suppose that $\max_k (t_{k+1}^{(n)} - t_k^{(n)}) < 1$, $H = 2(m+3)L^2 + 1$,

$$\begin{aligned}\alpha_n^{(k)} = \mathbf{M}|a(t_k^{(n)}, \bar{\xi}_k^{(n)}) - a^{(n)}(t_k^{(n)}, \bar{\xi}_k^{(n)})|^2 \\ + \sum_{i=1}^m \mathbf{M}|b_i(t_k^{(n)}, \bar{\xi}_k^{(n)}) - b_i^{(n)}(t_k^{(n)}, \bar{\xi}_k^{(n)})|^2 \\ + \int \mathbf{M}|f(t_k^{(n)}, \bar{\xi}_k^{(n)}, u) - f^{(n)}(t_k^{(n)}, \bar{\xi}_k^{(n)}, u)|^2 \frac{du}{|u|^{m+1}}.\end{aligned}$$

Then, from Condition 2 and formula (4.9), we obtain the inequality

$$\begin{aligned}\mathbf{M}|\bar{\xi}_{k+1}^{(n)} - \bar{\eta}_{k+1}^{(n)}|^2 &\le \mathbf{M}|\bar{\xi}_k^{(n)} - \bar{\eta}_k^{(n)}|^2(1 + H(t_{k+1}^{(n)} - t_k^{(n)})) \\ &\quad + H\alpha_n^{(k)}(t_{k+1}^{(n)} - t_k^{(n)}) \\ &\le \mathbf{M}|\bar{\xi}_k^{(n)} - \bar{\eta}_k^{(n)}|^2 e^{H(t_{k+1}^{(n)} - t_k^{(n)})} \\ &\quad + H\alpha_n^{(k)}(t_{k+1}^{(n)} - t_k^{(n)}).\end{aligned}$$

Since $\mathbf{M}|\bar{\xi}_0^{(n)} - \bar{\eta}_0^{(n)}|^2 = 0$, we can establish the following inequality by induction:

$$\mathbf{M}|\bar{\xi}_{k+1}^{(n)} - \bar{\eta}_{k+1}^{(n)}|^2 \le H \sum_{i=0}^k \alpha_n^{(i)} e^{H(t_{i+1}^{(n)} - t_0^{(n)})}(t_{i+1}^{(n)} - t_i^{(n)}).$$

From Condition 3, $\sum_{i=0}^n \alpha_n^{(i)}(t_{i+1}^{(n)} - t_i^{(n)}) \to 0$. Thus, from the preceding inequality, we obtain the proof of the lemma.

LEMMA 3. *Suppose that there exists a number* K *independent of* n *such that*

$$|a^{(n)}(t, x)|^2 + \sum_{i=1}^m |b_i^{(n)}(t, x)|^2 + \int |f^{(n)}(t, x, u)|^2 \frac{du}{|u|^{m+1}} \le K(1 + |x|^2) \tag{4.10}$$

and that $\bar{\xi}_k^{(n)}$ *is defined by the relations (4.7). Then if* $\sup_n \mathbf{M}|\bar{\xi}_0^{(n)}|^2 < \infty$, *there exist* H_1 *and* H_2 *such that for all n,*

$$\mathbf{M}|\bar{\xi}_i^{(n)}|^2 \leq H_1, \qquad \mathbf{M}|\bar{\xi}_i^{(n)} - \bar{\xi}_j^{(n)}|^2 \leq H_2|t_i^{(n)} - t_j^{(n)}|.$$

Proof. From condition (4.10), we can, as in the preceding lemma, derive the inequality

$$\mathbf{M}|\bar{\xi}_{k+1}^{(n)}|^2 \leq \mathbf{M}|\bar{\xi}_k^{(n)}|^2(1 + H(t_{k+1}^{(n)} - t_k^{(n)})) \leq |\mathbf{M}\bar{\xi}_k^{(n)}|^2 e^{H(t_{k+1}^{(n)} - t_k^{(n)})},$$

where $H = K(m + 3) + 1$. Consequently

$$\mathbf{M}|\bar{\xi}_{k+1}^{(n)}|^2 \leq \mathbf{M}|\bar{\xi}_0^{(n)}|^2 e^{H(T-t_0)} \tag{4.11}$$

For $k < j$,

$$\mathbf{M}|\bar{\xi}_j^{(n)} - \bar{\xi}_k^{(n)}|^2 \leq \mathbf{M}\left|\sum_{r=k}^{f-1}\left[a^{(n)}(t_r^{(n)}, \bar{\xi}_r^{(n)})(t_{r+1}^{(n)} - t_r^{(n)}) + \sum_{i=1}^{m} b_i^{(n)}(t_r^{(n)}, \bar{\xi}_r^{(n)})(\bar{\omega}_r^{(n)}(t_r^{(n)})) + \int f^{(n)}(t_r^{(n)}, \bar{\xi}_r^{(n)}, u)\bar{\mu}_r^{(n)}(du)\right]\right|^2.$$

By using Lemma 1 and inequalities (4.10) and (4.11), we find that for $|t_j - t_k| < 1$,

$$\begin{aligned}
\mathbf{M}|\bar{\xi}_j^{(n)} - \bar{\xi}_k^{(n)}|^2 &\leq (m + 2)\left\{\mathbf{M}\left|\sum_{r=k}^{j-1} a^{(n)}(t_r^{(n)}, \bar{\xi}_r^{(n)})(t_{r+1}^{(n)} - t_r^{(n)})\right|^2 \right.\\
&\quad + \sum_{i=1}^{m}\mathbf{M}\left|\sum_{r=k}^{j-1} b_i^{(n)}(t_r^{(n)}, \bar{\xi}_r^{(n)})\bar{\omega}_i^{(n)}(t_r^{(n)})\right|^2 \\
&\quad \left. + \mathbf{M}\left|\sum_{r=k}^{j-1}\int f^{(n)}(t_r^{(n)}, \bar{\xi}_r^{(n)}, u)\bar{\mu}_r^{(n)}(du)\right|^2\right\} \\
&\leq (m + 2)\sum_{r=k}^{j-1}\left(\mathbf{M}|a^{(n)}(t_r^{(n)}, \bar{\xi}_r^{(n)})|^2 \right.\\
&\quad + \sum_{i=1}^{m}\mathbf{M}|b_i^{(n)}(t_r^{(n)}, \bar{\xi}_r^{(n)})|^2 \\
&\quad \left. + \int \mathbf{M}|f^{(n)}(t_r^{(n)}, \bar{\xi}_r^{(n)}, u)|^2 \frac{du}{|u|^{m+1}}\right)(t_{r+1}^{(n)} - t_r^{(n)}) \\
&\leq (m + 2)K\sum_{r=k}^{j-1}(\mathbf{M}|\bar{\xi}_r^{(n)}|^2 + 1)(t_{r+1}^{(n)} - t_r^{(n)}) \\
&\leq (m + 2)K[\mathbf{M}|\bar{\xi}_0^{(n)}|^2 He^{H(T-t_0)} + 1](t_j^{(n)} - t_k^{(n)}).
\end{aligned} \tag{4.12}$$

This proves the lemma.

LEMMA 4. *Suppose that*

$$\max_k (t_{k+1}^{(n)} - t_k^{(n)}) \to 0, \qquad \delta_n \to 0, \qquad \frac{\max_k (t_{k+1}^{(n)} - t_k^{(n)})}{\delta_n^2} \to 0$$

and

$$\sup_{i,k} |w_i^{(n)}(t_k)| \leq \epsilon_n, \quad \text{where} \quad \epsilon_n \to 0.$$

Define

$$w_i^{(n)}(t) = \sum_{t_{k-1}^{(n)} \leq t} w_i^{(n)}(t_k^{(n)}), \qquad i = 1, 2, \ldots, m,$$

$$\zeta^{(n)}(t) = \sum_{t_{k-1}^{(n)} \leq t} \left(\int_{\delta_n < |u| \leq 1} u\bar{\mu}_k^{(n)}(du) + \int_{|u|>1} u\mu_k^{(n)}(du) \right).$$

Suppose further that $w_1(t), w_2(t), \ldots, w_m(t)$ are mutually independent Brownian processes such that $w(t_0) = 0$ and

$$\zeta(t) = \int_{t_0}^t \int_{|u| \leq 1} uq\,(ds \times du) + \int_{t_0}^t \int_{|u|>1} up\,(ds \times du),$$

where p and q are Poisson measures that are independent of $w_1(t), \ldots, w_m(t)$ with independent values, as defined in Section 4, Chapter 2. Then the joint distributions of the processes

$$w_1^{(n)}(t), w_2^{(n)}(t), \ldots, w_m^{(n)}(t), \zeta^{(n)}(t)$$

will converge to the joint distributions of the processes

$$w_1(t), w_2(t), \ldots, w_m(t), \zeta(t).$$

Proof. To prove the lemma, it is sufficient to show that no matter what the value of $C > 0$ or of the real numbers $\lambda_j^{(n)}(k)$, $j = 1, \ldots, m$, $k = 0, 1, \ldots, n$, satisfying the inequality $|\lambda_j^{(n)}(k)| \leq C$ may be, or what the vectors $z_k^{(n)}$ for which $|z_k^{(n)}| \leq C$ may be, the relation

$$\lim_{n\to\infty} \Bigg[\mathbf{M} \exp \Bigg\{ i \sum_{k=0}^n \Bigg(\sum_{j=1}^m \lambda_j^{(n)}(k) \omega_j^{(n)}(t_k^{(n)}) + \int_{\delta_n < |u| \leq 1} (z_k^{(n)}, u)\bar{\mu}_k^{(n)}(du) + \int_{|u|>1} (z_k^{(n)}, u)\mu_k^{(n)}(du) \Bigg) \Bigg\}$$
$$- \mathbf{M} \exp \Bigg\{ i \sum_{k=0}^n \Bigg(\sum_{j=1}^m \lambda_j^{(n)}(k) [\omega_j(t_{k+1}^{(n)}) - \omega_j(t_k^{(n)})] + (z_k^{(n)}, \zeta(t_{k+1}^{(n)}) - \zeta(t_k^{(n)})) \Bigg) \Bigg\} \Bigg] = 0 \qquad (4.13)$$

holds. To derive Formula (4.13), we need to introduce the following notations:

$$\varphi_k^{(n)} = \sum_{j=1}^{m} \lambda_j^{(n)}(k)\omega_j^{(n)}(t_k^{(n)}).$$

$$\psi_k^{(n)} = \int_{\delta_n<|u|\leq 1} (z_k^{(n)}, u)\bar{\mu}_k^{(n)}(du) + \int_{|u|>1} (u, z_k^{(n)})\mu_k^{(n)}(du),$$

$$\chi_k^{(n)} = \mu_k^{(n)}(\{|u| > \delta_n\}),$$

$$\bar{\chi}_k^{(n)} = \mu_k^{(n)}(\{|u| \leq \delta_n\})$$

(where $\{|u| > \delta_n\}$ and $\{|u| \leq \delta_n\}$ are sets of those u for which the inequalities shown within the parentheses are true). Suppose that $\mathbf{F}_k^{(n)}$ is the σ-algebra defined in Lemma 1. The variables $\psi_k^{(n)}$, $\chi_k^{(n)}$, and $\bar{\chi}_k^{(n)}$ do not depend on $\mathbf{F}_k^{(n)}$. We shall denote the conditional mathematical expectation with respect to $\mathbf{F}_k^{(n)}$ by M_k. It follows from the definition of $\omega_j^{(n)}(t_k^{(n)})$ that for $\chi_k^{(n)} = 1$, $\varphi_k^{(n)}$ is a function only of $\bar{\xi}_k^{(n)}$. We denote the value of $\varphi_k^{(n)}$ for $\chi_k^{(n)} = 1$ by $\alpha_k^{(n)}$. Then

$$\alpha_k^{(n)} = \frac{\mathbf{M}_k\varphi_k^{(n)}\chi_k^{(n)}}{\mathbf{M}\chi_k^{(n)}}.$$

For $\bar{\chi}_k^{(n)} = 1$,

$$\psi_k^{(n)} = -(t_{k+1}^{(n)} - t_k^{(n)})\int_{\delta_n<|u|\leq 1} (u, z_k^{(n)}) \frac{du}{|u|^{m+1}},$$

and we shall denote this quantity by $\beta_k^{(n)}$. Then

$$\mathbf{M}_k(e^{i\psi_k^{(n)}}\varphi_k^{(n)}) = \alpha_k^{(n)}\mathbf{M}_k\chi_k^{(n)}e^{i\psi_k^{(n)}} + e^{i\beta_k^{(n)}}\mathbf{M}_k\bar{\chi}_k^{(n)}\varphi_k^{(n)}. \tag{4.14}$$

Since

$$\mathbf{M}_k\varphi_k^{(n)} = \alpha_k^{(n)}\mathbf{M}x_k^{(n)} + \mathbf{M}_k\bar{\chi}_k^{(n)}\varphi_k^{(n)} = 0,$$

we have

$$\begin{aligned} 0 = \mathbf{M}_k(\chi_k^{(n)}e^{i\psi_k^{(n)}})\alpha_k^{(n)}\mathbf{M}\chi_k^{(n)} &+ \mathbf{M}_k(\chi_k^{(n)}e^{i\psi_k^{(n)}})\mathbf{M}_k(\bar{\chi}_k^{(n)}\varphi_k^{(n)}) \\ &+ e^{i\beta_k^{(n)}}\mathbf{M}\bar{\chi}_k^{(n)} \cdot \mathbf{M}_k\bar{\chi}_k^{(n)}\varphi_k^{(n)} \\ &+ \alpha_k^{(n)}e^{i\beta_k^{(n)}}\mathbf{M}\bar{\chi}_k^{(n)} \cdot \mathbf{M}\chi_k^{(n)}. \end{aligned}$$

Subtracting the last inequality from (4.14), we obtain

$$\begin{aligned} \mathbf{M}_k(e^{i\psi_k^{(n)}}\varphi_k^{(n)}) &= \alpha_k^{(n)}\mathbf{M}\chi_k^{(n)} \cdot \mathbf{M}_k(\chi_k^{(n)}e^{i\psi_k^{(n)}}) \\ &\quad + \mathbf{M}\chi_k^{(n)} \cdot e^{i\beta_k^{(n)}} \cdot \mathbf{M}_k(\bar{\chi}_k^{(n)}\varphi_k^{(n)}) - \mathbf{M}_k\chi_k^{(n)}e^{i\psi_k^{(n)}} \cdot \mathbf{M}_k\bar{\chi}_k^{(n)}\varphi_k^{(n)} \\ &\quad - \alpha_k^{(n)} \cdot e^{i\beta_k^{(n)}}\mathbf{M}\bar{\chi}_k^{(n)} \cdot \mathbf{M}\chi_k^{(n)} \\ &= -\mathbf{M}_k\big(\chi_k^{(n)}(e^{i\psi_k^{(n)}} - e^{i\beta_k^{(n)}})\big)\mathbf{M}_k\bar{\chi}_k^{(n)}(\varphi_k^{(n)} - \alpha_k^{(n)}). \end{aligned}$$

Remembering that

$$\alpha_k^{(n)} = \frac{\mathbf{M}_k \chi_k^{(n)} \varphi_k^{(n)}}{\mathbf{M}\chi_k^{(n)}} = -\frac{\mathbf{M}_k \bar{\chi}_k^{(n)} \varphi_k^{(n)}}{\mathbf{M}\chi_k^{(n)}}$$

$(\mathbf{M}_k(\bar{\chi}_k^{(n)} + \bar{\chi}_k^{(n)})\varphi_k^{(n)} = 0)$, from the last relation we see that

$$\mathbf{M}_k e^{i\psi_k^{(n)}} \varphi_k^{(n)} = -\mathbf{M}\chi_k^{(n)}(e^{i\psi_k^{(n)}} - e^{i\beta_k^{(n)}})\mathbf{M}_k \bar{\chi}_k^{(n)} \varphi_k^{(n)} \left(1 + \frac{\mathbf{M}\bar{\chi}_k^{(n)}}{\mathbf{M}\chi_k^{(n)}}\right)$$

$$= \frac{\mathbf{M}_k \chi_k^{(n)} \varphi_k^{(n)}}{\mathbf{M}\chi_k^{(n)}} \mathbf{M}_k\big(\chi_k^{(n)}(e^{i\psi_k^{(n)}} - e^{i\beta_k^{(n)}})\big).$$

Applying Cauchy's inequality and Lemma 1, we obtain

$$\left|\frac{\mathbf{M}_k \chi_k^{(n)} \varphi_k^{(n)}}{\mathbf{M}\chi_k^{(n)}}\right| \le \sqrt{\frac{\mathbf{M}_k |\varphi_k^{(n)}|^2}{\mathbf{M}\chi_k^{(n)}}}$$

$$= \sqrt{\sum_{j=1}^{m} |\lambda_j^{(n)}(k)|^2 (t_{k+1}^{(n)} - t_k^{(n)})} \left(\sqrt{(t_{k+1}^{(n)} - t_k^{(n)}) \int_{|u|>\delta_n} \frac{du}{|u|^{m+1}}}\right)^{-1} \cdot$$

Changing to polar coordinates in the integral, we see that for some C_1,

$$\left|\frac{\mathbf{M}_k \chi_k^{(n)} \varphi_k^{(n)}}{\mathbf{M}\chi_k^{(n)}}\right| \le C_1 \sqrt{\delta_n}.$$

We also see that

$$|\mathbf{M}_k \chi_k^{(n)}(e^{i\psi_k^{(n)}} - e^{i\beta_k^{(n)}})|$$

$$= |\mathbf{M}_k \chi_k^{(n)} \Bigg(\exp\left\{i\int_{\delta_n<|u|<\infty} (u, z_k^{(n)}) \mu_k^{(n)}(du) - i\int_{\delta_n<|u|\le 1} (u, z_k^{(n)}) \frac{du}{|u|^{m+1}} (t_{k+1}^{(n)} - t_k^{(n)})\right\}$$

$$- \exp\left\{- i\int_{\delta_n<|u|\le 1} (u, z_k^{(n)}) \frac{du}{|u|^{m+1}} (t_{k+1}^{(n)} - t_k^{(n)})\right\}$$

$$= \left|\mathbf{M}_k \chi_k^{(n)} \left(\exp\left\{i\int_{\delta_n<|u|<\infty} (u, z_k^{(n)}) \mu_k^{(n)}(du)\right\} - 1\right)\right|$$

$$= (t_{k+1}^{(n)} - t_k^{(n)}) \left|\int_{\delta_n<|u|<\infty} (e^{i(u, z_k^{(n)})} - 1) \frac{du}{|u|^{m+1}}\right| \le (t_{k+1}^{(n)} - t_k^{(n)})$$

$$\times \left[\int_{\delta_n<|u|\le 1} (u, z_k^{(n)}) \frac{du}{|u|^{m+1}} + 2\int_{|u|>1} \frac{du}{|u|^{m+1}}\right] \cdot$$

By changing to polar coordinates, we see that for some $H > 0$,

$$\int_{\delta_n<|u|\le 1} |u| \frac{du}{|u|^{m+1}} \le H \int_{\delta_n}^{1} \frac{dr}{r} = H \ln \frac{1}{\delta_n} \cdot$$

Therefore there exists a constant C_2 such that

$$|\mathbf{M}_k(\chi_k^{(n)}(e^{i\psi_k^{(n)}} - e^{i\beta_k^{(n)}}))| \leq C_2 \ln (t_{k+1}^{(n)} - t_k^{(n)}) \frac{1}{\delta_n}.$$

Thus

$$|\mathbf{M}_k e^{i\psi_k^{(n)}} \varphi_k^{(n)}| = O\left(\ln \frac{1}{\delta_n} \sqrt{\delta_n}\right) (t_{k+1}^{(n)} - t_k^{(n)}).$$

In an analogous fashion, we can show that

$$|\mathbf{M}_k e^{i\psi_k^{(n)}} [(\varphi_k^{(n)})^2 - \mathbf{M}_k(\varphi_k^{(n)})^2]| = O\left(\ln \frac{1}{\delta_n} \sqrt{\delta_n}\right) (t_{k+1}^{(n)} - t_k^{(n)}).$$

Finally, we note that

$$\begin{aligned}|\mathbf{M}_k e^{i\psi_k^{(n)}} (\varphi_k^{(n)})^3| \leq \mathbf{M}_k|\varphi_k^{(n)}|^3 &\leq \sup |\varphi_k^{(n)}| \cdot \mathbf{M}_k|\varphi_k^{(n)}|^2 \\ &= O(\epsilon_n)(t_{k+1}^{(n)} - t_k^{(n)}).\end{aligned}$$

By using all the evaluations that we have just made, we see that

$$\begin{aligned}
&\mathbf{M}_k e^{i\psi_k^{(n)}} \cdot e^{i\varphi_k^{(n)}} \\
&\quad = \mathbf{M}_k e^{i\psi_k^{(n)}} (1 + i\varphi_k^{(n)} - \tfrac{1}{2}(\varphi_k^{(n)})^2 + O(|\varphi_k^{(n)}|^3)) \\
&\quad = \mathbf{M}_k e^{i\psi_k^{(n)}} (1 - \tfrac{1}{2}\mathbf{M}_k(\varphi_k^{(n)})^2) + O\left(\epsilon_n + \sqrt{\delta_n} \ln \frac{1}{\delta_n}\right) (t_{k+1}^{(n)} - t_k^{(n)}) \\
&\quad = \mathbf{M}_k e^{i\psi_k^{(n)}} \left(1 - \tfrac{1}{2} \sum_{j=1}^{m} (\lambda_j^{(n)}(k))^2 (t_{k+1}^{(n)} - t_k^{(n)})\right) \\
&\qquad + O\left(\epsilon_n + \sqrt{\delta_n} \ln \frac{1}{\delta_n}\right) \times (t_{k+1}^{(n)} - t_k^{(n)}) \\
&\quad = \mathbf{M} e^{i\psi_k^{(n)}} \exp\left\{- \tfrac{1}{2} \sum_{j=1}^{m} (\lambda_j^{(n)}(k))^2 (t_{k+1}^{(n)} - t_k^{(n)})\right\} \\
&\qquad + O\left(\epsilon_n + \sqrt{\delta_n} \ln \frac{1}{\delta_n} + \max_k (t_{k+1}^{(n)} - t_k^{(n)})\right) (t_{k+1}^{(n)} - t_k^{(n)}).
\end{aligned}$$

Let us calculate $\mathbf{M} e^{i\psi_k^{(n)}}$:

$$\begin{aligned}
\mathbf{M} e^{i\psi_k^{(n)}} &= \exp\left\{- i(t_{k+1}^{(n)} - t_k^{(n)}) \int_{\delta_n < |u| \leq 1} (u, z_k^{(n)}) \frac{du}{|u|^{m+1}}\right\} \\
&\qquad \times \mathbf{M} \exp\left\{i \int_{\delta_n < |u| < \infty} (u, z_k^{(n)}) \mu_k^{(n)} (du)\right\} \\
&= \exp\left\{- i(t_{k+1}^{(n)} - t_k^{(n)}) \int_{\delta_n < |u| \leq 1} (u, z_k^{(n)}) \frac{du}{|u|^{m+1}}\right\} \\
&\qquad \times \left(1 + (t_{k+1}^{(n)} - t_k^{(n)}) \int_{\delta_n < |u| < \infty} (e^{i(u, z_k^{(n)})} - 1) \frac{du}{|u|^{m+1}}\right).
\end{aligned}$$

Since

$$\int_{\delta_n<|u|<\infty} \frac{du}{|u|^{m+1}} = O\left(\frac{1}{\delta_n}\right),$$

we have

$$1 + (t_{k+1}^{(n)} - t_k^{(n)}) \int_{\delta_n<|u|} (e^{i(u,z_k^{(n)})} - 1) \frac{du}{|u|^{m+1}}$$

$$= \exp\left\{(t_{k+1}^{(n)} - t_k^{(n)}) \int_{\delta_n<|u|<\infty} (e^{i(u,z_k^{(n)})} - 1) \frac{du}{|u|^{m+1}}\right\}$$

$$+ O\left(\frac{\max_k (t_{k+1}^{(n)} - t_k^{(n)})}{\delta_n^2}\right) \times (t_{k+1}^{(n)} - t_k^{(n)}).$$

Therefore

$$\mathbf{M}e^{i\psi_k^{(n)}} = \exp\left\{(t_{k+1}^{(n)} - t_k^{(n)})\right.$$

$$\times \left[\int_{\delta_n<|u|\le 1} (e^{i(z_k^{(n)},u)} - 1 - i(z_k^{(n)}, u)) \frac{du}{|u|^{m+1}}\right.$$

$$\left.\left. + \int_{|u|>1} (e^{i(z_k^{(n)},u)} - 1) \frac{du}{|u|^{m+1}}\right]\right\}$$

$$+ O\left(\frac{\max_k (t_{k+1}^{(n)} - t_k^{(n)})}{\delta_n^2}\right) (t_{k+1}^{(n)} - t_k^{(n)})$$

$$= \exp\left\{(t_{k+1}^{(n)} - t_k^{(n)}) \left[\int_{0<|u|\le 1} (e^{i(z_k^{(n)},u)} - 1 - i(z_k^{(n)}, u)) \frac{du}{|u|^{m+1}}\right.\right.$$

$$\left.\left. + \int_{|u|>1} (e^{i(z_k^{(n)},u)} - 1) \frac{du}{|u|^{m+1}}\right]\right\}$$

$$+ O\left(\delta_n + \frac{\max_k (t_{k+1}^{(n)} - t_k^{(n)})}{\delta_k^2}\right) (t_{k+1}^{(n)} - t_k^{(n)}),$$

because

$$\int_{|u|\le\delta_n} |e^{i(z_k^{(n)},u)} - 1 - i(z_k^{(n)}, u)| \frac{du}{|u|^{m+1}}$$

$$= O\left(\int_{|u|\le\delta_n} \frac{du}{|u|^{m-1}}\right) = O(\delta_n).$$

Thus

$$\mathbf{M}_k e^{i\psi_k^{(n)} + i\varphi_k^{(n)}} = \exp\left\{(t_{k+1}^{(n)} - t_k^{(n)}) \left[-\frac{1}{2} \sum_{j=1}^{m} (\lambda_j^{(n)}(k))^2 + \right.\right.$$

(cont.)

$$+ \int_{0<|u|\leq 1} \left(e^{i(z_k^{(n)},u)} - 1 - i(z_k^{(n)}, u)\right) \frac{du}{|u|^{m+1}}$$

$$\left. \left. + \int_{|u|>1} \left(e^{i(z_k^{(n)},u)} - 1\right) \frac{du}{|u|^{m+1}} \right] \right\}$$

$$+ O\left(\epsilon_n + \delta_n + \sqrt{\delta_n} \ln \frac{1}{\delta_n}\right.$$

$$\left. + \frac{\max_k (t_{k+1}^{(n)} - t_k^{(n)})}{\delta_n^2} + \max_k (t_{k+1}^{(n)} - t_k^{(n)})\right) (t_{k+1}^{(n)} - t_k^{(n)}). \qquad (4.15)$$

Remembering the relation

$$\mathbf{M} \exp\left[i \sum_{k=0}^{r} (\psi_k^{(n)} + \varphi_k^{(n)})\right] = \mathbf{M} \exp\left[i \sum_{k=0}^{r-1} (\psi_k^{(n)} + \varphi_k^{(n)})\right] \mathbf{M}_r e^{i(\varphi_r^{(n)} + \psi_r^{(n)})},$$

we easily derive from (4.15) the inequality

$$\left| \mathbf{M} \exp\left[i \sum_{k=0}^{n} (\psi_k^{(n)} + \varphi_k^{(n)})\right] - \exp\left\{\sum_{k=0}^{n} (t_{k+1}^{(n)} - t_k^{(n)}) \left[-\frac{1}{2} \sum_{j=1}^{m} (\lambda_j^{(n)}(k))^2 \right.\right.\right.$$

$$+ \int_{0<|u|\leq 1} \left(e^{i(u,z_k^{(n)})} - 1 - i(u, z_k^{(n)})\right) \frac{du}{|u|^{m+1}}$$

$$\left.\left.\left. + \int_{|u|>1} \left(e^{i(u,z_k^{(n)})} - 1\right) \frac{du}{|u|^{m+1}}\right]\right\}\right|$$

$$\leq O\left(\epsilon_n + \sqrt{\delta_n} \ln \frac{1}{\delta_n} + \frac{\max_k (t_{k+1}^{(n)} - t_k^{(n)})}{\delta_n^2}\right) (T - t_0). \quad (4.16)$$

Since by the condition of the lemma the right-hand side of (4.16) approaches zero, to prove (4.13), and hence the lemma, it remains to note that

$$\mathbf{M} \exp\left\{i \sum_{k=0}^{n} \left(\sum_{j=1}^{m} \lambda_j^{(n)}(k)[w_j(t_{k+1}^{(n)}) - w_j(t_k^{(n)})] + \left(z_k^{(n)}, \zeta(t_{k+1}^{(n)} - \zeta(t_k^{(n)})\right)\right)\right\}$$

$$= \exp\left\{\sum_{k=0}^{n} (t_{k+1}^{(n)} - t_k^{(n)}) \left[-\frac{1}{2} \sum_{j=1}^{m} (\lambda_j^{(n)}(k))^2\right.\right.$$

$$+ \int_{0<|u|\leq 1} \left(e^{i(z_k^{(n)},u)} - 1 - i(z_k^{(n)}, u)\right) \frac{du}{|u|^{m+1}}$$

$$\left.\left. + \int_{|u|>1} \left(e^{i(z_k^{(n)},u)} - 1\right) \frac{du}{|u|^{m+1}}\right]\right\}.$$

LEMMA 5. *If the conditions of Lemma 4 are satisfied and $\zeta^{(n)}(t)$ is the process defined in Lemma 4, then for every $\epsilon > 0$,*

$$\lim_{C\to 0} \varlimsup_{n\to\infty} \mathbf{P}\,\{\Delta_C(\zeta^{(n)}(t)) > \epsilon\} = 0$$

(the definition of the quantity $\Delta_C(x(t))$ being given in Section 2).

Proof. We note that $\zeta^{(n)}(t)$ is a process with independent increments and hence is a Markov process. Also,

$$\mathbf{P}\{|\zeta^{(n)}(t+h) - \zeta^{(n)}(t)| > \epsilon/\zeta^{(n)}(t)\} = \mathbf{P}\{|\zeta^{(n)}(t+h) - \zeta^{(n)}(t)| > \epsilon\}$$

$$= \mathbf{P}\left\{\left|\sum_{t_k^{(n)}\in(t,t+h)}\left(\int_{\delta_n<|u|\le 1} u\bar{\mu}_k^{(n)}\,(du) + \int_{|u|>1} u\mu_k^{(n)}\,(du)\right)\right| > \epsilon\right\}$$

$$\le \frac{1}{\epsilon^2}\mathbf{M}\left|\sum_{t_k^{(n)}\in(t,t+h)}\int_{\delta_n<|u|\le 1} u\bar{\mu}_k^{(n)}\,(du)\right|^2$$

$$+ \sum_{t_k^{(n)}\in(t,t+h)} \mathbf{P}\{\mu_k^{(n)}(\{|u| > 1\}) = 1\}$$

$$< \sum_{t_k^{(n)}\in(t,t+h)} (t_{k+1}^{(n)} - t_k^{(n)})\left[\frac{1}{\epsilon^2}\int_{0<|u|\le 1}\frac{du}{|u|^{m+1}} + \int_{|u|>1}\frac{du}{|u|^{m+1}}\right].$$

Thus for the process $\zeta^{(n)}(t)$ a relation analogous to (5.4) in Section 5 of this chapter is satisfied. Therefore for this process the relation (5.1) of Section 5 of this chapter will be valid, and hence we have proof of the lemma. (The conclusions of Section 5 referred to do not depend on the conclusions of this section.)

LEMMA 6. *Suppose that Conditions 1 and 2 of Lemma 2 and the conditions of Lemma 4 are fulfilled. Suppose also that $a(t, x)$, $b_j(t, x)$, and $f(t, x, u)$ satisfy the conditions of the existence theorem of Section 3, Chapter 3, and that $f(t, x, u)$ is continuous with respect to all t and x and almost all u. We define $\eta_n(t)$ as in Lemma 2. Then if the distribution of $\eta_n(t_0)$ converges to the distribution of $\xi(t_0)$, and if $\sup_n \mathbf{M}|\eta_n(t_0)|^2 < \infty$, the finite-dimensional distributions of the processes $\eta_n(t)$ will converge to the finite-dimensional distributions of the process $\xi(t)$ that is a solution to Eq. (1.8) of Chapter 3.*

Proof. Suppose that $\zeta^{(n)}(t)$ and $w_j^{(n)}(t)$ are defined as in Lemma 4. On the basis of Corollary 2 to the theorem of Section 6 of Chapter 1, we can choose a sequence of natural numbers n' and construct processes $\tilde{\eta}_{n'}(t)$,

$\tilde{w}_1^{(n')}(t), \ldots, \tilde{w}_m^{(n')}(t), \tilde{\zeta}^{(n')}(t)$ the joint distributions of which coincide with the joint distributions of the processes $\eta_{n'}(t), w_1^{(n')}(t), \ldots, w_m^{(n')}(t), \zeta^{(n')}(t)$; also, $\tilde{\eta}_{n'}(t), \tilde{w}_1^{(n')}(t), \ldots, \tilde{w}_m^{(n')}(t), \tilde{\zeta}^{(n')}(t)$ converge in probability as $n' \to \infty$ to certain processes $\tilde{\eta}(t), \tilde{w}_1(t), \ldots, \tilde{w}_m(t), \tilde{\zeta}(t)$. It is obvious that the joint distributions of the processes $\tilde{w}_1(t), \ldots, \tilde{w}_m(t), \tilde{\zeta}(t)$ are the same as the joint distributions of the processes $w_1(t), \ldots, w_m(t), \zeta(t)$ examined in Lemma 4.

Let $\tilde{p}^{(n')}(A)$ denote for every Borel set $A \subset [t_0, T] \times R^{(m)}$ the number of jumps of the process $\tilde{\zeta}^{(n')}(t)$ for which

$$(t, \tilde{\zeta}^{(n')}(t+0) - \tilde{\zeta}^{(n')}(t-0)) \in A, \ \tilde{q}^{(n')}(A) = \tilde{p}^{(n')}(A) - \tilde{\mathbf{M}}p^{(n')}(A).$$

Then from the coincidence of the joint distributions of the processes $\tilde{\eta}_{n'}(t), \tilde{w}_1^{(n')}(t), \ldots, \tilde{w}_m^{(n')}(t), \tilde{\zeta}^{(n')}(t)$ and the processes $\eta_{n'}(t), w_1^{(n')}(t), \ldots, w_m^{(n')}(t), \zeta^{(n')}(t)$ and also from formula (4.8), it follows that

$$\begin{aligned}\tilde{\eta}^{(n')}(t_{k+1}^{(n')}) = \tilde{\eta}^{(n')}(t_k^{(n')}) &+ a(t_k^{(n')}, \tilde{\eta}^{(n')})(t_k^{(n')})(t_{k+1}^{(n')} - t_k^{(n')}) \\ &+ \sum_{j=1}^{m} b(t_k^{(n')}, \tilde{\eta}(t_k^{(n')}))[\tilde{w}_j^{(n')}(t_{k+1}^{(n')}) - \tilde{w}_j^{(n')}(t_k^{(n')})] \\ &+ \int_{\substack{t_k^{(n')} < s \le t_{k+1}^{(n')} \\ u \in R^{(m)}}} f(t_k^{(n')}, \tilde{\eta}(t_k^{(n')}), u)\tilde{q}^{(n')}(ds \times du).\end{aligned}$$

From this relation, we obtain

$$\begin{aligned}\tilde{\eta}^{(n')}(t) = \tilde{\eta}^{(n')}(0) &+ \sum_{t_{i+1}^{(n')} \le t} a(t_i^{(n')}, \tilde{\eta}(t_i^{(n')}))(t_{i+1}^{(n')} - t_i^{(n')}) \\ &+ \sum_{t_{i+1}^{(n)} \le t} \sum_{j=1}^{m} b(t_i^{(n')}, \tilde{\eta}(t_i^{(n')}))[\tilde{w}_j^{(n')}(t_{i+1}^{(n')}) - \tilde{w}_j^{(n')}(t_i^{(n')})] \\ &+ \sum_{t_{i+1}^{(n)} \le t} \int_{\substack{t_i^{(n')} < s \le t_{i+1}^{(n')} \\ u \in R^{(m)}}} f(t_i^{(n')}, \tilde{\eta}^{(n')}(t_i^{(n')}), u)\tilde{q}^{(n')}(ds \times du).\end{aligned} \tag{4.17}$$

This formula is analogous to Formula (3.7) of Section 3, Chapter 3, which was obtained in the proof of the existence theorem of that section. The remainder of the proof of this lemma is almost the same as the proof of the existence theorem. Just as in the proof of the existence theorem referred to, we can show that

$$\sum_{t_{i+1}^{(n')} \le t} a(t_i^{(n')}, \tilde{\eta}^{(n')}(t_i^{(n')}))(t_{i+1}^{(n')} - t_i^{(n')}) \to \int_{t_0}^{t} a(s, \tilde{\eta}(s))\, ds \qquad (4.18)$$

in probability and that

$$\sum_{t_{i+1}^{(n')} \le t} b_j(t_i^{(n')}, \tilde{\eta}^{(n')}(t_i^{(n')}))(\tilde{w}_j^{(n')}(t_{i+1}^{(n')}) - \tilde{w}_j^{(n')}(t_i^{(n')}))$$

$$\to \int_{t_0}^{t} b_j(s, \tilde{\eta}(s))\, d\tilde{w}_j(s) \qquad (4.19)$$

in probability. We now show that

$$\sum_{t_{i+1}^{(n')} \le t} \int_{\substack{t_i^{(n)} < s \le t_{i+1}^{(n')} \\ u \in R^{(m)}}} f(t_i^{(n')}, \tilde{\eta}^{(n')}(t_i^{(n')}), u)\tilde{q}^{(n')}\,(ds \times du)$$

$$\to \int_{t_0}^{t} \int_{u \in R^{(m)}} f(s, \tilde{\eta}(s), u)\tilde{q}\,(ds \times du) \qquad (4.20)$$

in probability, where the measures $\tilde{q}$ (and also the $\tilde{p}$, which we shall meet below) are defined with respect to the process $\tilde{\zeta}(t)$ just as q_0 (and also p_0) were defined with respect to the process $\zeta_0(t)$ in Lemma 4 of Section 3, Chapter 3.

The proof of (4.20) can be obtained in the same way as the proof of the analogous relations in the existence theorem of Section 3, Chapter 3 (by using Lemma 5 of that section), provided that we establish the assertion analogous to the assertion of Lemma 4 of the same section, namely, that for every almost-everywhere-continuous function $\varphi(u)$ that is bounded in every bounded region and that becomes zero in some neighborhood of zero,

$$\int_{t}^{t+h} \int_{R^{(m)}} \varphi(u)\tilde{p}^{(n')}\,(ds \times du) \to \int_{t}^{t+h} \int_{R^{(m)}} \varphi(u)\tilde{p}\,(ds \times du) \qquad (4.21)$$

in probability. To establish (4.21), we note that the functional

$$F^*(x(s)) = \sum_{s \in [t, t+h]} \varphi(x(s+0) - x(s-0))$$

is defined on $D_m[t_0, T]$, that it is measurable with respect to $\mathbf{F}(D_m[t_0, T]$, and that it belongs to $F(\mathbf{J}, \mu_0)$, where μ_0 is the measure corresponding to the process $\zeta(s)$. Further,

$$F^*(\tilde{\zeta}^{(n')}(s)) = \int_{t}^{t+h} \int_{R^{(m)}} \varphi(u)\tilde{p}^{(n')}\,(ds \times du),$$

$$F^*(\tilde{\zeta}(s)) = \int_{t}^{t+h} \int_{R^{(m)}} \varphi(u)\tilde{p}\,(ds \times du).$$

It follows from Lemma 5 and from the proof of the theorem of Section 3 of this chapter that $F^*(\tilde{\zeta}^{(n')}(s)) \to F^*(\tilde{\zeta}(s))$ in probability; we can establish the validity of (4.21) in the same way in which (3.8) of Chapter 3 was derived from Lemma 4 of Section 3 of that chapter in the proof of the existence theorem. If we now go to the limit as $n' \to \infty$ in (4.17), we obtain

$$\tilde{\eta}(t) = \tilde{\eta}(t_0) + \int_{t_0}^{t} a(s, \tilde{\eta}(s))\, ds + \sum_{j=1}^{m} \int_{t_0}^{t} b_j(s, \tilde{\eta}(s))\, d\tilde{w}_j(s)$$
$$+ \int_{t_0}^{t} \int_{R^{(m)}} f(s, \tilde{\eta}(s), u)\tilde{q}\,(ds \times du).$$

Since the joint distributions of the variables $\tilde{\eta}(t_0)$, $\tilde{w}_1(s), \ldots, \tilde{w}(s)$, $\tilde{q}$ coincide with the joint distributions of the variables $\xi(t_0)$, $w_1(s), \ldots,$ $w_m(s)$, q that appear in Eq. (1.8) of Chapter 3, it follows that $\eta(t)$ has the same finite-dimensional distributions as does $\xi(t)$ in the solution of Eq. (1.8) of Chapter 3. This proves the lemma.

By combining the results of Lemmas 2 and 6, we can assert the validity of the following theorem.

THEOREM. *Let $\xi^{(n)}(t)$ be formed by a Markov chain as indicated in Section 1 and let $a^{(n)}(t, x)$, $b_j^{(n)}(t, x)$, and $f^{(n)}(t, x, u)$ be defined as in this section; also, let there exist a constant K for which*

$$\text{(a)}\quad |a^{(n)}(t, x)|^2 + \sum_{j=1}^{m} |b_j^{(n)}(t, x)|^2$$
$$+ \int_{R^{(m)}} |f^{(n)}(t, x, u)|^2 \frac{du}{|u|^{m+1}} \le K(1 + |x|^2).$$

Then suppose there exist

$$a(t, x), \quad b_j(t, x), \qquad \text{and} \qquad f(t, x, u)$$

satisfying the following conditions:

(b) *$a(t, x)$ and $b_j(t, x)$ are continuous over the set of variables.*

(c) *$f(t, x, u)$ is bounded in every bounded region of variation of x and u, is continuous for almost all u with respect to the set of variables t and x, and for all t and x is continuous with respect to u for almost all u; also, for all t_1 and x_1,*

$$\lim_{\substack{t \to t_1 \\ x \to x_1}} \int |f(t, x, u) - f(t_1, x_1, u)|^2 \frac{du}{|u|^{m+1}} = 0.$$

(d) *There exists K_1 and L such that*

$$|a(t, x)|^2 + \sum_{j=1}^{m} |b_j(t, x)|^2 + \int_{R^{(m)}} |f(t, x, u)|^2 \frac{du}{|u|^{m+1}} \leq K_1(1 + |x|^2),$$

$$|a(t, x) - a(t, y)|^2 \sum_{j=1}^{m} |b_j(t, x) - b_j(t, y)|^2$$

$$+ \int_{R^{(m)}} |f(t, x, u) - f(t, y, u)|^2 \frac{du}{|u|^{m+1}} \leq L|x - y|^2$$

for all x and y and, in addition, the following conditions are fulfilled:

(e)
$$\lim_{n\to\infty} \sum_{k=0}^{n-1} \Bigg[\mathbf{M}|a(t_k^{(n)}, \xi_k^{(n)}(t_k^{(n)})) - a^{(n)}(t_k^{(n)}, \xi_n^{(n)}(t_k^{(n)}))|^2$$

$$+ \sum_{j=1}^{m} \mathbf{M}|b_j(t_k^{(n)}, \xi^{(n)}(t_k^{(n)})) - b_j^{(n)}(t_k^{(n)}, \xi^{(n)}(t_k^{(n)}))|^2$$

$$+ \int_{R^{(m)}} \mathbf{M}|f(t_k^{(n)}, \xi_n^{(n)}(t_k^{(n)}), u)$$

$$- f^{(n)}(t_k^{(n)}, \xi^{(n)}(t_k^{(n)}), u)|^2 \frac{du}{|u|^{m+1}} \Bigg] (t_{k+1}^{(n)} - t_k^{(n)}) = 0.$$

(f) *The distribution of $\xi^{(n)}(t_0)$ converges to the distribution of $\xi(t_0)$ and*

$$\sup_n \mathbf{M}|\xi^{(n)}(t_0)|^2 < \infty.$$

Under these conditions, the finite-dimensional distributions of the process $\xi^{(n)}(t)$ converge to the finite-dimensional distributions of the process $\xi(t)$ which is a solution of Eq. (1.8) of Chapter 3.

Remark. Condition (e) can be replaced by the following condition:

(e′) For every $C > 0$,

$$\lim_{n\to\infty} \sum_{k=0}^{n} \sup_{|x|\leq C} \Bigg[|a(t_k^{(n)}, x) - a^{(n)}(t_k^{(n)}, x)|^2 + \sum_{j=1}^{m} |b_j(t_k^{(n)}, x) - b_j^{(n)}(t_k^{(n)}, x)|^2$$

$$+ \int_{u\in R^{(m)}} |f(t_k^{(n)}, x, u) - f^{(n)}(t_k^{(n)}, x, u)|^2 \frac{du}{|u|^{m+1}} \Bigg] (t_{k+1}^{(n)} - t_k^{(n)}) = 0.$$

Clearly, the fulfillment of (e′) implies the fulfillment of (e). The condition (e′) is sometimes more convenient for testing.

5. A limit theorem for the distributions of functionals of a sequence of Markov processes. In this section, the general theorem of Section 2 will be applied to the special case of convergence of the sequence of processes constructed in Section 1 for the solution of (1.8) of Chapter 3. In doing so, we shall use to a considerable degree the conditions for convergence of finite-dimensional distributions that we obtained in the preceding paragraph.

THEOREM. *Suppose that the sequence $\xi^{(n)}(t)$, $a(t, x)$, $b_j(t, x)$, $f(t, x, u)$ satisfies the conditions of the theorem of Section 4, that $\xi(t)$ is a solution to Eq. (1.8) of Chapter 3, and that μ is the measure corresponding to the process $\xi(t)$ on $\mathbf{F}(D_m[t_0, T])$. Then for every functional $F(x(t))$ in the class $F(\mathbf{J}, \mu)$, the distribution of $F(\xi^{(n)}(t))$ converges to the distribution of $F(\xi(t))$.*

Proof. Since (by the theorem of Section 4) the finite-dimensional distributions of the processes $\xi^{(n)}(t)$ converge to the finite-dimensional distribution of the process $\xi(t)$, it follows from the theorem of Section 2 that to prove this theorem, it is sufficient to show that

$$\lim_{c\to 0} \overline{\lim_{N\to\infty}} \mathbf{P}\,\{\Delta_c(\bar{\xi}_n^{(n)}(t)) > \epsilon\} = 0, \tag{5.1}$$

where $\bar{\xi}^{(n)}(t)$ is the process defined by Formula (4.2) and having the same finite-dimensional distributions as does the process $\xi^{(n)}(t)$. To prove formula (5.1), we shall prove some auxiliary propositions.

LEMMA 1. *Suppose that $\xi_1, \xi_2, \ldots, \xi_l$ form a Markov chain and that there exists $\alpha < 1$ such that for all $i > j$ with probability 1 the relation*

$$\mathbf{P}\{|\xi_j - \xi_i| > \epsilon/\xi_i\} \le \alpha$$

is satisfied. Then

$$\mathbf{P}\,\{\sup_{i,j} |\xi_j - \xi_i| > 4\epsilon\} < \frac{2}{1-\alpha}\mathbf{P}\{|\xi_1 - \xi_l| > \epsilon\}.$$

Proof. First we show that

$$\mathbf{P}\,\{\sup |\xi_j - \xi_1| > 2\epsilon\} < \frac{1}{1-\alpha}\mathbf{P}\{|\xi_1 - \xi_l| > \epsilon\}. \tag{5.2}$$

In fact,

$$\begin{aligned}
\mathbf{P}\{|\xi_1 - \xi_l| > \epsilon\} &\ge \mathbf{P}\{|\xi_1 - \xi_l| > \epsilon, \sup_j |\xi_j - \xi_1| > 2\epsilon\} \\
&= \sum_{k=2}^{l} \mathbf{P}\{|\xi_1 - \xi_2| \le 2\epsilon, \ldots, |\xi_1 - \xi_{k-1}| \\
&\qquad \le 2\epsilon, |\xi_1 - \xi_k| > 2\epsilon, |\xi_1 - \xi_l| > \epsilon\} \\
&\ge \sum_{k=2}^{l} \mathbf{P}\{|\xi_1 - \xi_2| \le 2\epsilon, \ldots, |\xi_1 - \xi_{k-1}| \\
&\qquad \le 2\epsilon, |\xi_1 - \xi_k| > 2\epsilon, |\xi_k - \xi_l| \le \epsilon\} \\
&\ge (1-\alpha) \sum_{k=2}^{l} \mathbf{P}\{|\xi_1 - \xi_2| \le 2\epsilon, \ldots, |\xi_1 - \xi_{k-1}| \\
&\qquad \le 2\epsilon, |\xi_1 - \xi_k| > 2\epsilon\},
\end{aligned}$$

because $\mathbf{P}\{|\xi_k - \xi_l| \leq \epsilon/\xi_k\} \geq 1 - \alpha$. Thus

$$\frac{1}{1-\alpha}\mathbf{P}\{|\xi_1 - \xi_l| > \epsilon\}$$
$$\geq \sum_{k=2}^{l} \mathbf{P}\{|\xi_1 - \xi_2| \leq 2\epsilon, \ldots, |\xi_1 - \xi_{k-1}| \leq 2\epsilon, |\xi_1 - \xi_k| > 2\epsilon\}$$
$$= \mathbf{P}\,\{\sup_k |\xi_1 - \xi_k| > 2\epsilon\}.$$

The inequality (5.2) follows from this inequality. We now note that

$$\mathbf{P}\,\{\sup_{i,j} |\xi_i - \xi_j| > 4\epsilon\} \leq \mathbf{P}\,\{\sup_i |\xi_i - \xi_1| > 2\epsilon\} + \mathbf{P}\,\{\sup_j |\xi_j - \xi_1| > 2\epsilon\}$$
$$\leq \frac{2}{1-\alpha}\mathbf{P}\{|\xi_l - \xi_1| > \epsilon\}.$$

This proves the lemma.

LEMMA 2. *Under the conditions of Lemma 1,*

$$\mathbf{P}\,\{\sup_{i<j<k} \min\,(|\xi_i - \xi_j|; |\xi_j - \xi_k|) > 4\epsilon\} < \frac{\alpha}{(1-\alpha)^2}\mathbf{P}\{|\xi_1 - \xi_l| > \epsilon\}. \tag{5.3}$$

Proof.

$$\mathbf{P}\,\{\sup_{i<j<k} \min\,(|\xi_i - \xi_j|; |\xi_j - \xi_k|) > 4\epsilon\}$$
$$\leq \sum_{k<j} \mathbf{P}\{|\xi_2 - \xi_1| \leq 2\epsilon, \ldots, |\xi_{k-1} - \xi_1| \leq 2\epsilon,$$
$$|\xi_k - \xi_1| > 2\epsilon, |\xi_{k+1} - \xi_k| \leq 2\epsilon, \ldots, |\xi_{j-1} - \xi_k| \leq 2\epsilon,$$
$$|\xi_j - \xi_k| > 2\epsilon\}$$
$$\leq \sum_k \mathbf{M}(\mathbf{P}\{|\xi_2 - \xi_1|) \leq 2\epsilon, \ldots, |\xi_{k-1} - \xi_1| \leq 2\epsilon,$$
$$|\xi_k - \xi_1| > 2\epsilon/\xi_k\}\,\mathbf{P}\,\{\sup_{j=k+1,\ldots,l} |\xi_j - \xi_k| > 2\epsilon/\xi_k\}$$
$$\leq \frac{\alpha}{1-\alpha}\,\{\sup_k |\xi_k - \xi_1| > 2\epsilon\} \leq \frac{\alpha}{(1-\alpha)^2}\mathbf{P}\{|\xi_l - \xi_1| > \epsilon\}.$$

[Here we use the inequality (5.2) twice.] This proves the lemma.

LEMMA 3. *For every $\epsilon > 0$, there exists an H_ϵ such that for $t_k^{(n)} - t_i^{(n)} > 0$,*

$$\mathbf{P}\{|\bar{\xi}^{(n)}(t_k^{(n)}) - \bar{\xi}^{(n)}(t_i^{(n)})| > \epsilon/\bar{\xi}^{(n)}(t_i^{(n)})\}$$
$$\leq (H_\epsilon(|\bar{\xi}_n(t_i^{(n)})|^2 + 1) \times (t_k^{(n)} - t_i^{(n)})).$$

Proof. We have

$$\mathbf{P}\{|\bar{\xi}^{(n)}(t_k^{(n)}) - \bar{\xi}^{(n)}(t_i^{(n)})| > \epsilon|\bar{\xi}^{(n)}(t_i^{(n)})|\}$$
$$\leq \frac{1}{\epsilon^2}\mathbf{M}(|\bar{\xi}^{(n)}(t_k^{(n)}) - \bar{\xi}^{(n)}(t_i^{(n)})|^2/(\bar{\xi}^{(n)}(t_i^{(n)})).$$

In exactly the same way as we obtained (4.12) in the proof of Lemma 3 of Section 4 (taking only the conditional mathematical expectations), we can show that

$$\mathbf{M}(|\bar{\xi}^{(n)}(t_k^{(n)}) - \bar{\xi}^{(n)}(t_i^{(n)})|^2/\bar{\xi}^{(n)}(t_i^{(n)}))$$
$$\leq (m+2)K[|\bar{\xi}^{(n)}(t_i^{(n)})|^2 e^{H(T-t_0)} + 1](t_k^{(n)} - t_i^{(n)}).$$

The proof of the lemma follows from these two inequalities.

LEMMA 4. *The quantity* $\sup_t |\bar{\xi}^{(n)}(t)|$ *is uniformly bounded in probability with respect to* n.

The proof of this lemma follows from the proof of Lemma 2 of Section 3, Chapter 3.

Let us now prove relation (5.1). Suppose that $\bar{\xi}_n(t) = \bar{\xi}_k^{(n)}(t)$ for

$$\sup_{s\in[t_0,t]} |\bar{\xi}^{(n)}(s)| > R$$

and that $\bar{\bar{\xi}}_n(t) = \bar{\bar{\xi}}_n(t_1)$ for $t_1 < t$ if

$$\sup_{s\in[t_0,t]} |\bar{\bar{\xi}}^{(n)}(s)| > R.$$

Then it follows from Lemma 3 that

$$\mathbf{P}\{\bar{\bar{\xi}}_n(t_k^{(n)}) - \bar{\bar{\xi}}_n(t_i^{(n)})\ | > \epsilon/\bar{\bar{\xi}}_n(t_i^{(n)})\} \leq H_\epsilon(R^2+1)(t_k^{(n)} - t_i^{(n)}). \quad (5.4)$$

Also, for an arbitrary set of functions A in $\mathbf{F}(D_m[t_0, T])$, the inequality

$$|\mathbf{P}\{\bar{\bar{\xi}}_n(t) \in \mathbf{A}\} - \mathbf{P}\{\bar{\xi}^{(n)}(t) \in \mathbf{A}\}| \leq \mathbf{P}\{\sup_t |\bar{\xi}^{(n)}(t)| > R\}$$

is valid. If $H_\epsilon(R^2+1)(c + \max_i (t_{i+1}^{(n)} - t_i^{(n)})) < 1$, we conclude on the basis of Lemma 1 that

$$\mathbf{P}\left\{\sup_{t_0\leq t\leq t_0+c} |\bar{\bar{\xi}}_n(t) - \bar{\bar{\xi}}_n(t_0)| > 2\epsilon\right\} \leq \frac{(R^2+1)H_\epsilon c}{1-(R^2+1)H_\epsilon c}$$

and that

$$\mathbf{P}\left\{\sup_{T-c\leq t\leq T} |\bar{\bar{\xi}}_n(t) - \bar{\bar{\xi}}_n(T)| > 4\epsilon\right\}$$
$$\leq 2\frac{(R^2+1)H_\epsilon(c + \max_k (t_{k+1}^{(n)} - t_k^{(n)}))}{1 - H_\epsilon(c + \max_k (t_{k+1}^{(n)} - t_k^{(n)}))(R^2+1)}.$$

In just the same manner, by using Lemma 2, we see that for

$$(R^2 + 1)H_\epsilon(2c + \max_i (t_{i+1}^{(n)} - t_i^{(n)})) < 1,$$

the inequality

$$\mathbf{P}\Big\{\sup_{t-2c<t'<t''<t'''\le t} \min [|\bar{\bar{\xi}}_n(t') - \bar{\bar{\xi}}_n(t'')|, |\bar{\bar{\xi}}_n(t'') - \bar{\bar{\xi}}_n(t''')|] > 4\epsilon\Big\} \le \frac{H_\epsilon^2(R^2+1)^2(2c + \max_i (t_{i+1}^{(n)} - t_i^{(n)}))^2}{[1 - H_\epsilon(R^2+1)(2c + \max_i (t_{i+1}^{(n)} - t_i^{(n)}))]^2}$$

holds. Let us choose n sufficiently large so that $\max_i (t_{i+1}^{(n)} - t_i^{(n)}) < C$. Then

$$\mathbf{P}\Big\{\sup_{\substack{t_0\le t'<t<t''\le T\\ |t'-t''|<C}} \min [|\bar{\bar{\xi}}_n(t') - \bar{\bar{\xi}}_n(t)|; |\bar{\bar{\xi}}_n(t) - \bar{\bar{\xi}}_n(t'')|] > 4\epsilon\Big\}$$

$$\le \sum_{0\le k<(T-t_0)/C} \mathbf{P}\Big\{\sup_{t_0+kC\le t'<t<t''\le t_0+(k+2)C} \min [|\bar{\bar{\xi}}_n(t') - \bar{\bar{\xi}}_n(t)|; |\bar{\bar{\xi}}_n(t) - \bar{\bar{\xi}}_n(t'')|] > 4\epsilon\Big\}$$

$$\le \frac{T - t_0}{C} \frac{9C^2(R^2+1)^2H_\epsilon^2}{(1 - 3C(R^2+1)H_\epsilon)^2}.$$

Thus for $\max (t_{i+1}^{(n)} - t_i^{(n)}) < C$ and $3C(R^2 + 1)H_{\epsilon/10} < 1$, we have

$$\mathbf{P}\{\Delta_C(\bar{\bar{\xi}}_n(t)) > \epsilon\} \le \mathbf{P}\Big\{\sup_{t_0<t\le t_0+c} |\bar{\bar{\xi}}_n(t) - \bar{\bar{\xi}}_n(t_0)| > \frac{2\epsilon}{10}\Big\}$$

$$+ \mathbf{P}\Big\{\sup_{\substack{t_0\le t'<t<t''\le T\\ |t'-t''|<C}} \min [|\bar{\bar{\xi}}_n(t') - \bar{\bar{\xi}}_n(t)|; |\bar{\bar{\xi}}_n(t) - \bar{\bar{\xi}}_n(t'')|] > \frac{4\epsilon}{10}\Big\}$$

$$+ \mathbf{P}\Big\{\sup_{T-c\le t<T} |\bar{\bar{\xi}}_n(t) - \bar{\bar{\xi}}_n(T)| > \frac{4\epsilon}{10}\Big\}$$

$$\le \frac{H_{(\epsilon/10)}(R^2+1)C}{1 - H_{(\epsilon/10)}(R^2+1)C} + \frac{4H_{(\epsilon/10)}(R^2+1)C}{1 - 2H_{(\epsilon/10)}(R^2+1)C}$$

$$+ \frac{9(H_{(\epsilon/10)}(R^2+1))^2(T - t_0)C}{(1 - 3H_{(\epsilon/10)}(R^2+1)C)^2}. \tag{5.5}$$

But since

$$\mathbf{P}\{\Delta_C(\bar{\xi}^{(n)}(t)) > \epsilon\} \le \mathbf{P}\{\Delta_C(\bar{\bar{\xi}}\ (t)) > \epsilon\} + \mathbf{P}\{\sup_t |\bar{\xi}^{(n)}(t)| > R\},$$

we have

$$\overline{\lim_{n\to\infty}} \mathbf{P} \left\{\Delta_C(\bar{\xi}^{(n)}(t)) > \epsilon\right\}$$

$$\leq (R^2 + 1)H_{(\epsilon/10)}C \frac{5 + 9(R^2 + 1)H_{(\epsilon/10)}(T - t_0)}{(1 - 3H_{(\epsilon/10)}(R^2 + 1)C)^2}$$

$$+ \overline{\lim_{n\to\infty}} \mathbf{P} \{\sup_t |\bar{\xi}^{(n)}(t)| > R\}.$$

Taking the limit in this inequality as $C \to 0$ and then as $R \to \infty$ and using Lemma 4, we obtain formula (5.1). This proves the theorem.

CHAPTER 7

SOME LIMIT THEOREMS FOR SUMS OF INDEPENDENT RANDOM VARIABLES

1. Preliminary remarks. In this chapter, we shall study a special form of the convergence of a discrete sequence to a continuous process.

Suppose that $\xi_{n1}, \xi_{n2}, \ldots, \xi_{nn}$ is a sequence for which each series of random variables is independent and that $S_{nk} = \sum_{i=1}^{k} \xi_{ni}$, $k = 1, 2, \ldots, n$, $S_{n0} = 0$. Let us examine the random variables satisfying the equations $\mathbf{M}\xi_{ni} = 0$ and $D\xi_{ni} = 1/n$, and let us investigate the case of the convergence of S_{nk} to a Brownian process. Let us set $\xi_n(t) = S_{nk}$ if $k/n \leq t < (k+1)/n$ and $t \in [0, 1]$. Under certain conditions (for example, if $\sup_k \mathbf{M}|\xi_{kn}|^3 = o(1/n)$), the finite-dimensional distributions of $\xi_n(t)$ will converge to finite-dimensional distributions of a Brownian process. Since the S_{nk} form a Markov chain, it follows from the theorem of Section 5, Chapter 6, that for every functional $F(x(t))$ in $F(\mathbf{J}, \mu_0)$, where μ_0 is the measure corresponding to a Brownian process on $D_m[t_0, T]$, the distribution of $F(\xi_n(t))$ will converge to the distribution of $F(w(t))$.

This chapter is devoted to a precise statement of a limit theorem for certain specific classes of functionals. We shall confine ourselves to finding the first term of an asymptotic expansion of the mathematical expectation of certain smooth functionals of $\xi_n(t)$ and to evaluating the rapidity of convergence of the distributions of certain functionals to a limiting distribution.

Such investigations for various special cases have been conducted before. However, the method suggested for the solution of these problems is quite different from the methods that have been applied for similar purposes. The usual method is based on an analytic technique (characteristic functions, difference and differential equations, etc.) by means of which one can obtain in explicit or implicit form the distributions of the corresponding functionals, which can then be subjected to asymptotic analysis.

The straightforward probabilistic methods that are used in this chapter are based on the construction of variables whose distributions coincide with the distributions of the corresponding functionals. In the solution of a problem, we investigate the asymptotic behavior of the same variables as $n \to \infty$. Such methods give results only in the case in which one can successfully construct variables whose distributions coincide with the distribution of the functional. To construct such random variables, we make use of the concept of a sequence of sums of independent random variables and their images under a Brownian process. The possibility of there being such a sequence is studied in Section 2. In Section 3, we shall evaluate the rapidity of the convergence of the probability

$$\mathbf{P}\{g_1(t) < \xi_n(t) < g_2(t), \quad t \in [0, 1]\}$$

to the probability

$$\mathbf{P}\{g_1(t) < w(t) < g_2(t), \quad t \in [0, 1]\},$$

where $w(t)$ is the process of Brownian motion, and $g_1(t)$ and $g_2(t)$ are continuous functions. In Section 4, we shall examine functionals of the integral form $\int_0^1 f(s, \xi_n(s))\, ds$ and shall study the convergence of such functionals to $\int_0^1 f(s, w(s))\, ds$.

2. The concept of a sequence of sums of independent random variables which are the images of a Brownian process. The basic result obtained in this section can be formulated in the following theorem.

THEOREM. *Suppose that $\xi_1, \xi_2, \ldots, \xi_n$ are independent random variables such that $\mathbf{M}\xi_i = 0$ and $D\xi_i < \infty$ and that $w(t)$ is a Brownian process. Then there exist nonnegative independent random variables $\tau_1, \tau_2, \ldots, \tau_n$ for which the variables*

$$w(\tau_1), w(\tau_1 + \tau_2) - w(\tau_1), \ldots, w(\tau_1 + \tau_2 + \cdots + \tau_n) - w(\tau_1 + \tau_2 + \cdots + \tau_{n-1})$$

have the same joint distribution as do $\xi_1, \xi_2, \ldots, \xi_n$. Also:

(a) $\mathbf{M}\tau_k = D\xi_k$.

(b) *There exist L_m such that*

$$\mathbf{M}(\tau_k)^m \leq L_m \mathbf{M}(\xi_k)^{2m}.$$

(c) *If $|\xi_i| \leq h$, then*

$$\left| w(s) - w\left(\sum_1^k \tau_i\right)\right| \leq h \quad \text{for} \quad s \in \left[\sum_1^k \tau_i, \sum_1^{k+1} \tau_i\right].$$

This theorem allows us to examine the sequence of values of the process $w(\sum_1^k \tau_i)$ instead of the sequence of sums $S_k = \sum_{i=1}^k \xi_i$ because the joint distributions of the terms of both sequences coincide. To prove the theorem, we shall establish several auxiliary propositions.

LEMMA 1. *Suppose that $w(t)$ is a Brownian process defined for $t \geq 0$ and $w(0) = 0$, and that τ is the smallest root of the equation*

$$(w(t) - a)(w(t) - b) = 0,$$

where $a < 0 < b$. *We define*

$$\varphi_1(t) = \mathbf{P}\{w(\tau) = a, \tau < t\}, \qquad \varphi_2(t) = \mathbf{P}\{w(\tau) = b, \tau < t\}.$$

Then for $\lambda > 0$,

$$\int_0^\infty e^{-\lambda t}\, d\varphi_1(t) = \frac{\sinh(b\sqrt{2\lambda})}{\sinh((b-a)\sqrt{2\lambda})}, \tag{2.1}$$

$$\int_0^\infty e^{-\lambda t}\, d\varphi_2(t) = -\frac{\sinh(a\sqrt{2\lambda})}{\sinh((b-a)\sqrt{2\lambda})}. \tag{2.2}$$

where $\sinh x = (e^x - e^{-x})/2$.

Proof. Suppose that $\xi_0^{(n)}, \xi_1^{(n)}, \ldots, \xi_n^{(n)}, \ldots,$ is a Markov chain and that the $\xi_k^{(n)}$ take only values of the form $a + l((b-a)/[\sqrt{n}])$, where the l are integers and

$$\mathbf{P}\left\{\xi_{k+1}^{(n)} = a + (l \pm 1)\frac{b-a}{[\sqrt{n}]} \mid \xi_k^{(n)} = a + l\frac{b-a}{[\sqrt{n}]}\right\} = \frac{1}{2}.$$

Suppose that $\xi^{(n)}(t) = \xi_k^{(n)}$ for $t \in [(k/n)(b-a)^2, ((k+1)/n)(b-a)^2]$. Then the conditions of the theorem of Section 5, Chapter 6, will be fulfilled provided δ_n is such that $(\sqrt{n}/\delta_n) \to \infty$, $f(t, x, u) = 0$, $a(t, x) = 0$, $b(t, x) = 1$. Therefore, if $\xi_0^{(n)} \to w(0)$ in probability, $F(x(t)) \in F(\mathbf{J}, \mu_0)$, and μ_0 is the measure corresponding to the process $w(t)$ on $D_1[0, T]$, then the distribution of $F(\xi_n(t))$ will converge to the distribution of $F(w(t))$. Let $F_T^{(1)}(x(t)) = t$ if for some $t \leq T$, the function $x(t)$ first intersects the straight line $x = a$ and does not intersect the straight line $x = b$ for $s < t$. In all remaining cases, $F_T^{(1)}(x(t)) = 0$. It is easy to see that $F_T^{(1)}(x(t)) \in F(\mathbf{J}, \mu_0)$; therefore

$$\mathbf{M}e^{-\lambda F_T^{(1)}(\xi_n(t))} \to Me^{-\lambda F_T^{(1)}(w(t))}.$$

Since the sequence $F_T^{(1)}(\xi_n(t))$ is uniformly bounded in probability with respect to T, this convergence is uniform with respect to T. Suppose that τ_n is the first zero of the equation $(\xi_n(t) - a)(\xi_n(t) - b) = 0$, that $\epsilon(x) = 1$ for $x = 0$, and that $\epsilon(x) = 0$ for $x \neq 0$. Then $F_T^{(1)}(\xi_n(t)) \to \tau_n\epsilon(\xi_n(\tau_n) - a)$ as $T \to \infty$ and $F_T^{(1)}(w(t)) \to \tau\epsilon(w(\tau) - a)$ as $T \to \infty$; therefore

$$\lim_{n\to\infty} \mathbf{M}e^{-\lambda\tau_n\epsilon(\xi_n(\tau_n)-a)} = \mathbf{M}e^{-\lambda\tau\epsilon(w(\tau)-a)} = \int_0^\infty e^{-\lambda t}\, d\varphi_1(t).$$

Define

$$g_n^{(k)}(l) = \mathbf{P}\left\{\tau_n\epsilon(\xi_n(\tau_n) - a) = \frac{k}{n}(b-a)^2/\xi_0 = a + l\frac{b-a}{[\sqrt{n}]}\right\}.$$

Then for $0 < l < [\sqrt{n}]$,

$$g_n^{(k)}(l) = \tfrac{1}{2}[g_n^{(k-1)}(l-1) + g_n^{(k-1)}(l+1)];$$

also, $g_n^{(k)}[\sqrt{n}]) = 0$, $g_n^{(k)}(0) = 0$ for $k > 0$, $g_n^{(0)}(l) = 0$ for $l \neq 0$, and $g_n^{(0)}(0) = 1$. Let us define

$$G_n(l) = \sum_{k=0}^{\infty} g_n^{(k)}(l) e^{-\lambda(k/n)(b-a)^2}.$$

Then

$$G_n(l) = \tfrac{1}{2} e^{-(\lambda/n)(b-n)^2}[G_n(l-1) + G_n(l+1)], \qquad 0 < l < [\sqrt{n}], \tag{2.3}$$

$$G_n(0) = 1, \qquad G_n([\sqrt{n}]) = 0. \tag{2.4}$$

From the relation (2.3), we see that

$$G_n(l) = C_1\left[e^{(\lambda/n)(b-a)^2} + \sqrt{e^{(2\lambda/n)(b-a)^2} - 1}\right]^l + C_2\left[e^{(\lambda/n)(b-a)^2} - \sqrt{e^{(2\lambda/n)(b-a)^2} - 1}\right],$$

where C_1 and C_2 are defined by the relations (2.4):

$$C_1 = -C_2 = \frac{1}{\left\{e^{(\lambda/n)(b-a)^2} + \sqrt{e^{(2\lambda/n)(b-a)^2} - 1}\right\}^{[\sqrt{n}]} - \left\{e^{(\lambda/n)(b-a)^2} - \sqrt{e^{(2\lambda/n)(b-a)^2} - 1}\right\}^{[\sqrt{n}]}}.$$

Thus

$$G_n(l) = \frac{\left\{e^{(\lambda/n)(b-a)^2} + \sqrt{e^{(2\lambda/n)(b-a)^2} - 1}\right\}^{l} - \left\{e^{(\lambda/n)(b-a)^2} - \sqrt{e^{(2\lambda/n)(b-a)^2} - 1}\right\}^{l}}{\left\{e^{(\lambda/n)(b-a)^2} + \sqrt{e^{(2\lambda/n)(b-a)^2} - 1}\right\}^{[\sqrt{n}]} - \left\{e^{(\lambda/n)(b-a)^2} - \sqrt{e^{(2\lambda/n)(b-a)^2} - 1}\right\}^{[\sqrt{n}]}}. \tag{2.5}$$

It is clear that

$$G_n(l) = \mathbf{M}\left(e^{-\lambda\tau_n\epsilon(\xi_n(\tau_n)-a)}/\xi_n(0) = a + \frac{l}{[\sqrt{n}]}(b-a)\right).$$

Choosing $l = [(b/(b-a))[\sqrt{n}]]$ and taking the limit as $n \to \infty$, we obtain Formula (2.1) from (2.5). Formula (2.2) is derived in an analogous manner. This proves the lemma.

COROLLARY. *If τ is the variable defined in Lemma 1, then*

1.

$$\mathbf{M}e^{-\lambda\tau} = \frac{\sinh b\sqrt{2\lambda} - \sinh a\sqrt{2\lambda}}{\sinh (b-a)\sqrt{2\lambda}}. \tag{2.6}$$

2. *Assuming that*

$$\mathbf{P}\{w(\tau) = a\} = \int_0^\infty d\varphi_1(t) = \lim_{\lambda\to 0} \int_0^\infty e^{-\lambda t}\, d\varphi_1(t),$$

we obtain

$$\mathbf{P}\{w(\tau) = a\} = \frac{b}{b-a}, \quad \mathbf{P}\{w(\tau) = b\} = \frac{-a}{b-a}.$$

3. *Since* $\mathbf{M}\tau = -(d/d\lambda)\mathbf{M}e^{-\lambda\tau}|_{\lambda=0}$, we have $\mathbf{M}\tau = -ab$.

4.
$$\begin{aligned}\mathbf{M}\tau w(\tau) &= a\int_0^\infty t\, d\varphi_1(t) + b\int_0^\infty t\, d\varphi_2(t) \\ &= -\frac{d}{d\lambda}\left.\frac{a \sinh b\sqrt{2\lambda} - b \sinh a\sqrt{2\lambda}}{\sinh (b-a)\sqrt{2\lambda}}\right|_{\lambda=0} \\ &= \tfrac{1}{3}ab(a+b) = -\tfrac{1}{3}\mathbf{M}\big(w(\tau)\big)^3.\end{aligned}$$

5. *By using formula (2.6) and the relation*

$$(-1)^k \frac{d^k}{d\lambda^k}\mathbf{M}e^{-\lambda\tau}\bigg|_{\lambda=0} = \mathbf{M}\tau^k,$$

we find that for any k there exists a C_k such that $\mathbf{M}\tau_k \le C_k ab(b-a)^{2k-2}$.

LEMMA 2. *If $w(t)$ is a Brownian process, $w(0) = 0$, and τ is defined as in Lemma 1, then the process $w_1(t) = w(t+\tau) - w(\tau)$ will also be a Brownian process, and for all k and $t_1 < t_2 < \cdots < t_k$, the random variables $w_1(t_1), w_1(t_2), \ldots, w_1(t_n)$ will be independent of the two random variables τ and $w(\tau)$.*

Proof. Suppose that τ_n is the smallest number of the form k/n for which

$$\left(w\left(\frac{k}{n}\right) - a\right)\left(w\left(\frac{k}{n}\right) - b\right) \ge 0.$$

Then $\tau_n \to \tau$, $w(\tau_n) \to w(\tau)$ with probability 1, and consequently $w_1^{(n)}(t) = w(\tau_n + t) - w(\tau_n)$ for every t with probability 1 converges to $w_1(t) = w(\tau + t) - w(\tau)$. Therefore to prove the lemma it is sufficient to show that the process $w_1^{(n)}(t)$ is a Brownian process and that $w_1^{(n)}(t_1), \ldots, w_1^{(n)}(t_k)$ are independent of the two random variables $w_1^{(n)}(\tau_n)$ and τ_n. But for arbitrary $x_1, x_2, \ldots, x_k$ in an interval Δ and for any natural number l, the following relation is valid:

$$\mathbf{P}\left\{w_1^{(n)}(t_1) < x_1, \ldots, w_1^{(n)}(t_k) < x_k, w(\tau_n) \in \Delta, \tau_n = \frac{l}{n}\right\}$$

$$= \mathbf{P}\left\{w\left(t_1 + \frac{l}{n}\right) - w\left(\frac{l}{n}\right) < x_1, \ldots, w\left(t_k + \frac{l}{n}\right) - w\left(\frac{l}{n}\right) < x_k, w\left(\frac{l}{n}\right) \in \Delta\right\};$$

$$\left\{\left(w\left(\frac{l}{n}\right) - a\right)\left(w\left(\frac{l}{n}\right) - b\right) < 0, \ldots, \left(w\left(\frac{l-1}{n}\right) - a\right)\left(w\left(\frac{l-1}{n}\right) - b\right) < 0, \left(w\left(\frac{l}{n}\right) - a\right)\left(w\left(\frac{l}{n}\right) - b\right) \geq 0\right\}$$

$$= \mathbf{P}\left\{w\left(t_1 + \frac{l}{n}\right) - w(t_1) < x_1, \ldots, w\left(t_k + \frac{l}{n}\right) - w\left(\frac{l}{n}\right) < x_k\right\}$$

$$\times \mathbf{P}\left\{w\left(\frac{l}{n}\right) \in \Delta, a < w\left(\frac{j}{n}\right) < b,\ j = 1, 2, \ldots, l-1, \left(w\left(\frac{l}{n}\right) - a\right)\left(w\left(\frac{l}{n}\right) - b\right) \geq 0\right\}$$

$$= \mathbf{P}\{w(t_1) < x_1, \ldots, w(t_k) < x_k\}\mathbf{P}\left\{w(\tau_n) \in \Delta, \tau_n = \frac{l}{n}\right\}.$$

This proves the lemma.

LEMMA 3. *Let the variable ξ be independent of $w(t)$ and have a continuous distribution function $F(x)$, let $M\xi = 0$ and $G(x) > 0$ for $x < 0$ so that $\int_x^{G(x)} y\, dF(y) = 0$, and for $x > 0$, let $G(x) < 0$ so that $\int_{G(x)}^{x} y\, dF(y) = 0$. Then if τ is the smallest root of the equation $(w(t) - \xi)(w(t) - G(\xi)) = 0$, the variable $w(\tau)$ has the same distribution as ξ.*

Proof. The existence of the function $G(x)$ follows from the continuity of $F(y)$ and $\int_{-\infty}^{\infty} y\, dF(y) = 0$. Clearly $G(0) = 0$ and $G(x)$ is nonincreasing so that $G(x)$ can be defined in such a way that $G(G(x)) = x$. On the basis of Corollary 2 of Lemma 1, we may write

$$\mathbf{P}\{w(\tau) = \xi/\xi\} = \frac{|G(\xi)|}{|\xi| + |G(\xi)|}, \qquad \mathbf{P}\{w(\tau)G(\xi)/\xi\} = \frac{|\xi|}{|\xi| + G(\xi)|}.$$

Therefore

$$\mathbf{P}\{0 < w(\tau) < x\} = \int_0^x \frac{-G(y)}{y - G(y)}\, dF(y) + \int_{G(x)}^0 \frac{-y}{-y + G(y)}\, dF(y)$$

$$= \int_0^x \frac{-G(y)}{y - G(y)}\, dF(y) - \int_0^x \frac{-G(z)}{z - G(z)}\, dF(G(z)).$$

[In the second integral, make the substitution $y = G(z)$ and use the relation $G(G(z)) = z$.] We now note that the relation $\int_{G(x)}^x y\, dF(y) = 0$ follows from the fact that $z\, dF(z) = G(z)\, dF(G(z))$. Consequently

$$\mathbf{P}\{0 < w(\tau) < x\} = \int_0^x \frac{-G(y)}{y - G(y)}\, dF(y) + \int_0^x \frac{z\, dF(z)}{z - G(z)} = \int_0^x dF(y).$$

In an analogous fashion, we may show that

$$\mathbf{P}\{x < w(\tau) < 0\} = \int_x^0 dF(y).$$

This proves the lemma.

Remark 1. It follows from Corollary 3 of Lemma 1 that $\mathbf{M}(\tau|\xi) = |\xi|\,|G(\xi)|$. Therefore

$$\mathbf{M}\tau = \mathbf{M}|\xi|\,|G(\xi)| = -\int_{-\infty}^{\infty} xG(x)\, dF(x) = -\int_{-\infty}^{\infty} (G(x))^2\, dF(G(x))$$

$$= \int_{-\infty}^{\infty} x^2\, dF(x) = D\xi.$$

[Again, we make use of the equality $x\, dF(x) = G(x)\, dF(G(x))$.]

Remark 2.

$$\mathbf{M}\tau^k \leq C_k \mathbf{M}|G(\xi)| \cdot |\xi| (|G(\xi)| + |\xi|)^{2k-2}$$

$$\leq C_k 2^{2k-1} \mathbf{M} \left(|G(\xi)| \cdot |\xi| \frac{|G(\xi)|^{2k-1} + |\xi|^{2k-1}}{|\xi| + |G(\xi)|} \right)$$

$$= C_k 2^{2k-1} \mathbf{M}|w(\tau)|^{2k} = 2^{2k-1} C_k \mathbf{M}|\xi|^{2k}.$$

Remark 3. Clearly, if $|\xi| \leq C$, then $|G(\xi) \leq C$, which means that $|w(\tau)| \leq C$ and $\sup_{s \leq \tau} |w(s)| \leq C$.

We now proceed to prove the theorem for the case in which the variables $\xi_1, \xi_2, \ldots, \xi_n$ have continuous distributions. Suppose that the process $w(t)$ is independent of the variables $\xi_1, \xi_2, \ldots, \xi_n$, that $F_1(x), F_2(x), \ldots, F_n(x)$ are the distribution functions of these variables, and that $G_k(x)$ satisfies the conditions of Lemma 3 for $F_k(x)$. We denote by τ_1 the first zero of the equation $(w(t) - \xi_1)(w(t) - G_1(\xi_1)) = 0$. We set $w_1(t) = w(t + \tau_1) -$

$w(\tau_1)$. $w_1(t)$ does not depend on the pair $w(\tau_1)$, τ_1. Suppose that τ_2 is the first zero of the equation $(w_1(t) - \xi_2)(w_1(t)) - G(\xi_2)) = 0$; then $w_1(\tau_2)$ and τ_2 do not depend on $w_1(\tau_1)$ and τ_1. We define $w_2(t) = w_1(t + \tau_2) - w_1(\tau_2)$. This process does not depend either on $w_1(\tau_2)$ and τ_2 or on $w(\tau_1)$ and τ_1. Suppose that τ_3 is the first zero of the equation $(w_2(t) - \xi_3)(w_2(t) - G_3(\xi_3)) = 0$. Then the pairs of variables $w(\tau_1), \tau_1$; $w_1(\tau_2), \tau_2$; $w_2(\tau_3), \tau_3$ are mutually independent, and the variables $w(\tau_1)$, $w_1(\tau_2)$, $w_2(\tau_3)$ are distributed in the same way as the variables ξ_1, ξ_2, ξ_3. Continuing this process, we form the processes $w_1(t), w_2(t), \ldots, w_{n-1}(t)$ and the variables $\tau_1, \tau_2, \ldots, \tau_n$ in such a way that $w_k(t) = w_{k-1}(\tau_k + t) - w_{k-1}(t_k)$; thus $\tau_1, \tau_2, \ldots, \tau_n$ will be independent and $w(\tau_1), w_1(\tau_2), \ldots, w_{n-1}(\tau_n)$ will have the same joint distribution as the $\xi_1, \xi_2, \ldots, \xi_n$. It follows from the definition of the processes that

$$w_k(\tau_{k+1}) = w\left(\sum_1^{k+1} \tau_i\right) - w\left(\sum_1^{k} \tau_i\right).$$

To show that assertions (a)—(c) are satisfied, it is sufficient to use the remarks following Lemma 3. The theorem is proved for the case in which the ξ_k have continuous distributions. The proof for the general case can be obtained by taking the limit of variables with continuous distributions.

3. The probability of determining a sequence of sums between two bounds. Suppose that the variables $\xi_1^{(n)}, \xi_2^{(n)}, \ldots, \xi_n^{(n)}$ are independent and have identical distributions, that $\mathbf{M}\xi_i^{(n)} = 0$, that $D\xi_i^{(n)} = 1/n$ and that there exists a C such that $\mathbf{P}\{|\xi_i^{(n)}| \leq C(1/\sqrt{n})\} = 1$. We define $S_{nk} = \sum_{i=1}^k \xi_i^{(n)}$. Suppose that the functions $g_1(t)$ and $g_2(t)$ are defined for $t \geq 0$ and that they satisfy the following conditions:

1. $g_1(0) < 0 < g_2(0)$.
2. There exists a K such that

$$|g_1(t+h) - g_1(t)| \leq Kh, \; |g_2(t+h) - g_2(t)| \leq Kh.$$

We denote by Q_n the probability

$$Q_n = \mathbf{P}\left\{g_1\left(\frac{k}{n}\right) < S_{nk} < g_2\left(\frac{k}{n}\right), \; k = 1, 2, \ldots, n\right\} \tag{3.1}$$

and by Q the probability

$$Q = \mathbf{P}\{g_1(t) < w(t) < g_2(t), \; t \in [0, 1]\}, \tag{3.2}$$

where $w(t)$ is a Brownian process. Then we have the following theorem.

THEOREM. *There exists a constant L, depending only on K, C, $g_1(0)$ and $g_2(0)$ such that*

$$|Q_n - Q| \leq L \frac{\ln n}{\sqrt{n}}. \tag{3.3}$$

To prove this theorem, we need several auxiliary propositions.

LEMMA 1. *Suppose that $\eta_1, \eta_2, \ldots, \eta_n$ are independent random variables such that $\mathbf{M}\eta_i = 0$, $D\eta_i = 1/n$, and $\mathbf{M}\eta_i^4 \leq H/n^2$. Then*

$$\mathbf{P}\left\{\sup_{1\leq k\leq n}\left|\sum_{i=1}^{k}\eta_i\right| > 2\ln n\right\} \leq \frac{1}{n}\left(H + 2\exp\left(\sqrt{H} + \frac{e}{2}\right)\right).$$

Proof. We define $\eta_k' = \eta_k$ if $|\eta_k| \leq 1$, $\eta_k' = 0$ and $\eta_k'' = \eta_k - \eta_k'$ if $|\eta_k| > 1$. Then

$$\mathbf{P}\left\{\sup_k\left|\sum_{i=1}^{k}\eta_i\right| > 2\ln n\right\} \leq \mathbf{P}\left\{\sup_k\left|\sum_{i=1}^{k}\eta_i'\right| > 2\ln n\right\}$$

$$+ \mathbf{P}\left\{\sup_k |\eta_k''| > 0\right\} \leq \sum_{k=1}^{n}\mathbf{P}\left\{\left|\sum_{i=1}^{k}\eta_i'\right| > 2\ln n\right\} + \sum_{k=1}^{n}\mathbf{P}\{|\eta_k| > 1\}$$

$$\leq \sum_{k=1}^{n}\mathbf{P}\left\{\sum_{i=1}^{k}\eta_k' > 2\ln n\right\} + \sum_{k=1}^{n}\mathbf{P}\left\{-\sum_{i=1}^{k}\eta_k' > 2\ln n\right\}$$

$$+ \sum_{k=1}^{n}\mathbf{M}|\eta_k|^4 \leq \sum_{k=1}^{n}\left(\frac{\mathbf{M}e^{\sum_{i=1}^{k}\eta_i'}}{n^2} + \frac{\mathbf{M}e^{-\sum_{i=1}^{k}\eta_i'}}{n^2}\right) + \frac{H}{n}.$$

(Here, we used Chebyshev's inequality.) Then

$$\mathbf{M}e^{\sum_{i=1}^{k}\eta_i'} = \prod_{i=1}^{k}\mathbf{M}e^{\eta_i'} = \prod_{i-1}^{k}\mathbf{M}(1 + \eta_i' + \tfrac{1}{2}(\eta_i')^2 e^{\Theta_i'\eta_i'}),$$

where $0 < \Theta_i < 1$. Since $|\eta_i'| \leq 1$, we obtain

$$\mathbf{M}e^{\sum_{i=1}^{k}\eta_i'} \leq \prod_{i=1}^{k}(1 + |\mathbf{M}\eta_i'| + \tfrac{1}{2}e\mathbf{M}\eta_i^2).$$

But

$$|\mathbf{M}\eta_i'| = |\mathbf{M}\eta_i''| \leq \mathbf{M}|\eta_i''| \leq \sqrt{\mathbf{M}|\eta_i''|^2 \cdot \mathbf{P}\{|\eta_i''| > 0\}}$$

$$\leq \sqrt{\mathbf{M}|\eta_i|^2\mathbf{M}|\eta_i|^4} \leq \frac{1}{n}\sqrt{H}.$$

Therefore

$$\mathbf{M}e^{\sum_{i=1}^{k}\eta_i'} \leq \prod_{i=1}^{k}\left(1 + \frac{\sqrt{H}}{n} + \frac{e}{2n}\right) \leq \exp\left\{\frac{k}{n}\left(\sqrt{H} + \frac{e}{2}\right)\right\}$$

$$\leq \exp\left\{\sqrt{H} + \frac{e}{2}\right\}.$$

In an analogous fashion, we could show that

$$\mathbf{M}e^{-\Sigma_{i=1}^{k}\eta_i'} \leq \exp\left(\sqrt{H} + \frac{l}{2}\right).$$

Thus

$$\mathbf{P}\left\{\sup_k \left|\sum_{i=1}^{k} \eta_i\right| > 2\ln n\right\} \leq \frac{1}{n}\left(H + 2\exp\left(\sqrt{H} + \frac{e}{2}\right)\right).$$

This proves the lemma.

LEMMA 2. *For $h > 0$ and $k > 0$,*

$$\mathbf{P}\left\{\sup_{0\leq t\leq T}(w(t) - kt) < h\right\} = \frac{1}{\sqrt{2\pi T}}\int_{-h}^{h}\exp\left\{-\frac{(z+kT)^2}{2T}\right\}dz + \frac{1-e^{-2kh}}{\sqrt{2\pi T}}\int_{-\infty}^{-h}\exp\left\{-\frac{(z+kT)^2}{2T}\right\}dz. \tag{3.4}$$

Proof. We note that the conditional distributions $w(t) - kt$ under the condition $w(T) - kT = z$ do not depend on k. In fact, the distributions of the process $w(t) - kt - (t/T)[w(T) - kT] = w(t) - (t/T)w(T)$ do not depend on the parameter k, and the process $w(t) - (t/T)w(T)$ does not depend on the value of $w(T)$. To prove this last assertion, it is sufficient (since $w(t) - (t/T)w(T)$ and $w(T)$ have common Gaussian distributions) to note that

$$\mathbf{M}\left(w(t) - \frac{t}{T}w(T)\right)w(T) = \mathbf{M}w(t)w(T) - \frac{t}{T}\mathbf{M}w(T)^2$$
$$= \mathbf{M}w(t)^2 + \mathbf{M}w(t)[w(T) - w(t)] - t = 0.$$

Thus the distributions of $w(t) - kt$ under the condition that $w(T) - kT = z$ coincide with the distributions of $w(t)$ under the condition that $w(T) = z$. Therefore

$$\mathbf{P}\left\{\sup_{0\leq t\leq T}(w(t) - kt) < h\right\}$$
$$= \int_{-\infty}^{h}\mathbf{P}\left\{\sup_{0\leq t\leq T}(w(t) - kt) < h/w(T) = z + kt\right\}\frac{1}{\sqrt{2\pi T}}e^{-(z+kT)^2/2T}dz$$
$$= \int_{-\infty}^{h}\mathbf{P}\left\{\sup_{0\leq t\leq T}w(t) < h/w(T) = z\right\}\frac{1}{\sqrt{2\pi T}}e^{-(z+kT)^2/2T}\,dz.$$

In order to evaluate

$$\mathbf{P}\left\{\sup_{0\leq t\leq T}w(t) < h/w(T) = z\right\},$$

we find the joint distribution of the variables $\sup_{0\leq t\leq T} w(t)$ and $w(T)$. If $z < h$, then

$$\mathbf{P}\left\{\sup_{0\leq t\leq T} w(t) < h, w(T) < z\right\}$$

$$= \mathbf{P}\,\{w(T) < z\} - \mathbf{P}\left\{\sup_{0\leq t\leq T} w(t) \geq h, w(T) < z\right\}.$$

Suppose that τ is the time of first passage through the point h by the process $w(t)$. Then

$$\begin{aligned}\mathbf{P}\left\{\sup_{0\leq t\leq T} w(t) \geq h, w(T) < z\right\} &= \mathbf{P}\{\tau < T, w(T) < z\}\\ &= \mathbf{P}\{\tau < T, w(T) - w(\tau) + w(\tau) < z\}\\ &= \mathbf{P}\{\tau < T, w(T) - w(\tau) < z - h\}\\ &= \mathbf{P}\{\tau < T, w(T) - w(\tau) > h - z\},\end{aligned}$$

because $w(T) - w(\tau)$ for fixed τ has a symmetrical distribution. Remembering that $w(\tau) = h$, we obtain

$$\begin{aligned}\mathbf{P}\left\{\sup_{0\leq t\leq T} w(t) \geq h, w(T) < z\right\} &= \mathbf{P}\{\tau < T, w(T) > 2h - z\}\\ &= \mathbf{P}\{w(T) > 2h - z\},\end{aligned}$$

because the event $\{w(\mathbf{T}) > 2h - z\}$ implies the event $\{\tau < T\}$. Therefore

$$\mathbf{P}\left\{\sup_{0\leq t\leq T} w(t) < h; w(T) < z\right\} = \mathbf{P}\{w(T) < z\} - \mathbf{P}\{w(T) > 2h - z\}.$$

Hence

$$\mathbf{P}\left\{\sup_{0\leq t\leq T} w(t) < h/w(T) = z\right\}$$

$$= \frac{\dfrac{\partial}{\partial z}\left(\mathbf{P}\{w(T) < z\} - \mathbf{P}\{w(T) > 2h - z\}\right)}{\dfrac{\partial}{\partial z}\mathbf{P}\{w(T) < z\}}$$

$$= 1 - e^{-(2h-z)^2/2T}\cdot e^{(z^2/2T)} = 1 - e^{-2h(h-z)/T}.$$

Thus

$$\mathbf{P}\left\{\sup_{0\leq t\leq T} (w(t) - kt) < h\right\}$$

$$= \frac{1}{\sqrt{2\pi T}}\int_{-\infty}^{h} e^{-(z+kT)^2/T}(1 - e^{-2h(h-z)/T})\,dz$$

(cont.)

$$= \frac{1}{\sqrt{2\pi T}} \int_{-\infty}^{h} e^{-(z+kT)^2/2T}\, dz - \frac{1}{\sqrt{2\pi T}} \int_{-\infty}^{h} e^{-((z+kT-2h)^2/2T)-2kh}\, dz$$

$$= \frac{1}{\sqrt{2\pi T}} \int_{-\infty}^{h} e^{-(z+kT)^2/2T}\, dz - \frac{e^{-2kh}}{\sqrt{2\pi T}} \int_{-\infty}^{-h} e^{-(z+kT)^2/2T}\, dz$$

$$= \frac{1}{\sqrt{2\pi T}} \int_{-\infty}^{h} e^{-(z+kT)^2/2T}\, dz - \frac{1 - e^{-2kh}}{\sqrt{2\pi T}} \int_{-\infty}^{-h} e^{-(z+kT)^2/2T}\, dz.$$

This proves the lemma.

COROLLARY. *For $k \geq 0$, $h \geq 0$,*

$$\mathbf{P}\left\{\sup_{0 \leq t \leq T} (w(t) - kt) < h\right\} \leq \frac{2h}{\sqrt{2\pi T}} + 2kh.$$

LEMMA 3. *Suppose that $\psi(x)$ is the characteristic function of the interval $(0, 1)$ and that τ is the time of first passage through the point a by the process $w(t)$. Then*

$$\mathbf{M}\left(\frac{\psi(\tau)}{\sqrt{1-\tau}}\right)^{3/2} \leq \frac{4}{a^2\sqrt{2\pi}}.$$

Proof. Because of the symmetry of the process $w(t)$, it will be sufficient to examine the case $a > 0$. By using Lemma 2, we obtain for $k = 0$

$$\mathbf{P}\{\tau > s\} = \mathbf{P}\left\{\sup_{0 \leq t \leq s} w(t) < a\right\} = \frac{2}{\sqrt{2\pi s}} \int_{0}^{a} e^{-(x^2/2s)}\, dx$$

$$= \frac{2}{\sqrt{2\pi}} \int_{0}^{(a/\sqrt{s})} e^{-(x^2/2)}\, dx.$$

Therefore the density function of the variable τ will be

$$p_\tau(s) = \frac{a}{\sqrt{2\pi}} s^{-3/2} e^{-(a^2/2s)};$$

consequently

$$p_\tau(s) \leq \frac{1}{a^2\sqrt{2\pi}}.$$

Therefore

$$\mathbf{M}\left(\frac{\psi(\tau)}{\sqrt{1-\tau}}\right)^{3/2} = \int_0^1 \left(\frac{1}{\sqrt{1-s}}\right)^{3/2} p_\tau(s)\, ds$$

$$\leq \frac{1}{a^2\sqrt{2\pi}} \int_0^1 (1-s)^{-3/4}\, ds = \frac{4}{a^2\sqrt{2\pi}}.$$

This proves the lemma.

LEMMA 4. *Suppose that $g(t)$ is defined for $t \in [0, 1]$ and that it satisfies the following conditions:*

1. $g(0) > 0$.
2. *There exists a K such that $|g(t_2) - g(t_1)| < K(t_2 - t_1)$. Then*

$$\mathbf{P}\left\{0 < \sup_{0\le t\le 1} (w(t) - g(t)) < h\right\} \le 2h\left[\frac{e^{K^2}}{\sqrt{2\pi}}\left[\frac{4}{g(0)^2\sqrt{2\pi}}\right]^{2/3} + K\right].$$

Proof. Suppose that τ' is the first zero of the difference $w(t) - g(t)$. Then, if $\psi^{(x)}$ is the density function on $(0, 1)$,

$$\begin{aligned}\mathbf{P}&\left\{0 < \sup_{0\le t\le 1} [w(t) - g(t)] < h\right\}\\ &= \mathbf{MP}\left\{\sup_{\tau'<t\le 1} [w(t) - g(t)] < h/\tau'\right\}\psi(\tau')\\ &= \mathbf{MP}\left\{\sup_{\tau'<t\le 1} [w(t) - w(\tau') - (g(t) - g(\tau'))] < h/\tau'\right\}\psi(\tau')\\ &\le \mathbf{MP}\left\{\sup_{\tau'<t\le 1} [w(t) - w(\tau') - k(t - \tau')] < h/\tau'\right\}\psi(\tau').\end{aligned}$$

But the process $w(t) - w(\tau')$, for fixed τ', has the same distributions as does a Brownian process. (This can be shown in the same way as in Lemma 2 of Section 2.) On the basis of the corollary to Lemma 2, we may assert that

$$\mathbf{P}\left\{\sup_{\tau<t<1} [w(t) - w(\tau') - k(t - \tau')] < h/\tau'\right\} \le \frac{2h}{\sqrt{2\pi(1-\tau')}} + 2Kh.$$

Thus

$$\mathbf{P}\left\{0 < \sup_{0\le t\le 1} (w(t) - g(t)) < h\right\} \le \frac{2h}{\sqrt{2\pi}}\mathbf{M}\frac{\psi(\tau')}{\sqrt{1-\tau'}} + 2Kh.$$

We note now that the measure μ_1 corresponding to the process $w(t) - (g(t) - g(0))$ is absolutely continuous with respect to the measure μ corresponding to the process $w(t)$. (This follows from Theorem 3 of Section 1, Chapter 5.) Then

$$\frac{d\mu_1}{d\mu}(w(t)) = \exp\left\{-\int_0^1 g(t)\,dw(t) - \tfrac{1}{2}\int_0^1 g'(t)^2\,dt\right\}.$$

Therefore for an arbitrary functional $F(x(t))$, we have the following relation:

$$\mathbf{M}F(w(t) - g(t) + g(0)) = \mathbf{M}F(w(t))\frac{d\mu_1}{d\mu}(w(t));$$

consequently

$$\mathbf{M}\frac{\psi(\tau_k')}{\sqrt{1-\tau'}} = \mathbf{M}\frac{\psi(\tau)}{\sqrt{1-\tau}} \times \exp\left\{-\int_0^1 g'(t)\,dw(t) - \tfrac{1}{2}\int_0^1 g'(t)^2\,dt\right\},$$

where τ is the first zero of $w(t) - g(0)$. By using Hölder's inequality, we obtain

$$\mathbf{M}\frac{\psi(\tau')}{\sqrt{1-\tau'}} \leq \left[\mathbf{M}\left(\frac{\psi(\tau)}{\sqrt{1-\tau}}\right)^{3/2}\right]^{2/3}$$

$$\times\left[\mathbf{M}\exp\left\{-3\int_0^1 g'(t)\,dw(t) - \frac{3}{2}\int_0^1 g'(t)^2\,dt\right\}\right]^{1/3}$$

$$\leq \left[\frac{4}{g(0)^2\sqrt{2\pi}}\right]^{2/3}\exp\left\{-\tfrac{1}{2}\int_0^1 g'(t)^2\,dt\right\}$$

$$\times\left[\mathbf{M}\exp\left\{-3\int_0^1 g'(t)\,dw(t)\right\}\right]^{1/3}.$$

The variable $3\int_0^1 g'(t)\,dw(t)$ has a Gaussian distribution with mean zero and with dispersion $9\int_0^1 g'(t)^2\,dt$; therefore

$$\mathbf{M}\exp\left\{-3\int_0^1 g'(t)\,dw(t)\right\} = \exp\left\{\tfrac{9}{2}\int_0^1 g'(t)^2\,dt\right\}.$$

Since $g'(t)^2 \leq K^2$, we have

$$\mathbf{M}\frac{\psi(\tau')}{\sqrt{1-\tau'}} \leq \left[\frac{4}{g(0)^2\sqrt{2\pi}}\right]^{2/3} e^{K^2}.$$

This proves the lemma.

We can now prove the theorem. It follows from the theorem of Section 2 that it is possible to construct variables $\tau_1, \tau_2, \ldots, \tau_n$ such that $w(\tau_1)$, $w(\tau_1+\tau_2)$, $\ldots$, $w(\tau_1+\tau_2+\cdots+\tau_n)$ will have the same distributions as do $s_{n1}, s_{n2}, \ldots, s_{nn}$; then $\tau_1, \tau_2, \ldots, \tau_n$ will be independent and have identical distributions, $\mathbf{M}\tau_i = D\xi_{ni} = 1/n$, and $\mathbf{M}\tau_i^m < L_m\mathbf{M}\xi_{ni}^{2m} \leq L_m(c^{2m}/n^m)$ since, by hypothesis, $|\xi_{ni}| \leq (c/\sqrt{n})$. Also, for $s \in (\sum_{i=1}^k \tau_i, \sum_{i=1}^{k+1}\tau_i)$, we have

$$s \in \left(\sum_{i=1}^k \tau_i, \sum_{i=1}^{k+1}\tau_i\right),$$

$$\left|w(s) - w\left(\sum_{i=1}^k \tau_i\right)\right| \leq \frac{c}{\sqrt{n}}. \tag{3.5}$$

We define $D\tau_i = b^2/n^2$, $\eta_i^{(n)} = (\sqrt{n}/b)[\tau_i - (1/n)]$, and $\zeta_{nk} = \sum_{i=1}^k \eta_i^{(n)}$. Then

$$w(\tau_1 + \tau_2 + \cdots + \tau_k) = w[(k/n) + (b/\sqrt{n})\,\zeta_{nk}].$$

Thus

$$Q_n = \mathbf{P}\left\{g_1\left(\frac{k}{n}\right) < w\left(\frac{k}{n} + \frac{b}{\sqrt{n}}\zeta_{nk}\right) < g_2\left(\frac{k}{n}\right), \quad k = 1, 2, \ldots, n\right\}.$$

From (3.5), we have

$$\mathbf{P}\left\{g_1\left(\frac{k}{n}\right) + \frac{c}{\sqrt{n}} < w(s) < g_2\left(\frac{k}{n}\right) - \frac{c}{\sqrt{n}},\right.$$

$$\left. s \in \left(\frac{k}{n} + \frac{b}{\sqrt{n}}\zeta_{nk}, \frac{k+1}{n} + \frac{b}{\sqrt{n}}\zeta_{nk+1}\right), \quad k = 0, 1, \ldots, n-1\right\}$$

$$\leq Q_n \leq \mathbf{P}\left\{g_1\left(\frac{k}{n}\right) - \frac{c}{\sqrt{n}} < w(s) < g_2\left(\frac{k}{n}\right) + \frac{c}{\sqrt{n}},\right.$$

$$\left. s \in \left(\frac{k}{n} + \frac{b}{\sqrt{n}}\zeta_{nk}, \frac{k+1}{n} + \frac{b}{\sqrt{n}}\zeta_{nk+1}\right), \quad k = 0, 1, \ldots, n-1\right\}.$$

We now note that for

$$s \in \left(\frac{k}{n} + \frac{b}{\sqrt{n}}\zeta_{nk}, \frac{k+1}{n} + \frac{b}{\sqrt{n}}\zeta_{nk+1}\right),$$

we have

$$\left|g_i\left(\frac{k}{n}\right) - g_1(s)\right| < K\left(\frac{1}{n} + \frac{b}{\sqrt{n}}\sup_k |\zeta_{nk}|\right).$$

Therefore if $\alpha_n = (c/\sqrt{n}) + K[(1/n) + (b/\sqrt{n})\sup_k |\zeta_{nk}|]$, we have the inequality

$$\mathbf{P}\left\{g_1(s) + \alpha_n < w(s) < g_2(s) - \alpha_n, \quad 0 \leq s \leq 1 + \frac{b}{\sqrt{n}}\zeta_{nn}\right\}$$

$$\leq Q_n \leq \mathbf{P}\left\{g_1(s) - \alpha_n < w(s) < g_2(s) + \alpha_n, \quad 0 \leq s \leq 1 + \frac{b}{\sqrt{n}}\zeta_{nn}\right\}.$$

The variables $\eta_i^{(n)}$ satisfy the conditions of Lemma 1 because

$$\mathbf{P}\left\{\sup_k |\zeta_{nk}| > 2\ln n\right\} \leq A_1\frac{1}{n},$$

where A_1 depends only on C. Consequently

$$\mathbf{P}\left\{g_1(s) + \alpha_n < w(s) < g_2(s) - \alpha_n,\ s \in \left[0, 1 + \frac{b}{\sqrt{n}}\zeta_{nn}\right]\right\}$$

$$> \mathbf{P}\left\{g_1(s) + \frac{c + 2b\ln n + K\sqrt{n^{-1}}}{\sqrt{n}} < w(s) < g_2(s) - \frac{c + 2b\ln n + K\sqrt{n^{-1}}}{\sqrt{n}},\ s \in \left[0, 1 + \frac{b\ln n}{\sqrt{n}}\right]\right\} - A_1\frac{1}{n}$$

$$= \mathbf{P}\left\{g_1\left(u\left(1 + \frac{2b\ln n}{\sqrt{n}}\right)\right) + \beta_n < w\left(u\left(1 + \frac{2b\ln n}{\sqrt{n}}\right)\right) < g_2\left(u\left(1 + \frac{2b\ln n}{\sqrt{n}}\right)\right) - \beta_n,\quad 0 \le u \le 1\right\} - A_1 \cdot \frac{1}{n},$$

where $\beta_n = (c + 2b\ln n + K\sqrt{n^{-1}})/\sqrt{n}$. But

$$\left|g_i\left(u\left(1 + \frac{2b\ln n}{\sqrt{n}}\right)\right) - g_i(u)\right| \le K\frac{2b\ln n}{\sqrt{n}}.$$

Therefore for $\gamma_n = \beta_n + K(2b\ln n/\sqrt{n})$,

$$\mathbf{P}\left\{g_1(s) + \alpha_n < w(s) < g_2(s) - \alpha_n,\quad 0 \le s \le 1 + \frac{b}{\sqrt{n}}\zeta_{nn}\right\}$$

$$> \mathbf{P}\left\{g_1(u) + \gamma_n < w\left(u\left(1 + \frac{2b\ln n}{\sqrt{n}}\right)\right) < g_2(u) - \gamma_n,\quad 0 \le u \le 1\right\} - A_1 \cdot \frac{1}{n}.$$

The process $w_1(u) = w(u(1 + (2b\ln n)/\sqrt{n}))/\sqrt{1 + (2b\ln n)/\sqrt{n}}$ will also be a Brownian process; consequently

$$\mathbf{P}\left\{g_1(s) + \alpha_n < w(s) < g_2(s) - \alpha_n,\ 0 \le s \le 1 + \frac{b}{\sqrt{n}}\zeta_{nn}\right\}$$

$$> \mathbf{P}\left\{\frac{g_1(u) + \gamma_n}{\sqrt{1 + (2b\ln n)/\sqrt{n}}} < w(u) < \frac{g_2(u) - \gamma_n}{\sqrt{1 + (2b\ln n)/\sqrt{n}}},\ 0 \le u \le 1\right\} - A_1 \cdot \frac{1}{n}.$$

In an analogous fashion, we can show that

$$\mathbf{P}\left\{g_1(s) - \alpha_n < w(s) < g_2(s) + \alpha_n, \quad 0 \le s < 1 + \frac{b}{\sqrt{n}}\zeta_{nn}\right\}$$

$$< \mathbf{P}\left\{\frac{g_1(u) - \gamma_n}{\sqrt{1 - (2b \ln n)/\sqrt{n}}} < w(u)\right.$$

$$\left. < \frac{g_2(u) + \gamma_n}{\sqrt{1 - (2b \ln n)/\sqrt{n}}}, \quad 0 \le u \le 1\right\} + A_1 \cdot \frac{1}{n}.$$

From the definition of γ_n, we may assert that there exists a constant H, dependent only on $g_1(0)$, $g_2(0)$, K, and c, such that

$$\left| g_1(u) - \frac{g_1(u) + \gamma_n}{\sqrt{1 + (2b \ln n)/\sqrt{n}}} \right| \le H \frac{\ln n}{\sqrt{n}},$$

$$\left| \frac{g_2(u) - \gamma_n}{\sqrt{1 + (2b \ln n)/\sqrt{n}}} - g_2(u) \right| \le H \frac{\ln n}{\sqrt{n}},$$

$$\left| g_1(u) - \frac{g_1(u) - \gamma_n}{\sqrt{1 - (2b \ln n)/\sqrt{n}}} \right| \le H \frac{\ln n}{\sqrt{n}},$$

$$\left| \frac{g_2(u) + \gamma_n}{\sqrt{1 - (2b \ln n)/\sqrt{n}}} - g_2(u) \right| \le H \frac{\ln n}{\sqrt{n}}.$$

Then

$$\mathbf{P}\left\{g_1(s) + \frac{H \ln n}{\sqrt{n}} < w(s) < g_2(s) - \frac{H \ln n}{\sqrt{n}}, \quad s \in [0, 1]\right\} - A_1 \cdot \frac{1}{n}$$

$$< Q_n < \mathbf{P}\left\{g_1(s) - \frac{H \ln n}{\sqrt{n}} < w(s)\right.$$

$$\left. < g_2(s) + \frac{H \ln n}{\sqrt{n}}, \quad s \in [0, 1]\right\} + A_1 \cdot \frac{1}{n}.$$

Also,

$$\mathbf{P}\left\{g_1(s) + \frac{H \ln n}{\sqrt{n}} < w(s) < g_2(s) - \frac{H \ln n}{\sqrt{n}}, \quad s \in [0, 1]\right\}$$

$$< Q < \mathbf{P}\left\{g_1(s) - \frac{H \ln n}{\sqrt{n}} < w(s) < g_2(s) + \frac{H \ln n}{\sqrt{n}}, \quad s \in [0, 1]\right\}.$$

Therefore

$$|Q_n - Q| \le \frac{2A_1}{n}$$

$$+ \mathbf{P}\left\{g_1(s) - \frac{H \ln n}{\sqrt{n}} < w(s) < g_2(s) + \frac{H \ln n}{\sqrt{n}},\ 0 \le s \le 1\right\}$$

$$- \mathbf{P}\left\{g_1(s) + \frac{H \ln n}{\sqrt{n}} < w(s) < g_2(s) - \frac{H \ln n}{\sqrt{n}},\ 0 \le s \le 1\right\} \le \frac{2A_1}{n}$$

$$+ \mathbf{P}\left\{- \frac{H \ln n}{\sqrt{n}} < \sup_{0 \le s \le 1} (g_1(s) - w(s)) < \frac{H \ln n}{\sqrt{n}}\right\}$$

$$+ \mathbf{P}\left\{- \frac{H \ln n}{\sqrt{n}} < \sup_{0 \le s \le 1} (w(s) - g_2(s)) < \frac{H \ln n}{\sqrt{n}}\right\}.$$

It follows from Lemma 4 that there exist constants H_1 (depending only on $|g_1(0)|$ and K) and H_2 (depending only on $|g_2(0)|$ and K) such that

$$\mathbf{P}\left\{- \frac{H \ln n}{\sqrt{n}} < \sup_s [g_1(s) - w(s)] < \frac{H \ln n}{\sqrt{n}}\right\} \le H_1 \cdot \frac{H \ln n}{\sqrt{n}},$$

$$\mathbf{P}\left\{- \frac{H \ln n}{\sqrt{n}} < \sup_s [w(s) - g_2(s)] < \frac{H \ln n}{\sqrt{n}}\right\} \le H_2 \cdot \frac{H \ln n}{\sqrt{n}}.$$

Thus

$$|Q_n - Q| \le \frac{2A_1}{n} + H(H_1 + H_2) \frac{\ln n}{\sqrt{n}}.$$

The proof of the theorem follows from this inequality.

4. The distribution of certain additive functionals of a sequence of sums of independent random variables. As in the preceding section, we shall examine a sequence of identically distributed independent variables $\xi_{n1}, \ldots, \xi_{nn}$ such that $\mathbf{M}\xi_{ni} = 0$, $D\xi_{ni} = 1/n$, and for some $C > 0$, $\mathbf{P}\{\xi_{ni} | > c/\sqrt{n}\} = 0$. Suppose that $s_{nk} = \sum_{k=1}^{k} \xi_{ni}$. Let us consider a sufficiently smooth function $f(t, x)$ such that all its derivatives with which we shall be concerned exist and are continuous. We denote by φ_n the random variable defined by the equation

$$\varphi_n = \frac{1}{n} \sum_{k=1}^{n} f\left(\frac{k}{n}, s_{nk}\right), \tag{4.1}$$

and consider the behavior of the distribution of φ_n as $n \to \infty$. Suppose that $\tau_1, \tau_2, \ldots, \tau_n$ are independent random variables for which the variables

$w(\tau_1), w(\tau_1 + \tau_2), \ldots, w(\tau_1 + \tau_2 + \cdots + \tau_n)$ have the same joint distribution as do $s_{n1}, s_{n2}, \ldots, s_{nn}$. As in the preceding section, we introduce the notation

$$D\tau_i = \frac{b^2}{n^2}, \qquad \eta_i^{(n)} = \frac{\sqrt{n}}{b}\left(\tau_i - \frac{1}{n}\right), \qquad \zeta_{nk} = \sum_{i=1}^{k} \eta_i^{(n)}.$$

The distribution of the variable φ_n will coincide with the distribution of the variable

$$\overline{\varphi}_n = \frac{1}{n}\sum_{k=1}^{n} f\left(\frac{k}{n}, w\left(\frac{k}{n} + \frac{b}{\sqrt{n}}\zeta_{nk}\right)\right). \tag{4.2}$$

Let us rewrite the expression (4.2) in the following form:

$$\begin{aligned}\overline{\varphi}_n = {} & \frac{1}{n}\sum_{k=1}^{n} f\left(\frac{k}{n} + \frac{b}{\sqrt{n}}\zeta_{nk}, w\left(\frac{k}{n} + \frac{b}{\sqrt{n}}\zeta_{nk}\right)\right) \\ & + \frac{1}{n}\sum_{k=1}^{n}\left[f\left(\frac{k}{n}, w\left(\frac{k}{n} + \frac{b}{\sqrt{n}}\zeta_{nk}\right)\right)\right. \\ & \left. - f\left(\frac{k}{n} + \frac{b}{\sqrt{n}}\zeta_{nk}, w\left(\frac{k}{n} + \frac{b}{\sqrt{n}}\zeta_{nk}\right)\right)\right].\end{aligned}$$

Let us examine separately both sums:

$$\begin{aligned}& \frac{1}{n}\sum_{k=1}^{n} f\left(\frac{k}{n} + \frac{b}{\sqrt{n}}\zeta_{nk}, w\left(\frac{k}{n} + \frac{b}{\sqrt{n}}\zeta_{nk}\right)\right) \\ & \quad = \sum_{k=0}^{n-1}\int_{(k/n)+(b/\sqrt{n})\zeta_{nk}}^{(k+1)/n+(b/\sqrt{n})\zeta_{nk}} f\left(\frac{k}{n} + \frac{b}{\sqrt{n}}\zeta_{nk}, w\left(\frac{k}{n} + \frac{b}{\sqrt{n}}\zeta_{nk}\right)\right) ds \\ & \qquad - \frac{b}{\sqrt{n}}\sum_{k=0}^{n-1} f\left(\frac{k}{n} + \frac{b}{\sqrt{n}}\zeta_{nk}, w\left(\frac{k}{n} + \frac{b}{\sqrt{n}}\zeta_{nk}\right)\right)[\zeta_{nk+1} - \zeta_{nk}] \\ & \qquad + \frac{f(0, 0)}{n} - \frac{f\left(1, w\left(1 + \frac{b}{\sqrt{n}}\zeta_{nn}\right)\right)}{n}.\end{aligned}$$

But

$$\begin{aligned}& \int_{(k/n)+(b/\sqrt{n})\zeta_{nk}}^{(k+1)/n+(b/\sqrt{n})\zeta_{nk+1}} f\left(\frac{k}{n} + \frac{b}{\sqrt{n}}\zeta_{nk}, w\left(\frac{k}{n} + \frac{b}{\sqrt{n}}\zeta_{nk}\right)\right) ds \\ & \quad = \int_{(k/n)+(b/\sqrt{n})\zeta_{nk}}^{(k+1)/n+(b/\sqrt{n})\zeta_{nk+1}} f(s)w(s)\, ds +\end{aligned}$$

(cont.)

$$+ \int_{(k/n)+(b/\sqrt{n})\zeta_{nk}}^{(k+1)/n+(b/\sqrt{n})\zeta_{nk+1}} \left[f\left(\frac{k}{n} + \frac{b}{\sqrt{n}}\zeta_{nk}, w\left(\frac{k}{n} + \frac{b}{\sqrt{n}}\zeta_{nk}\right)\right) - f(s, w(s)) \right] ds$$

$$= \int_{(k/n)+(b/\sqrt{n})\zeta_{nk}}^{(k+1)/n+(b/\sqrt{n})\zeta_{nk+1}} f(s, w(s))\, ds$$

$$- \int_{(k/n)+(b/\sqrt{n})\zeta_{nk}}^{(k+1)/n+(b/\sqrt{n})\zeta_{nk+1}} f'_t\left(\Theta_s, w\left(\frac{k}{n} + \frac{b}{\sqrt{n}}\zeta_{nk}\right)\right)\left[s - \frac{k}{n} - \frac{b}{\sqrt{n}}\zeta_{nk}\right] ds$$

$$- \int_{(k/n)+(b/\sqrt{n})\zeta_{nk}}^{(k+1)/n+(b/\sqrt{n})\zeta_{nk+1}} \left[f(s, w(s)) - f\left(s, w\left(\frac{k}{n} + \frac{b}{\sqrt{n}}\zeta_{nk}\right)\right)\right] ds.$$

[Here, Θ_s is some point in the interval $(k/n + (b/\sqrt{n})\zeta_{nk}, s)$.] If

$$|f'_t| < C_1, |f'_x| < C_2, |f''_{xx}| < C_3,$$

then

$$\left| \int_{(k/n)+(b/\sqrt{n})\zeta_{nk}}^{(k+1)/n+(b/\sqrt{n})\zeta_{nk+1}} f'_t\left(\Theta_s, w\left(\frac{k}{n} + \frac{b}{\sqrt{n}}\zeta_{nk}\right)\right)\left(s - \frac{k}{n} - \frac{b}{\sqrt{n}}\zeta_{nk}\right) ds \right|$$

$$\leq C_1 \left[\frac{1}{n} + \frac{b}{\sqrt{n}}(\zeta_{nk+1} - \zeta_{nk})\right]^2.$$

Further,

$$\int_{(k/n)+(b/\sqrt{n})\zeta_{nk}}^{(k+1)/n+(b/\sqrt{n})\zeta_{nk+1}} \left[f(s, w(s)) - f\left(s, w\left(\frac{k}{n} + \frac{b}{\sqrt{n}}\zeta_{nk}\right)\right)\right] ds = \beta_{nk} + \gamma_{nk},$$

where

$$\beta_{nk} = \int_{(k/n)+(b/\sqrt{n})\zeta_{nk}}^{(k+1)/n+(b/\sqrt{n})\zeta_{nk+1}} f'_x\left(s, w\left(\frac{k}{n} + \frac{b}{\sqrt{n}}\zeta_{nk}\right)\right) \times \left[w(s) - w\left(\frac{k}{n} + \frac{b}{\sqrt{n}}\zeta_{nk}\right)\right] ds,$$

$$\gamma_{nk} = \frac{1}{2}\int_{(k/n)+(b/\sqrt{n})\zeta_{nk}}^{(k+1)/n+(b/\sqrt{n})\zeta_{nk+1}} O\left(f''_{xx}\left(s, w\left(\frac{k}{n} + \frac{b}{\sqrt{n}}\zeta_{nk}\right)\right)\right) \times \left[w(s) - w\left(\frac{k}{n} + \frac{b}{\sqrt{n}}\zeta_{nk}\right)\right]^2 ds.$$

Clearly, $\mathbf{M}\beta_{ni}\beta_{nk} = 0$ for $i \neq k$, and

$$\mathbf{M}\beta_{nk}^2 \leq C_2^2\, \mathbf{M}\left[\frac{1}{n} + \frac{b}{\sqrt{n}}(\zeta_{nk+1} - \zeta_{nk})\right]$$

$$\times \int_{(k/n)+(b/\sqrt{n})\zeta_{nk}}^{(k+1)/n+(b/\sqrt{n})\zeta_{nk+1}} \left[w(s) - w\left(\frac{k}{n} + \frac{b}{\sqrt{n}}\zeta_{nk}\right)\right]^2 ds$$

$$\leq \frac{c^2}{n} C_2^2\, \mathbf{M}\left[\frac{1}{n} + \frac{b}{\sqrt{n}}(\zeta_{nk+1} - \zeta_{nk})\right]^2 = c^2 \frac{c_2^2}{n^3}(b^2 + 1),$$

because for $s \in [(k/n) + (b/\sqrt{n})\zeta_{nk}, ((k+1)/n) + (b/\sqrt{n})\zeta_{nk+1}]$, the inequality $|w(s) - w((k/\sqrt{n}) + (b/\sqrt{n})\zeta_{nk})| \leq c/\sqrt{n}$ is fulfilled since $|\xi_{ni}| \leq c/\sqrt{n}$. Also,

$$|\gamma_{nk}| \leq \frac{1}{2} C_3 \frac{c^2}{n}\left[\frac{1}{n} + \frac{b}{\sqrt{n}}(\zeta_{nk+1} - \zeta_{nk})\right].$$

Thus

$$\sum_{k=0}^{n-1} \int_{(k/n)+(b/\sqrt{n})\zeta_{nk}}^{(k+1)/n+(b/\sqrt{n})\zeta_{nk+1}} f\left(\frac{k}{n} + \frac{b}{\sqrt{n}}\zeta_{nk}, w\left(\frac{k}{n} + \frac{b}{\sqrt{n}}\zeta_{nk}\right)\right) ds$$

$$= \int_0^{1+(b/\sqrt{n})\zeta_{nn}} f(s, w(s))\, ds + O\left(\sum_{k=0}^{n-1}\left(\left[\frac{1}{n} + \frac{b}{\sqrt{n}}(\zeta_{nk+1} - \zeta_{nk})\right]^2\right.\right.$$

$$\left.\left. + \frac{1}{n}\left[\frac{b}{\sqrt{n}}(\zeta_{nk+1} - \xi_{nk}) + \frac{1}{n}\right]\right)\right) - \sum_{k=1}^{n} \beta_{nk}$$

$$= \int_0^{1+(b/\sqrt{n})\zeta_{nn}} f(s, w(s))\, ds + \frac{\epsilon_n}{n},$$

where $\mathbf{M}\epsilon_n^2$ is bounded by a constant independent of n.

Now let us examine the second sum:

$$\frac{1}{n}\sum_{k=1}^{n}\left[f\left(\frac{k}{n}, w\left(\frac{k}{n} + \frac{b}{\sqrt{n}}\zeta_{nk}\right)\right) - f\left(\frac{k}{n} + \frac{b}{\sqrt{n}}\zeta_{nk}, w\left(\frac{k}{n} + \frac{b}{\sqrt{n}}\zeta_{nk}\right)\right)\right]$$

$$= -\frac{1}{n}\sum_{k=1}^{n} f_t'\left(\frac{k}{n}, w\left(\frac{k}{n} + \frac{b}{\sqrt{n}}\zeta_{nk}\right)\right)\frac{b}{\sqrt{n}}\zeta_{nk}$$

$$- \frac{1}{2n}\sum_{k=1}^{n} f_t''\left(\Theta_k^{(n)}, w\left(\frac{k}{n} + \frac{b}{\sqrt{n}}\zeta_{nk}\right)\right)\frac{b^2}{n}\zeta_{nk}^2,$$

where $\Theta_k^{(n)}$ lies between k/n and $k/n + (b/\sqrt{n})\zeta_{nk}$. Consequently

$$\bar{\varphi}_n = \int_0^1 f(s, w(s))\,ds + \frac{b}{\sqrt{n}}\left[\frac{\sqrt{n}}{b}\int_1^{1+(b/\sqrt{n})\zeta_{nn}} f(s, w(s))\,ds\right.$$
$$- \sum_{k=0}^{n-1} f\left(\frac{k}{n} + \frac{b}{\sqrt{n}}\zeta_{nk}, w\left(\frac{k}{n} + \frac{b}{\sqrt{n}}\zeta_{nk}\right)\right)[\zeta_{nk+1} - \zeta_{nk}]$$
$$- \frac{1}{n}\sum_{k=1}^{n} f'_t\left(\frac{k}{n}, w\left(\frac{k}{n} + \frac{b}{\sqrt{n}}\zeta_{nk}\right)\right)\zeta_{nk} + \frac{\epsilon'_n}{n}, \tag{4.3}$$

where the $\mathbf{M}\epsilon_n'^2$ are bounded by a constant independent of n. For what follows, we shall need a lemma.

LEMMA. *Suppose that* $\mathbf{M}\,\xi_{ni}^3 = (1/n^{3/2})\mu$. *We denote by* $\zeta_n(t)$ *the process defined by* $\zeta_n(t) = \zeta_{nk}$ *for* $t \in ((k-1)/n, k/n)$ *and* $\zeta_n(0) = 0$. *Then the joint finite-dimensional distributions of the processes* $\zeta_n(t)$ *and* $w(t)$ *will converge to the finite-dimensional distributions of the pair of processes* $w_1(t)$ *and* $w(t)$, *each of which is a Brownian process, and* $\mathbf{M}w_1(t)w(t) = -t\mu/3b$.

Proof. Suppose that

$$w_n(t) = w\left(\frac{k}{n} + \frac{b}{\sqrt{n}}\zeta_{nk}\right) \quad \text{if} \quad t \in \left(\frac{k-1}{n}, \frac{k}{n}\right), \quad w_n(0) = 0.$$

It is sufficient for us to show that the joint finite-dimensional distributions of the pair of processes $\zeta_n(t)$ and $w_n(t)$ will converge to the joint finite-dimensional distributions of the processes $w_1(t)$ and $w(t)$ referred to in the statement of the lemma.

We note that the two-dimensional process $(\zeta_n(t); w_n(t))$ is a process with independent increments; $\zeta_n(t)$ and $w_n(t)$ are the sums of identically distributed independent terms $\zeta_{nk+1} - \zeta_{nk}$ and $w(\sum_{i=1}^{k+1}\tau_i) - w(\sum_{i=1}^{k}\tau_i)$. Since

$$\mathbf{M}(\zeta_{nk+1} - \zeta_{nk}) = 0, \qquad D(\zeta_{nk+1} - \zeta_{nk}) = \frac{1}{n},$$

$$\mathbf{M}\left(w\left(\sum_{i=1}^{k+1}\tau_i\right) - w\left(\sum_{i=1}^{k}\tau_i\right)\right) = 0,$$

$$D\left(w\left(\sum_{i=1}^{k+1}\tau_i\right) - w\left(\sum_{i=1}^{k}\tau_i\right)\right) = \frac{1}{n},$$

$$\mathbf{M}(\zeta_{nk+1} - \zeta_{nk})^4 = 0\left(\frac{1}{n^2}\right),$$

$$\mathbf{M}\left(w\left(\sum_{i=1}^{k+1}\tau_i\right) - w\left(\sum_{i=1}^{k}\tau_i\right)\right)^4 = 0\left(\frac{1}{n^2}\right),$$

the conditions under which the central limit theorem can be applied are satisfied; therefore the joint finite-dimensional distributions of $\zeta_n(t)$ and $w_n(t)$ will converge to the joint finite-dimensional distributions of the pair of processes $w_1(t)$ and $w(t)$, each of which is a Brownian process. To compute $\mathbf{M}w_1(t)w(t)$, we use the relation

$$\begin{aligned}\mathbf{M}w_1(t)w(t) &= t\mathbf{M}w_1(1)w(1) \\ &= t \lim_{n\to\infty} \cdot \mathbf{M} \sum_{k=0}^{n-1} \left[w\left(\frac{k+1}{n} + \frac{b}{\sqrt{n}} \zeta_{nk+1}\right)\right. \\ &\qquad \left. - w\left(\frac{k}{n} + \frac{b}{\sqrt{n}}\zeta_{nk}\right)\right] (\zeta_{nk+1} - \zeta_{nk}).\end{aligned}$$

But

$$\begin{aligned}&\mathbf{M}\left(w\left(\frac{k+1}{n} + \frac{b}{\sqrt{n}}\zeta_{nk+1}\right) - w\left(\frac{k}{n} + \frac{b}{\sqrt{n}}\zeta_{nk}\right)\right)(\zeta_{nk+1} - \zeta_{nk}) \\ &\qquad = \frac{\sqrt{n}}{b}\mathbf{M}\left[w\left(\frac{k+1}{n} + \frac{b}{\sqrt{n}}\zeta_{nk+1}\right) - w\left(\frac{k}{n} + \frac{b}{\sqrt{n}}\xi_{nk}\right)\right] \\ &\qquad\qquad \times \left(\frac{1}{n} + \frac{b}{\sqrt{n}}(\zeta_{nk+1} - \zeta_{nk})\right) \\ &\qquad = \frac{\sqrt{n}}{b}\mathbf{M}\left[w\left(\sum_{i=1}^{k+1}\tau_i\right) - w\left(\sum_{i=1}^{k}\tau_i\right)\right]\tau_{k+1} \\ &\qquad = \frac{\sqrt{n}}{b}\mathbf{M}\cdot\mathbf{M}(w(\tau_1)\tau_1/\xi_{n1}).\end{aligned}$$

(From the definition of τ_1 in Section 2, the variables τ_1 and $w(\tau_1)$ depend on ξ_{n1}.) From Corollary 4 of Lemma 1 of Section 2, we have

$$\mathbf{M}(w(\tau_1)\tau_1/\xi_{n1}) = -\tfrac{1}{3}\mathbf{M}(w(\tau_1)^3/\xi_{n1}).$$

Therefore $\mathbf{M}w_1(t)w(t) = -t\mu/3b$. This proves the lemma.

COROLLARY. *For any continuous bounded function $g(x)$,*

$$\begin{aligned}\mathbf{M}g\left(\int_0^1 f(s, w(s))\, ds\right)&\left[\frac{\sqrt{n}}{b}\int_1^{1+(b/\sqrt{n})\zeta_{nn}} f(s, w(s))\, ds\right. \\ &- \sum_{k=1}^{n-1} f\left(\frac{k}{n} + \frac{b}{\sqrt{n}}\zeta_{nk}, w\left(\frac{k}{n} + \frac{b}{\sqrt{n}}\zeta_{nk}\right)\right)[\zeta_{nk+1} - \zeta_{nk}]\end{aligned}$$

(cont.)

$$- \frac{1}{n} \sum_{k=1}^{n} f'_t \left(\frac{k}{n}, w \left(\frac{k}{n} + \frac{b}{\sqrt{n}} \zeta_{nk} \right) \right) \zeta_{nk} \Bigg]$$

$$\to \mathbf{M} g \left(\int_0^1 f(s, w(s))\, ds \right) \Bigg[f(1, w(1)) w_1(1)$$

$$- \int_0^1 f(s, w(s))\, dw_1(s) - \int_0^1 f'_t(s, w(s)) w_1(s)\, ds \Bigg], \qquad (4.4)$$

where $(w(s); w_1(s))$ form a two-dimensional Gaussian process with independent increments such that $w(s)$ and $w_1(s)$ are Brownian processes and

$$\mathbf{M} w(s) w_1(s) = -\frac{\mu t}{3b}.$$

To prove this assertion, we define the processes $\tilde{w}_n(s)$ and $\tilde{\zeta}_n(s)$ having the same joint distributions as $w(s)$ and $\zeta_n(s)$ and converging to the processes $\tilde{w}(s)$ and $\tilde{w}_1(s)$, whose joint distribution coincides with the joint distribution of the processes $w(s)$ and $w_1(s)$. (We may do this by the theorem of Section 6, Chapter 1.) Substituting $\tilde{w}_n(s)$ and $\tilde{\zeta}_n(s)$ in (4.4) in place of $w(s)$ and $\zeta_n(s)$, we may obtain a proof of the assertion by passing to the limit under the mathematical expectation sign and using the theorem of Section 3, Chapter 2, on passage to the limit under the stochastic integral sign. The results obtained above allow us to establish the following theorem on the asymptotic behavior of the distribution of the variable φ_n.

THEOREM. *Suppose that $g(x)$ and its first two derivatives are bounded. Then*

$$\mathbf{M} g(\varphi_n) = \mathbf{M} g \left(\int_0^1 f(s, w(s))\, ds \right)$$

$$+ \frac{\mu}{\sqrt{n}} \mathbf{M} g' \left(\int_0^1 f(s, w(s))\, ds \right) \Bigg[f(1, w(1)) w_1(1)$$

$$- \int_0^1 f(s, w(s))\, dw_1(s) - \int_0^1 f'_t(s, w(s)) w_1(s)\, ds \Bigg] + O\left(\frac{1}{\sqrt{n}} \right).$$

The proof of this theorem follows immediately if we use the expression (4.3) for $\overline{\varphi}_n$, apply Taylor's formula for $g(x)$ at the point $\int_0^1 f(s, w(s))\, ds$ with a second-order remainder term, Formula (4.4), and, finally, use the fact that the process $w_1(t) - (\mu/3b) w(t)$ does not depend on $w(t)$.

SUPPLEMENTARY REMARKS

CHAPTER 1

1. For the definitions of a random process, see the books by A. N. Kolmogorov (1) and J. L. Doob (4). Theorem 1 is a combination of Theorems 2.4 and 2.6 of Chapter 2 of Doob's book (4). Theorem 2 was first published in E. Slutskii's work (1); the proof is that of P. L. Dobrushin and A. M. Yaglom (1). The proof of Theorem 3 is contained in A. N. Kolmogorov's book (1), page 39.

2. The theory of conditional probabilities and mathematical expectations is developed by A. N. Kolmogorov in (1). A detailed exposition of the properties of conditional mathematical expectations and probabilities is found in Doob's book (4), Chapter 1, Sections 7–10.

3. The general picture of processes with independent increments is given by A. N. Kolmogorov in (5) (the case of finite dispersion) and by P. Lévy in (1). The theory of these processes is most completely expounded in the books of P. Lévy (3) and Doob (4).

4. The fundamentals of the general theory of Markov processes are given in A. N. Kolmogorov's work (2). A detailed exposition of the theory of Markov processes is contained in the books of Doob (4) and E. B. Dynkin (6).

5. The theory of martingales is contained in Doob's book (4), Chapter 7.

6. The theorem in this section is a minor modification of Theorem 3.1.2 in the article of A. V. Skorokhod (3).

CHAPTER 2

1. Stochastic integrals with respect to a Brownian process were introduced by N. Wiener (1) for the case in which an integrable function is not random. The definitions used by us are those of K. Itō (1–5).

2. Theorems 1–3 are contained in the works of Itō referred to. This seems to be the first presentation of Theorem 4.

3. The definition of a stochastic integral with respect to a martingale is contained in Doob's book (4), Chapter 9, Section 5. Apparently, the theorem of this section is presented here for the first time.

4. Stochastic integrals with respect to Poisson measures are defined in Itō's article (5). Random measures and integrals with respect to these measures were studied by Bochner (1) and A. Prekopa (1) for the case in which an integrable function is not random.

CHAPTER 3

Stochastic differential equations were first studied by S. N. Bernstein (1) [see also (2), supplement 4], who did not construct a trajectory of processes satisfying the equations, but studied only one-dimensional distributions of processes. The equation itself was regarded as a means of obtaining these distributions and was written in finite-difference form. Bernstein examined also the question of the existence of a limiting distribution for one-dimensional distributions of the solution of the finite-difference equation when the lengths of the intervals of a subdivision approach zero, and investigated the conditions under which this limiting distribution will satisfy the Fokker-Planck equation. These works of S. N. Bernstein are closely associated with his works on the distribution of a central limit theorem on dependent random processes.

On the other hand, physicists have approached similar questions by investigating systems with rapidly changing forces and by trying to determine the averaged characteristics of the behavior of a system without solving the equations which determine the system. Here we have a whole series of works devoted to Brownian motion and to the derivation of the Fokker-Planck equations. Of interest in this connection is the work of N. N. Bogolyubov and N. M. Krylov (1) in which a study is made of the question of the limiting behavior of a dynamical system under the influence of a random force which, in the limit, is converted into "white noise." Krylov and Bogolyubov showed that, in the limit, the motion of the system can be described by a Markov process whose transition probabilities satisfy the Fokker-Planck equation. However, the limiting transition in the equations of dynamics had not been rigorously proved. For proof of the limiting transition, I. I. Gikhman (continuing these investigations) succeeded in constructing a theory of stochastic differential equations [see (1–4)].

Gikhman's stochastic equations made it possible, not only to find the distribution of a process at a single point, but also to construct the trajectories of the process. He showed the differentiability of the solutions of the stochastic equations with given initial conditions, which permitted him to derive the inverse equations of A. N. Kolmogorov for mathematical expectations from smooth functions of a process.

Further significant progress in the theory of stochastic differential equations is associated with the name of K. Itō. He constructed a general theory of stochastic integrals, which allowed him to construct stochastic equations even for discontinuous processes (in contrast to the work of earlier authors, who in general examined continuous processes; however, he confined himself to the one-dimensional case) and for continuous processes in differentiable manifolds. Itō's works on stochastic differential equations (2–5) are also of great interest because he succeeded in constructing a rather wide class of Markov random processes by using only a Brownian process and the Poisson measure with independent increments.

We note further that the idea of a stochastic equation in the sense of S. N. Bernstein was made precise by P. Lévy in (3), who also succeeded in constructing a trajectory of the solution of the equation. Some generalizations of Bernstein's equation were examined by F. Zitek (1, 2). Itō's equations were studied also by Doob (3) and by G. Maruyama (1).

2. Theorem 1 generalizes Itō's theorem in (5) to the higher-dimensional case. The idea of the proof belongs to Itō.

3. The uniqueness theorem is a generalization of the corresponding theorem of Itō in (5) to the higher-dimensional case. The existence theorem and the reasoning used in proving it seemed to be new. This idea was used in my article (9) for proving the existence of a solution in the case of the equations of a one-dimensional diffusion process.

4. The definition of Itō's differential appears in his article (5).

5. The theorems on the differentiability of the solutions of stochastic differential equations with respect to a parameter in the case of diffusion processes were obtained by I. I. Gikhman (3, 4).

6. The idea of the derivation of integro-differential equations for mathematical expectations of smooth functions of a process was originated by I. I. Gikhman, who did it for the diffusion case. The integro-differential equation obtained is a generalization of the equations for the distributions of a Markov process which were obtained by A. N. Kolmogorov (2) and W. Feller (2). The differential equation method proposed by I. G. Petrovskii (1) is extensively used in the works of A. N. Kolmogorov, A. Ya. Khinchine, and I. I. Gikhman. The equation obtained is closely related to the results of the semigroup theory of Markov processes developed by W. Feller (3, 4) and E. B. Dynkin (3, 4). The connection between Markov processes and partial differential equations is examined in the works of M. Kac (1, 2), Dynkin (2), P. Z. Khas'minskii (1–3); an account of the questions associated with these results appears in the survey of I.M. Gel'fand and A. M. Yaglom (1).

Chapter 4

1. For the absolute continuity of measures and for the Radon-Nikodym theorem, see P. Halmos's book (1), Chapter 6.

3. The results of this section are contained in my article (6).

4. The results of this section in the case of one-dimensional diffusion processes with diffusion coefficient equal to one were obtained under different assumptions by Cameron and Martin (1, 2), by Cameron and Fagen (1), and by Yu. V. Prokhorov (3) (Appendix II); see also the survey by I. M. Gel'fand and A. M. Yaglom (1). I have examined the general case in (7).

Chapter 5

This chapter is a minor revision of my article (9).

Chapter 6

The limit theorems for the distributions of Markov sequences at a single point have been examined by S. N. Bernstein (1, 2) and A. Ya. Khinchine (1). (In both cases, convergence to a diffusion process was considered.) Convergence in the case of functionals depending on the complete flow of the trajectory of a process has been studied by A. N. Kolmogorov (3, 4) and A. Ya. Khinchine (1), who considered the limit theorems on the probabilities of a sequence of sums of independent random variables or a sequence of values of a Markov chain lying between two given curves. Erdös and Kac (1, 2) have examined the limit theorems for certain other functionals of a sequence of sums of independent random variables. The general limit theorem for the distributions of a broad class of functionals of a sequence of sums of independent random variables was discovered by M. Donsker (1). The case of convergence to discontinuous processes with independent increments and to diffusion processes was studied by I. I.

Gikhman (5, 6). Yu. V. Prokhorov (1, 2) generalized the results of Donsker to the case of convergence to arbitrary continuous processes. The survey article of A. N. Kolmogorov and Yu. V. Prokhorov (1) is devoted to an exposition of the processes mentioned above.

In my own works (1–3), I have introduced a topology in the space of functions with no discontinuities of the second kind and have proved limit theorems for those functionals of the sequence of processes that are continuous in this topology. This topology was then examined by A. N. Kolmogorov and Yu. V. Prokhorov (3), who showed that it could be generated by a certain metric and then formed this metric. Yu. V. Prokhorov (3) studied the convergence of the distributions of continuous functionals (with this metric) of a sequence of sums of differently distributed independent terms. [The case of identically distributed terms was examined by myself (1).] In my articles (4, 5), the general theorems (3) were used to obtain limit theorems for processes with independent increments and homogeneous Markov processes. Convergence to diffusion processes has also been examined by G. Maruyama (1). N. M. Chentsov (1) has obtained an interesting limit theorem for the probability of finding a process between two curves in the case of convergence to an arbitrary process with no discontinuities of the second kind.

The results of Sections 2 and 3 were published in my article (3). By and large, the results of Sections 4 and 5 are published here for the first time. It should be noted that estimates analogous to those obtained in Lemmas 2 and 3 of Section 5 were obtained earlier by E. B. Dynkin (1) and J. R. Kinney (1).

Chapter 7

The works of H. Cramér (1) and S. Esseen (1) are devoted to making precise the limit theorems for the distributions of sums of independent random variables. [See also the book by B. V. Gnedenko and A. N. Kolmogorov (1).] Estimates for the rapidity of convergence for distributions of certain specific functionals appear in the articles of K. L. Chung (1, 2). General methods for obtaining asymptotic analyses for the probability that a sequence of values of a Markov chain will lie between two straight lines have been developed by V. C. Korolyuk (1–3). I have published the results of Sections 2 and 3 in (8, 11). The problem examined in Section 3 has also been solved by Yu. V. Prokhorov (3) with no assumption as to the boundedness of the terms; he obtained the estimate

$$|Q_n - Q| = O(\log^2 n/n^{1/4}).$$

BIBLIOGRAPHY

BARTLETT, M. S.

1. *An introduction to stochastic processes.* Cambridge University Press (1955).

BERNSTEIN, S. N.

1. "Principes de la théorie des équations différentielles stochastiques," *Trudy Fiz.-Mat. Inst. Akad. Nauk,* **5,** 95–124 (1934).

2. *Teoriya veroyatnostei* [Probability theory]. 4th ed., Moscow and Leningrad (1946).

BOCHNER, S.

1. "Stochastic processes," *Ann. Math.,* **48,** 1014–1061 (1947).

BOGOLYUBOV, N. N., and N. M. KRYLOV.

1. "Pro rivnyannya Fokker-Planka shcho vivodit'sya v teorii perturbartsii metodom osnovanim na vlastivostyakh perturbatsiinogo gamil'toniana" [Derivation of the Fokker-Planck equation in perturbation theory by a method based on properties of the perturbation Hamiltonian], *Zapiski Kafedri Mat. Fiz., Akad. Nauk (Ukraine),* Kiev, **4,** 5–157 (1939).

CAMERON, R. H., and R. E. FAGEN.

1. "Nonlinear transformations of Volterra type in Wiener space," *Trans. Am. Math. Soc.,* **75,** 552–575 (1953).

CAMERON, R. H., and W. T. MARTIN.

1. "Transformation of Wiener integrals under translations," *Ann. Math.,* (2), **45,** 386–396 (1944).

2. "The transformation of Wiener integrals by nonlinear transformations," *Trans. Am. Math. Soc.,* **66,** 253–283 (1949).

CHENTSOV, N. N.

1. "Slabaya skhodimost' sluchainykh protsessov s traektoriyami bez razryvov vtorogo roda i tak nazyvaemyi 'evristichesii' podkhod k kriteriyam soglasiya tipa Kolmogorova-Smirnova" [Weak convergence of random processes whose trajectories have no discontinuities of the second kind and the so-called "heuristic" approach to the criteria of agreement of the Kolmogorov-Smirnov type], *Teor. Ver. i ee Prim.,* **1,** 155–161 (1956).

CHUNG, K. L.

1. "On the maximum partial sums of sequences of independent random variables," *Trans. Am. Math. Soc.,* **64,** 205–233 (1948).

2. "An estimate concerning the Kolmogoroff limit distribution," *Trans. Am. Math. Soc.,* **67,** 36–50 (1949).

CRAMÉR, H.

1. *Random variables and probability distributions.* Cambridge University Press (1937).

DOBRUSHIN, P. L., and A. M. YAGLOM.

1. Translators' notes to Russian edition (1956) of J. L. Doob's *Stochastic Processes.* (John Wiley, New York, 1953).

DONSKER, M. D.

1. "An invariance principle for certain probability limit theorems," *Mem. Am. Math. Soc.*, No. 6, 1–12 (1951).

2. "Justification and extension of Doob's heuristic approach to the Kolmogorov-Smirnov theorems," *Ann. Math. Stat.*, **23,** 277–281. (1952).

DOOB, J. L.

1. "The Brownian movement and stochastic equations," *Ann. Math.*, **43,** (2), 351–369 (1942).

2. "The elementary Gaussian processes," *Ann. Math. Stat.*, **15,** 229–282 (1944).

3. "Martingales and one-dimensional diffusion," *Trans. Am. Math. Soc.*, **78,** 168–208 (1955).

4. *Stochastic processes.* John Wiley, New York, 1953.

DYNKIN, E. B.

1. "Kriterii nepreryvnosti i otsutstviya razryvov vtorogo roda dlya traektorii markovskogo sluchainogo protsessa" [Criteria for continuity and for the absence of discontinuities of the second kind for trajectories of a Markov random process], *Izdat. Akad. Nauk, Ser. Mat.*, **16,** 563–572 (1952).

2. "Funktsionali ot traektorii markovskikh sluchainykh protsessov" [Functionals of the trajectories of Markov random processes], *Doklady Akad. Nauk SSSR,* **104,** 691–694 (1955).

3. "Infinitezimal'nye operatory markovskikh protsessov" [Infinitesimal operators of Markov processes], *Teor. Ver. i ee Prim.*, **1,** 38–60 (1956).

4. "Odnomernye nepreryvnye strogo markovskie protsessy" [One-dimensional continuous strong Markov processes], *Teor. Ver. i ee Prim.*, **4,** 3–54 (1959).

5. "Neodnorodnye strogo markovskie protsessy" [Nonhomogeneous strong Markov processes], *Doklady Akad. Nauk SSSR,* **113,** 261–263 (1957).

6. *Osnovaniya teorii markovskikh protsessov* [Fundamentals of the theory of Markov processes], Fizmatgiz, Moscow (1959).

DYNKIN, E. B., and A. A. YUSHKEVICH.

1. "Strogo markovskie protsessy" [Strong Markov processes], *Teor. Ver. i ee Prim.*, **1,** 149–155 (1956).

ERDÖS, P., and M. KAC.

1. "On certain limit theorems on the theory of probability," *Bull. Am. Math. Soc.*, **52**, 292–302 (1946).

2. "On the number of positive sums of independent random variables," *Bull. Am. Math. Soc.*, **53**, 1011–1020 (1947).

ESSEEN, S.

1. "Fourier analysis of distribution functions. A mathematical study of the Laplace-Gaussian law," *Acta Math.*, **77**, 1–125 (1945).

FELLER, W.

1. "Zur Theorie der stochastischen Prozesse (Existenz und Eindeutigkeitsätze)," *Ann. Math.*, **113**, 113–160 (1936).

2. "On the integro-differential equations of purely discontinuous Markoff processes," *Trans. Am. Math. Soc.*, **48**, 488–515 (1940).

3. "Diffusion processes in one dimension," *Trans. Am. Math. Soc.*, **77**, 1–31 (1954).

4. "The general diffusion operator and positivity preserving semigroups in one dimension," *Ann. Math.*, **60**, (2), 417–436 (1954).

FREIDLIN, M. I.

1. "O nekotorykh primeneniyakh stokhasticheskikh uravnenii Ito k differentsial'nym uravneniyam" [Some applications of Itō's stochastic equations to differential equations], *Teor. Ver. i ee Prim.*, **4**, 472–473 (1959).

GEL'FAND, I. M., and A. M. YAGLOM.

1. "Integrirovanie v funktsional'nykh prostranstvakh i ego primenenie v kvantovoi fizike" [Integration in function spaces and its application in quantum physics], *Usp. Mat. Nauk*, **11**, No. 1, 77–114 (1956).

GIKHMAN, I. I.

1. "Ob odnoi skheme obrazovaniya sluchainykh protsessov" [A method of constructing random processes], *Doklady Akad. Nauk SSSR*, **58**, 961–964 (1947).

2. "O nekotorykh differentsial'nykh uravneniyakh so sluchainymi funktsiyami" [Certain differential equations with random functions], *Ukr. Mat. Zh.*, **2**, No. 3, 45–69 (1950).

3. "K teorii differentsial'nykh uravnenii sluchainykh protsessov" [On the theory of differential equations of random processes], *Ukr. Mat. Zh.*, **2**, No. 4, 37–63 (1950).

4. "K teorii differential'nykh uravnenii sluchainykh protsessov, II" [On the theory of differential equations of random processes, II], *Ukr. Mat. Zh.*, **3**, 317–339 (1951).

5. "O nekotorykh predel'nykh teoremakh dlya uslovnykh raspredelenii i o svyazannykh s nimi zadachami matematicheskoi statistiki" [Some limit theorems on conditional distributions and related problems in mathematical statistics], *Ukr. Mat. Zh.*, **5**, 413–433 (1953).

6. "Ob odnoi teoreme A. N. Kolmogorova" [On a theorem of Kolmogorov], *Nauchn. Zap.* Kiev. Univ, **12**, 6, *Mat. b.*, No. 7, 75–94 (1953).

7. "Protsessy markova v zadachakh matematicheskoi statistiki" [Markov processes in mathematical statistical problems], *Ukr. Mat. Zh.*, **6**, 28–36 (1954).

GIRSANOV, I. V.

1. "Sil'no fellerovskie protsessy" [Strong Feller processes, I, General Properties] *Teor. Ver. i ee Prim.*, **5**, (1960).

2. "O preobrazovanii odnogo klassa sluchainykh protsessov s pomoshch'yu absolyutno-nepreryvnoi zameny mery" [Transformation of a certain class of random processes by means of an absolutely continuous change of measure], *Teor. Ver. i ee Prim.*, **5**, 314–330 (1960).

GNEDENKO, B. V.

1. *Kurs teorii veroyatnostei* [A course in probability theory]. 2nd ed., Moscow and Leningrad (1954).

GNEDENKO, B. V., and A. N. KOLMOGOROV.

1. *Predel'nye raspredeleniya dlya summ nezavisimykh sluchainykh velichin* [Limit distributions for sums of independent random variables]. Moscow and Leningrad (1949).

HALMOS, P.

1. *Measure theory*, D. Van Nostrand, Princeton, N.J. (1950).

ITŌ, K.

1. "Stochastic integral," *Proc. Imp. Acad. Tokyo*, **20**, 519–524 (1944).

2. "On a stochastic integral equation," *Proc. Japan Acad.*, No. 1–4, 32–35 (1946).

3. "Stochastic differential equations in a differentiable manifold," *Nagoya Math. J.*, **1**, 35–47 (1950).

4. "On a formula concerning stochastic differentials," *Nagoya Math. J.*, **3**, 55–65 (1951).

5. "On stochastic differential equations," *Mem. Am. Math. Soc.*, No. 4, 1–51 (1951).

KAC, M.

1. "On distributions of certain Wiener functionals," *Trans. Am. Math. Soc.*, **65**, 1–13 (1949).

2. "On some connections between probability theory and differential and integral equations," *Proc. 2nd Berkeley Sympos. Math. Stat. and Prob.*, 189–215 (1951).

KHAS'MINSKII, R. Z.

1. "Raspredelenie veroyatnostei dlya funktsionalov ot traektorii sluchainogo protsessa diffuzionnogo tipa" [The distribution of probabilities

for functionals of the trajectory of a random process of diffusion type], *Doklady Akad. Nauk SSSR*, **104**, 22–25 (1955).

2. "Diffuzionnye protsessy i ellipticheskie differentsial'nye uravneniya, vyrozhdayushchiesya na granitse oblasti" [Diffusion processes and elliptic differential equations that degenerate at the boundary of a region], *Teor. Ver. i ee Prim.*, **3**, 430–451 (1958).

3. "O polozhitel'nykh resheniyakh uravneniya $\mathfrak{A}u + \mathfrak{B}u = 0$" [Positive solutions to the equation $\mathfrak{A}u + \mathfrak{B}u = 0$], *Teor. Ver. i ee Prim.*, **4**, 332–341 (1959).

KHINCHIN, A. YA.

1. *Asimptoticheskie zakony teorii veroyatnostei* [Asymptotic laws of probability theory]. Moscow and Leningrad (1936).

2. *Predel'nye teoremy dlya summ nezavisimykh sluchainykh velichin* [Limit theorems for the sums of random variables]. Moscow and Leningrad (1938).

KINNEY, J. R.

1. "Continuity properties of sample functions of Markov processes," *Trans. Am. Math. Soc.*, **74**, 280–302 (1953).

KOLMOGOROV, A. N.

1. *Osnovnye ponyatiya teorii veroyatnostei* [Basic concepts of probability theory]. Moscow and Leningrad (1936).

2. "Analiticheskie metody v teorii veroyatnostei" [Analytic methods in probability theory], *Usp. Mat. Nauk*, **5**, 5–41 (1938).

3. *Eine Verallgemeinerung des Laplace-Liapunoffschen Satzes.* Izdat. Akad. Nauk (Otd. mat. i estestv. nauk) 959–962 (1931).

4. *Über die Grenzwertsätze der Wahrscheinlichkeitsrechnung.* Izdat. Akad. Nauk (Otd. mat. i estestv. nauk) 363–372 (1933).

5. "Sulla forma generale di un processo stocastico omogeneo," *Rend. Accad. Naz. dei Lincei*, **15**, (6), 805–808, 866 (1932).

6. "O skhodimosti A. V. Skorokhoda" [On the convergence of A. V. Skorokhod], *Teor. Ver. i ee Prim.*, **1**, 239–247 (1956).

KOLMOGOROV, A. N., and YU. V. PROKHOROV.

1. "Zufällige Funktionen und Grenzverteilungssätze," *Bericht über die Tagung Wahrscheinlichkeitsrechnung und mathematische Statistik in Berlin* (1954), pp. 113–126, Deutsche Verlag der Wissensch., Berlin (1956).

KOROLYUK, V. S.

1. "Asimptoticheskie razlozheniya dlya raspredeleniya maksimal'nykh uklonenii v skheme Bernulli" [Asymptotic expansions for distributions of maximum deviations in a Bernoulli scheme], *Doklady Akad. Nauk SSSR*, **108**, 183–186 (1956).

2. "Asimptotichnii analiz imovirnosti vbiraniya v odnomirnii skhemi vipadkovikhblukan' z reshitchastim rozpodilom imovirnostei perekhody"

[Asymptotic analysis of the probability of inclusion in a one-dimensional scheme of random variables from a lattice configuration of transition probabilities], *Dopouidi Akad. Nauk (Ukraine)*, **7** (1959).

3. "Asimptoticheskii analiz raspredelenii maksimal'nykh uklonenii v skheme Bernulli" [Asymptotic analysis of the distributions of maximum deviations in a Bernoulli scheme]. *Teor. Ver. i ee Prim.*, **4,** 369–397 (1959).

LÉVY, P.

1. "Sur les intégrales dont les éléments sont des variables aléatoires indépendantes," *Ann. Scuola Norm. Sup. Pisa*, (2), **3,** 337–366 (1934); "Observation sur un précédent mémoire de l'auteur," *ibid.*, **4,** 217–218 (1935).

2. *Théorie de l'addition des variables aléatoires.* Paris (1937).

3. *Processus stochastiques et mouvement Brownien.* Paris (1948).

MARUYAMA, G.

1. "Continuous Markov processes and stochastic equations," *Rend. Circ. Mat. Palermo*, (2), **4,** 48–90 (1955).

2. "On the strong Markov property," *Memoirs of the Faculty of Science, Kyushu University, Ser. A*, **13,** No. 1, 17–29 (1959).

PETROVSKII, I. G.

1. "Über das Irrfahrtproblem," *Ann. Math.*, **109,** 425 (1934).

PREKOPA, A.

1. "On stochastic set functions, I, II, III," *Acta Math. Acad. Sci. Hung.*, **7** 215–263 (1956); *ibid.* **8,** 337–374, 375–400 (1957).

PROKHOROV, YU. V.

1. "Raspredelenie veroyatnostei v funktsional'nykh prostranstvakh" [Probability distributions in functional spaces], *Usp. Mat. Nauk*, **8,** No. 3, 165–167 (1953).

2. "Metody funktsional'nogo analiza v predel'nykh teoremakh teorii veroyatnostei" [Methods of functional analysis in the limit theorems of probability theory], *Vest. Leningr. Univ.* No. II, 44 (1955).

3. "Skhodimosti' sluchainykh protsessov i predel'nye teoremy teorii veroyatnostei" [The convergence of random processes and limit theorems in probability theory], *Teor. Ver. i ee Prim.*, **1,** 177–238 (1956).

SKOROKHOD, A. V.

1. "O predel'nom perekhode ot posledovatel'nosti summ nezavisimykh velichin k odnomu sluchainomy protsessy s nezavisimymi prirashcheniyami" [Limit transition from a sequence of sums of independent variables to a homogeneous random process with independent increments], *Doklady Akad. Nauk SSSR*, **104,** 364–367 (1955).

2. "Ob odnom klasse predel'nykh teorem dlya tsepei Markova" [A class of limit theorems for Markov chains], *Doklady Akad. Nauk SSSR*, **106,** 781–784 (1956).

3. "Predel'nye teoremy dlya sluchainykh protsessov" [Limit theorems for random processes], *Teor. Ver. i ee Prim.*, **1**, 289–319, (1956).
4. "Predel'nye teoremy dlya protsessov s nezavisimymi prirashcheniyami" [Limit theorems for processes with independent increments], *Teor. Ver. i ee Prim.*, **2**, 145–177 (1957).
5. "Predel'nye teoremy dlya protsessov Markova" [Limit theorems for Markov processes], *Teor. Ver. i ee Prim.*, **3**, 217–264 (1958).
6. "O differentsiruemosti mer, sootvetstvuyushchikh sluchainym protsessam, I. Protsessy s nezavisimymi prirashcheniyami" [Differentiability of measures corresponding to random processes, I. Processes with independent increments], *Teor. Ver. i ee Prim.*, **2**, 417–443 (1957).
7. "O differentsiruemosti mer, sootvetstvuyushchikh sluchainym protsessam, II. Protsessy Markova" [Differentiability of measures corresponding to random processes, II. Markov processes], *Teor. Ver. i ee Prim.*, **5**, 45–53 (1960).
8. "Odna predel'naya teorema dlya summ nezavisimykh sluchainykh velichin" [A limit theorem for sums of independent random variables], *Doklady Akad. Nauk SSSR*, **133**, 34–35 (1960).
9. "O sushchestvovanii i edinstvennosti reshenii stokhasticheskikh diffuzionnykh uravenii" [Existence and uniqueness of solutions to stochastic diffusion equations], *Sibirsk. Mat. Zh.*, **2**, No. 1, 129–137 (1961).
10. "O stokhasticheskikh uravneniyakh dlya diffuzionnykh protsessov s otrazheniem na granitse" [Stochastic equations for diffusion processes with reflection at the boundary], *Teor. Ver. i ee Prim.*, **6**, 287–298 (1961).

SLUTSKII, E. E.
1. "Alcuni proposizioni sulla teoria degli funzioni aleatorie," *Giorn. Inst. Ital. Attuarie*, **8**, 183–199 (1937).

VENTZEL', A. D.
1. "Polugruppy operatorov, sootvetstvuyushchikh obobshchennomy differentsial'nomy operatory vtorogo poryadka" [Semigroups of operators corresponding to a generalized second-order differential operator], *Doklady Akad. Nauk SSSR*, **111**, 269–272 (1956).
2. "O granichnykh usloviyakh dlya mnogomernykh diffuzionnykh protsessov" [Boundary conditions for multi-dimensional diffusion processes], *Teor. Ver. i ee Prim.*, **4**, 172–185 (1956).

WIENER, N.
1. "Differential space," *J. Math. Phys.*, **2**, 131–179 (1923).
2. "Generalized harmonic analysis," *Acta Math.*, **55**, 117–258 (1930).
3. "The homogeneous chaos," *Am. J. Math.*, **60**, 897–936 (1938).

YUSHKEVICH, A. A.
1. "O strogo markovskikh protsessakh" [Strong Markov processes], *Teor. Ver. i ee Prim.*, **2**, 187–213 (1957).

Zitek, F.
1. "Equations différentielles stochastiques," *Czech. Math. J.*, **8**, (83), 465–472 (1958).
2. "Sur l'intégrabilité d'une équation différentielle stochastique," *Czech. Math. J.*, **8**, (83), 473–482 (1958).
3. "Fonctions aléatoires d'intervalle," *Czech. Math. J.*, **8**, (83), 583–609 (1958).

INDEX

A CATALOGUE OF SELECTED DOVER BOOKS IN ALL FIELDS OF INTEREST

A CATALOGUE OF SELECTED DOVER BOOKS IN ALL FIELDS OF INTEREST

THE SENSE OF BEAUTY, George Santayana. Masterfully written discussion of nature of beauty, materials of beauty, form, expression; art, literature, social sciences all involved. 168pp. 5⅜ x 8½. 20238-0 Pa. $2.50

ON THE IMPROVEMENT OF THE UNDERSTANDING, Benedict Spinoza. Also contains *Ethics, Correspondence,* all in excellent R. Elwes translation. Basic works on entry to philosophy, pantheism, exchange of ideas with great contemporaries. 402pp. 5⅜ x 8½. 20250-X Pa. $4.50

THE TRAGIC SENSE OF LIFE, Miguel de Unamuno. Acknowledged masterpiece of existential literature, one of most important books of 20th century. Introduction by Madariaga. 367pp. 5⅜ x 8½.
20257-7 Pa. $4.50

THE GUIDE FOR THE PERPLEXED, Moses Maimonides. Great classic of medieval Judaism attempts to reconcile revealed religion (Pentateuch, commentaries) with Aristotelian philosophy. Important historically, still relevant in problems. Unabridged Friedlander translation. Total of 473pp. 5⅜ x 8½. 20351-4 Pa. $6.00

THE I CHING (THE BOOK OF CHANGES), translated by James Legge. Complete translation of basic text plus appendices by Confucius, and Chinese commentary of most penetrating divination manual ever prepared. Indispensable to study of early Oriental civilizations, to modern inquiring reader. 448pp. 5⅜ x 8½. 21062-6 Pa. $4.00

THE EGYPTIAN BOOK OF THE DEAD, E. A. Wallis Budge. Complete reproduction of Ani's papyrus, finest ever found. Full hieroglyphic text, interlinear transliteration, word for word translation, smooth translation. Basic work, for Egyptology, for modern study of psychic matters. Total of 533pp. 6½ x 9¼. (Available in U.S. only) 21866-X Pa. $5.95

THE GODS OF THE EGYPTIANS, E. A. Wallis Budge. Never excelled for richness, fullness: all gods, goddesses, demons, mythical figures of Ancient Egypt; their legends, rites, incarnations, variations, powers, etc. Many hieroglyphic texts cited. Over 225 illustrations, plus 6 color plates. Total of 988pp. 6⅛ x 9¼. (Available in U.S. only)
22055-9, 22056-7 Pa., Two-vol. set $12.00

THE ENGLISH AND SCOTTISH POPULAR BALLADS, Francis J. Child. Monumental, still unsuperseded; all known variants of Child ballads, commentary on origins, literary references, Continental parallels, other features. Added: papers by G. L. Kittredge, W. M. Hart. Total of 2761pp. 6½ x 9¼.
21409-5, 21410-9, 21411-7, 21412-5, 21413-3 Pa., Five-vol. set $37.50

CORAL GARDENS AND THEIR MAGIC, Bronsilaw Malinowski. Classic study of the methods of tilling the soil and of agricultural rites in the Trobriand Islands of Melanesia. Author is one of the most important figures in the field of modern social anthropology. 143 illustrations. Indexes. Total of 911pp. of text. 5⅝ x 8¼. (Available in U.S. only)
23597-1 Pa. $12.95

HOLLYWOOD GLAMOUR PORTRAITS, edited by John Kobal. 145 photos capture the stars from 1926-49, the high point in portrait photography. Gable, Harlow, Bogart, Bacall, Hedy Lamarr, Marlene Dietrich, Robert Montgomery, Marlon Brando, Veronica Lake; 94 stars in all. Full background on photographers, technical aspects, much more. Total of 160pp. 8⅜ x 11¼. 23352-9 Pa. $6.00

THE NEW YORK STAGE: FAMOUS PRODUCTIONS IN PHOTOGRAPHS, edited by Stanley Appelbaum. 148 photographs from Museum of City of New York show 142 plays, 1883-1939. *Peter Pan, The Front Page, Dead End, Our Town,* O'Neill, hundreds of actors and actresses, etc. Full indexes. 154pp. 9½ x 10. 23241-7 Pa. $6.00

MASTERS OF THE DRAMA, John Gassner. Most comprehensive history of the drama, every tradition from Greeks to modern Europe and America, including Orient. Covers 800 dramatists, 2000 plays; biography, plot summaries, criticism, theatre history, etc. 77 illustrations. 890pp. 5⅜ x 8½. 20100-7 Clothbd. $10.00

THE GREAT OPERA STARS IN HISTORIC PHOTOGRAPHS, edited by James Camner. 343 portraits from the 1850s to the 1940s: Tamburini, Mario, Caliapin, Jeritza, Melchior, Melba, Patti, Pinza, Schipa, Caruso, Farrar, Steber, Gobbi, and many more—270 performers in all. Index. 199pp. 8⅜ x 11¼. 23575-0 Pa. $6.50

J. S. BACH, Albert Schweitzer. Great full-length study of Bach, life, background to music, music, by foremost modern scholar. Ernest Newman translation. 650 musical examples. Total of 928pp. 5⅜ x 8½. (Available in U.S. only) 21631-4, 21632-2 Pa., Two-vol. set $10.00

COMPLETE PIANO SONATAS, Ludwig van Beethoven. All sonatas in the fine Schenker edition, with fingering, analytical material. One of best modern editions. Total of 615pp. 9 x 12. (Available in U.S. only) 23134-8, 23135-6 Pa., Two-vol. set $15.00

KEYBOARD MUSIC, J. S. Bach. Bach-Gesellschaft edition. For harpsichord, piano, other keyboard instruments. English Suites, French Suites, Six Partitas, Goldberg Variations, Two-Part Inventions, Three-Part Sinfonias. 312pp. 8⅛ x 11. (Available in U.S. only) 22360-4 Pa. $6.95

FOUR SYMPHONIES IN FULL SCORE, Franz Schubert. Schubert's four most popular symphonies: No. 4 in C Minor ("Tragic"); No. 5 in B-flat Major; No. 8 in B Minor ("Unfinished"); No. 9 in C Major ("Great"). Breitkopf & Hartel edition. Study score. 261pp. 9⅜ x 12¼. 23681-1 Pa. $6.50

THE AUTHENTIC GILBERT & SULLIVAN SONGBOOK, W. S. Gilbert, A. S. Sullivan. Largest selection available; 92 songs, uncut, original keys, in piano rendering approved by Sullivan. Favorites and lesser-known fine numbers. Edited with plot synopses by James Spero. 3 illustrations. 399pp. 9 x 12. 23482-7 Pa. $7.95

THE COMPLETE BOOK OF DOLL MAKING AND COLLECTING, Catherine Christopher. Instructions, patterns for dozens of dolls, from rag doll on up to elaborate, historically accurate figures. Mould faces, sew clothing, make doll houses, etc. Also collecting information. Many illustrations. 288pp. 6 x 9. 22066-4 Pa. $4.50

THE DAGUERREOTYPE IN AMERICA, Beaumont Newhall. Wonderful portraits, 1850's townscapes, landscapes; full text plus 104 photographs. The basic book. Enlarged 1976 edition. 272pp. 8¼ x 11¼. 23322-7 Pa. $7.95

CRAFTSMAN HOMES, Gustav Stickley. 296 architectural drawings, floor plans, and photographs illustrate 40 different kinds of "Mission-style" homes from *The Craftsman* (1901-16), voice of American style of simplicity and organic harmony. Thorough coverage of Craftsman idea in text and picture, now collector's item. 224pp. 8⅛ x 11. 23791-5 Pa. $6.00

PEWTER-WORKING: INSTRUCTIONS AND PROJECTS, Burl N. Osborn. & Gordon O. Wilber. Introduction to pewter-working for amateur craftsman. History and characteristics of pewter; tools, materials, step-by-step instructions. Photos, line drawings, diagrams. Total of 160pp. 7⅞ x 10¾. 23786-9 Pa. $3.50

THE GREAT CHICAGO FIRE, edited by David Lowe. 10 dramatic, eyewitness accounts of the 1871 disaster, including one of the aftermath and rebuilding, plus 70 contemporary photographs and illustrations of the ruins—courthouse, Palmer House, Great Central Depot, etc. Introduction by David Lowe. 87pp. 8¼ x 11. 23771-0 Pa. $4.00

SILHOUETTES: A PICTORIAL ARCHIVE OF VARIED ILLUSTRATIONS, edited by Carol Belanger Grafton. Over 600 silhouettes from the 18th to 20th centuries include profiles and full figures of men and women, children, birds and animals, groups and scenes, nature, ships, an alphabet. Dozens of uses for commercial artists and craftspeople. 144pp. 8⅜ x 11¼. 23781-8 Pa. $4.00

ANIMALS: 1,419 COPYRIGHT-FREE ILLUSTRATIONS OF MAMMALS, BIRDS, FISH, INSECTS, ETC., edited by Jim Harter. Clear wood engravings present, in extremely lifelike poses, over 1,000 species of animals. One of the most extensive copyright-free pictorial sourcebooks of its kind. Captions. Index. 284pp. 9 x 12. 23766-4 Pa. $7.95

INDIAN DESIGNS FROM ANCIENT ECUADOR, Frederick W. Shaffer. 282 original designs by pre-Columbian Indians of Ecuador (500-1500 A.D.). Designs include people, mammals, birds, reptiles, fish, plants, heads, geometric designs. Use as is or alter for advertising, textiles, leathercraft, etc. Introduction. 95pp. 8¾ x 11¼. 23764-8 Pa. $3.50

SZIGETI ON THE VIOLIN, Joseph Szigeti. Genial, loosely structured tour by premier violinist, featuring a pleasant mixture of reminiscenes, insights into great music and musicians, innumerable tips for practicing violinists. 385 musical passages. 256pp. 5⅝ x 8¼. 23763-X Pa. $3.50

THE AMERICAN SENATOR, Anthony Trollope. Little known, long unavailable Trollope novel on a grand scale. Here are humorous comment on American vs. English culture, and stunning portrayal of a heroine/villainess. Superb evocation of Victorian village life. 561pp. 5⅜ x 8½.
23801-6 Pa. $6.00

WAS IT MURDER? James Hilton. The author of *Lost Horizon* and *Goodbye, Mr. Chips* wrote one detective novel (under a pen-name) which was quickly forgotten and virtually lost, even at the height of Hilton's fame. This edition brings it back—a finely crafted public school puzzle resplendent with Hilton's stylish atmosphere. A thoroughly English thriller by the creator of Shangri-la. 252pp. 5⅜ x 8. (Available in U.S. only)
23774-5 Pa. $3.00

CENTRAL PARK: A PHOTOGRAPHIC GUIDE, Victor Laredo and Henry Hope Reed. 121 superb photographs show dramatic views of Central Park: Bethesda Fountain, Cleopatra's Needle, Sheep Meadow, the Blockhouse, plus people engaged in many park activities: ice skating, bike riding, etc. Captions by former Curator of Central Park, Henry Hope Reed, provide historical view, changes, etc. Also photos of N.Y. landmarks on park's periphery. 96pp. 8½ x 11. 23750-8 Pa. $4.50

NANTUCKET IN THE NINETEENTH CENTURY, Clay Lancaster. 180 rare photographs, stereographs, maps, drawings and floor plans recreate unique American island society. Authentic scenes of shipwreck, lighthouses, streets, homes are arranged in geographic sequence to provide walking-tour guide to old Nantucket existing today. Introduction, captions. 160pp. 8⅞ x 11¾. 23747-8 Pa. $6.95

STONE AND MAN: A PHOTOGRAPHIC EXPLORATION, Andreas Feininger. 106 photographs by *Life* photographer Feininger portray man's deep passion for stone through the ages. Stonehenge-like megaliths, fortified towns, sculpted marble and crumbling tenements show textures, beauties, fascination. 128pp. 9¼ x 10¾. 23756-7 Pa. $5.95

CIRCLES, A MATHEMATICAL VIEW, D. Pedoe. Fundamental aspects of college geometry, non-Euclidean geometry, and other branches of mathematics: representing circle by point. Poincare model, isoperimetric property, etc. Stimulating recreational reading. 66 figures. 96pp. 5⅝ x 8¼.
63698-4 Pa. $2.75

THE DISCOVERY OF NEPTUNE, Morton Grosser. Dramatic scientific history of the investigations leading up to the actual discovery of the eighth planet of our solar system. Lucid, well-researched book by well-known historian of science. 172pp. 5⅜ x 8½. 23726-5 Pa. $3.00

THE DEVIL'S DICTIONARY. Ambrose Bierce. Barbed, bitter, brilliant witticisms in the form of a dictionary. Best, most ferocious satire America has produced. 145pp. 5⅜ x 8½. 20487-1 Pa. $2.00

THE COMPLETE WOODCUTS OF ALBRECHT DURER, edited by Dr. W. Kurth. 346 in all: "Old Testament," "St. Jerome," "Passion," "Life of Virgin," Apocalypse," many others. Introduction by Campbell Dodgson. 285pp. 8½ x 12¼. 21097-9 Pa. $7.50

DRAWINGS OF ALBRECHT DURER, edited by Heinrich Wolfflin. 81 plates show development from youth to full style. Many favorites; many new. Introduction by Alfred Werner. 96pp. 8⅛ x 11. 22352-3 Pa. $5.00

THE HUMAN FIGURE, Albrecht Dürer. Experiments in various techniques—stereometric, progressive proportional, and others. Also life studies that rank among finest ever done. Complete reprinting of *Dresden Sketchbook*. 170 plates. 355pp. 8⅜ x 11¼. 21042-1 Pa. $7.95

OF THE JUST SHAPING OF LETTERS, Albrecht Dürer. Renaissance artist explains design of Roman majuscules by geometry, also Gothic lower and capitals. Grolier Club edition. 43pp. 7⅞ x 10¾ 21306-4 Pa. $8.00

TEN BOOKS ON ARCHITECTURE, Vitruvius. The most important book ever written on architecture. Early Roman aesthetics, technology, classical orders, site selection, all other aspects. Stands behind everything since. Morgan translation. 331pp. 5⅜ x 8½. 20645-9 Pa. $4.00

THE FOUR BOOKS OF ARCHITECTURE, Andrea Palladio. 16th-century classic responsible for Palladian movement and style. Covers classical architectural remains, Renaissance revivals, classical orders, etc. 1738 Ware English edition. Introduction by A. Placzek. 216 plates. 110pp. of text. 9½ x 12¾. 21308-0 Pa. $8.95

HORIZONS, Norman Bel Geddes. Great industrialist stage designer, "father of streamlining," on application of aesthetics to transportation, amusement, architecture, etc. 1932 prophetic account; function, theory, specific projects. 222 illustrations. 312pp. 7⅞ x 10¾. 23514-9 Pa. $6.95

FRANK LLOYD WRIGHT'S FALLINGWATER, Donald Hoffmann. Full, illustrated story of conception and building of Wright's masterwork at Bear Run, Pa. 100 photographs of site, construction, and details of completed structure. 112pp. 9¼ x 10. 23671-4 Pa. $5.50

THE ELEMENTS OF DRAWING, John Ruskin. Timeless classic by great Viltorian; starts with basic ideas, works through more difficult. Many practical exercises. 48 illustrations. Introduction by Lawrence Campbell. 228pp. 5⅜ x 8½. 22730-8 Pa. $2.75

GIST OF ART, John Sloan. Greatest modern American teacher, Art Students League, offers innumerable hints, instructions, guided comments to help you in painting. Not a formal course. 46 illustrations. Introduction by Helen Sloan. 200pp. 5⅜ x 8½. 23435-5 Pa. $4.00

THE ANATOMY OF THE HORSE, George Stubbs. Often considered the great masterpiece of animal anatomy. Full reproduction of 1766 edition, plus prospectus; original text and modernized text. 36 plates. Introduction by Eleanor Garvey. 121pp. 11 x 14¾. 23402-9 Pa. $6.00

BRIDGMAN'S LIFE DRAWING, George B. Bridgman. More than 500 illustrative drawings and text teach you to abstract the body into its major masses, use light and shade, proportion; as well as specific areas of anatomy, of which Bridgman is master. 192pp. 6½ x 9¼. (Available in U.S. only) 22710-3 Pa. $3.00

ART NOUVEAU DESIGNS IN COLOR, Alphonse Mucha, Maurice Verneuil, Georges Auriol. Full-color reproduction of *Combinaisons ornementales* (c. 1900) by Art Nouveau masters. Floral, animal, geometric, interlacings, swashes—borders, frames, spots—all incredibly beautiful. 60 plates, hundreds of designs. 9⅜ x 8-1/16. 22885-1 Pa. $4.00

FULL-COLOR FLORAL DESIGNS IN THE ART NOUVEAU STYLE, E. A. Seguy. 166 motifs, on 40 plates, from *Les fleurs et leurs applications decoratives* (1902): borders, circular designs, repeats, allovers, "spots." All in authentic Art Nouveau colors. 48pp. 9⅜ x 12¼. 23439-8 Pa. $5.00

A DIDEROT PICTORIAL ENCYCLOPEDIA OF TRADES AND INDUSTRY, edited by Charles C. Gillispie. 485 most interesting plates from the great French Encyclopedia of the 18th century show hundreds of working figures, artifacts, process, land and cityscapes; glassmaking, papermaking, metal extraction, construction, weaving, making furniture, clothing, wigs, dozens of other activities. Plates fully explained. 920pp. 9 x 12. 22284-5, 22285-3 Clothbd., Two-vol. set $40.00

HANDBOOK OF EARLY ADVERTISING ART, Clarence P. Hornung. Largest collection of copyright-free early and antique advertising art ever compiled. Over 6,000 illustrations, from Franklin's time to the 1890's for special effects, novelty. Valuable source, almost inexhaustible.
Pictorial Volume. Agriculture, the zodiac, animals, autos, birds, Christmas, fire engines, flowers, trees, musical instruments, ships, games and sports, much more. Arranged by subject matter and use. 237 plates. 288pp. 9 x 12. 20122-8 Clothbd. $13.50

Typographical Volume. Roman and Gothic faces ranging from 10 point to 300 point, "Barnum," German and Old English faces, script, logotypes, scrolls and flourishes, 1115 ornamental initials, 67 complete alphabets, more. 310 plates. 320pp. 9 x 12. 20123-6 Clothbd. $15.00

CALLIGRAPHY (CALLIGRAPHIA LATINA), J. G. Schwandner. High point of 18th-century ornamental calligraphy. Very ornate initials, scrolls, borders, cherubs, birds, lettered examples. 172pp. 9 x 13. 20475-8 Pa. $6.00

A MAYA GRAMMAR, Alfred M. Tozzer. Practical, useful English-language grammar by the Harvard anthropologist who was one of the three greatest American scholars in the area of Maya culture. Phonetics, grammatical processes, syntax, more. 301pp. 5⅜ x 8½. 23465-7 Pa. $4.00

THE JOURNAL OF HENRY D. THOREAU, edited by Bradford Torrey, F. H. Allen. Complete reprinting of 14 volumes, 1837-61, over two million words; the sourcebooks for *Walden*, etc. Definitive. All original sketches, plus 75 photographs. Introduction by Walter Harding. Total of 1804pp. 8½ x 12¼. 20312-3, 20313-1 Clothbd., Two-vol. set $50.00

CLASSIC GHOST STORIES, Charles Dickens and others. 18 wonderful stories you've wanted to reread: "The Monkey's Paw," "The House and the Brain," "The Upper Berth," "The Signalman," "Dracula's Guest," "The Tapestried Chamber," etc. Dickens, Scott, Mary Shelley, Stoker, etc. 330pp. 5⅜ x 8½. 20735-8 Pa. $3.50

SEVEN SCIENCE FICTION NOVELS, H. G. Wells. Full novels. *First Men in the Moon, Island of Dr. Moreau, War of the Worlds, Food of the Gods, Invisible Man, Time Machine, In the Days of the Comet.* A basic science-fiction library. 1015pp. 5⅜ x 8½. (Available in U.S. only) 20264-X Clothbd. $8.95

ARMADALE, Wilkie Collins. Third great mystery novel by the author of *The Woman in White* and *The Moonstone.* Ingeniously plotted narrative shows an exceptional command of character, incident and mood. Original magazine version with 40 illustrations. 597pp. 5⅜ x 8½. 23429-0 Pa. $5.00

MASTERS OF MYSTERY, H. Douglas Thomson. The first book in English (1931) devoted to history and aesthetics of detective story. Poe, Doyle, LeFanu, Dickens, many others, up to 1930. New introduction and notes by E. F. Bleiler. 288pp. 5⅜ x 8½. (Available in U.S. only) 23606-4 Pa. $4.00

FLATLAND, E. A. Abbott. Science-fiction classic explores life of 2-D being in 3-D world. Read also as introduction to thought about hyperspace. Introduction by Banesh Hoffmann. 16 illustrations. 103pp. 5⅜ x 8½. 20001-9 Pa. $1.75

THREE SUPERNATURAL NOVELS OF THE VICTORIAN PERIOD, edited, with an introduction, by E. F. Bleiler. Reprinted complete and unabridged, three great classics of the supernatural: *The Haunted Hotel* by Wilkie Collins, *The Haunted House at Latchford* by Mrs. J. H. Riddell, and *The Lost Stradivarious* by J. Meade Falkner. 325pp. 5⅜ x 8½. 22571-2 Pa. $4.00

AYESHA: THE RETURN OF "SHE," H. Rider Haggard. Virtuoso sequel featuring the great mythic creation, Ayesha, in an adventure that is fully as good as the first book, *She*. Original magazine version, with 47 original illustrations by Maurice Greiffenhagen. 189pp. 6½ x 9¼. 23649-8 Pa. $3.50

THE DEPRESSION YEARS AS PHOTOGRAPHED BY ARTHUR ROTHSTEIN, Arthur Rothstein. First collection devoted entirely to the work of outstanding 1930s photographer: famous dust storm photo, ragged children, unemployed, etc. 120 photographs. Captions. 119pp. 9¼ x 10¾.
23590-4 Pa. $5.00

CAMERA WORK: A PICTORIAL GUIDE, Alfred Stieglitz. All 559 illustrations and plates from the most important periodical in the history of art photography, *Camera Work* (1903-17). Presented four to a page, reduced in size but still clear, in strict chronological order, with complete captions. Three indexes. Glossary. Bibliography. 176pp. 8⅜ x 11¼.
23591-2 Pa. $6.95

ALVIN LANGDON COBURN, PHOTOGRAPHER, Alvin L. Coburn. Revealing autobiography by one of greatest photographers of 20th century gives insider's version of Photo-Secession, plus comments on his own work. 77 photographs by Coburn. Edited by Helmut and Alison Gernsheim. 160pp. 8⅛ x 11. 23685-4 Pa. $6.00

NEW YORK IN THE FORTIES, Andreas Feininger. 162 brilliant photographs by the well-known photographer, formerly with *Life* magazine, show commuters, shoppers, Times Square at night, Harlem nightclub, Lower East Side, etc. Introduction and full captions by John von Hartz. 181pp. 9¼ x 10¾. 23585-8 Pa. $6.00

GREAT NEWS PHOTOS AND THE STORIES BEHIND THEM, John Faber. Dramatic volume of 140 great news photos, 1855 through 1976, and revealing stories behind them, with both historical and technical information. Hindenburg disaster, shooting of Oswald, nomination of Jimmy Carter, etc. 160pp. 8¼ x 11. 23667-6 Pa. $5.00

THE ART OF THE CINEMATOGRAPHER, Leonard Maltin. Survey of American cinematography history and anecdotal interviews with 5 masters—Arthur Miller, Hal Mohr, Hal Rosson, Lucien Ballard, and Conrad Hall. Very large selection of behind-the-scenes production photos. 105 photographs. Filmographies. Index. Originally *Behind the Camera*. 144pp. 8¼ x 11. 23686-2 Pa. $5.00

DESIGNS FOR THE THREE-CORNERED HAT (LE TRICORNE), Pablo Picasso. 32 fabulously rare drawings—including 31 color illustrations of costumes and accessories—for 1919 production of famous ballet. Edited by Parmenia Migel, who has written new introduction. 48pp. 9⅜ x 12¼. (Available in U.S. only) 23709-5 Pa. $5.00

NOTES OF A FILM DIRECTOR, Sergei Eisenstein. Greatest Russian filmmaker explains montage, making of *Alexander Nevsky*, aesthetics; comments on self, associates, great rivals (Chaplin), similar material. 78 illustrations. 240pp. 5⅜ x 8½. 22392-2 Pa. $4.50

THE CURVES OF LIFE, Theodore A. Cook. Examination of shells, leaves, horns, human body, art, etc., in "*the* classic reference on how the golden ratio applies to spirals and helices in nature . . . ."—Martin Gardner. 426 illustrations. Total of 512pp. 5⅜ x 8½. 23701-X Pa. $5.95

AN ILLUSTRATED FLORA OF THE NORTHERN UNITED STATES AND CANADA, Nathaniel L. Britton, Addison Brown. Encyclopedic work covers 4666 species, ferns on up. Everything. Full botanical information, illustration for each. This earlier edition is preferred by many to more recent revisions. 1913 edition. Over 4000 illustrations, total of 2087pp. 6⅛ x 9¼. 22642-5, 22643-3, 22644-1 Pa., Three-vol. set $24.00

MANUAL OF THE GRASSES OF THE UNITED STATES, A. S. Hitchcock, U.S. Dept. of Agriculture. The basic study of American grasses, both indigenous and escapes, cultivated and wild. Over 1400 species. Full descriptions, information. Over 1100 maps, illustrations. Total of 1051pp. 5⅜ x 8½. 22717-0, 22718-9 Pa., Two-vol. set $15.00

THE CACTACEAE,, Nathaniel L. Britton, John N. Rose. Exhaustive, definitive. Every cactus in the world. Full botanical descriptions. Thorough statement of nomenclatures, habitat, detailed finding keys. The one book needed by every cactus enthusiast. Over 1275 illustrations. Total of 1080pp. 8 x 10¼. 21191-6, 21192-4 Clothbd., Two-vol. set $35.00

AMERICAN MEDICINAL PLANTS, Charles F. Millspaugh. Full descriptions, 180 plants covered: history; physical description; methods of preparation with all chemical constituents extracted; all claimed curative or adverse effects. 180 full-page plates. Classification table. 804pp. 6½ x 9¼. 23034-1 Pa. $10.00

A MODERN HERBAL, Margaret Grieve. Much the fullest, most exact, most useful compilation of herbal material. Gigantic alphabetical encyclopedia, from aconite to zedoary, gives botanical information, medical properties, folklore, economic uses, and much else. Indispensable to serious reader. 161 illustrations. 888pp. 6½ x 9¼. (Available in U.S. only) 22798-7, 22799-5 Pa., Two-vol. set $12.00

THE HERBAL or GENERAL HISTORY OF PLANTS, John Gerard. The 1633 edition revised and enlarged by Thomas Johnson. Containing almost 2850 plant descriptions and 2705 superb illustrations, Gerard's *Herbal* is a monumental work, the book all modern English herbals are derived from, the one herbal every serious enthusiast should have in its entirety. Original editions are worth perhaps $750. 1678pp. 8½ x 12¼. 23147-X Clothbd. $50.00

MANUAL OF THE TREES OF NORTH AMERICA, Charles S. Sargent. The basic survey of every native tree and tree-like shrub, 717 species in all. Extremely full descriptions, information on habitat, growth, locales, economics, etc. Necessary to every serious tree lover. Over 100 finding keys. 783 illustrations. Total of 986pp. 5⅜ x 8½. 20277-1, 20278-X Pa., Two-vol. set $10.00